AF897945

FRONTIERS IN HUMAN GENETICS

DISEASES AND TECHNOLOGIES

FRONTIERS IN HUMAN GENETICS

DISEASES AND TECHNOLOGIES

With a Foreword from Lap-Chee Tsui
President of the Human Genome Organisation (HUGO)

Expanded and Updated from the Proceedings of the
International Symposium on Human Genetics and Gene Therapy, Singapore

Editors

Lai Poh San, Coral
National University of Singapore

Eric P H Yap
Defence Medical Research Institute, Singapore

World Scientific
Singapore • New Jersey • London • Hong Kong

Published by

World Scientific Publishing Co. Pte. Ltd.

P O Box 128, Farrer Road, Singapore 912805

USA office: Suite 1B, 1060 Main Street, River Edge, NJ 07661

UK office: 57 Shelton Street, Covent Garden, London WC2H 9HE

British Library Cataloguing-in-Publication Data
A catalogue record for this book is available from the British Library.

FRONTIERS IN HUMAN GENETICS
Diseases and Technologies

Copyright © 2001 by World Scientific Publishing Co. Pte. Ltd.

All rights reserved. This book, or parts thereof, may not be reproduced in any form or by any means, electronic or mechanical, including photocopying, recording or any information storage and retrieval system now known or to be invented, without written permission from the Publisher.

For photocopying of material in this volume, please pay a copying fee through the Copyright Clearance Center, Inc., 222 Rosewood Drive, Danvers, MA 01923, USA. In this case permission to photocopy is not required from the publisher.

ISBN 981-02-4458-4

Printed in Singapore.

LIST OF CORRESPONDING AUTHORS

Thuan D. Bui
Membrane Biology Laboratory
Institute of Molecular and Cell Biology
30 Medical Drive
Singapore 117609
E-mail: dtbui@imcb.nus.edu.sg

Peh Yean Cheah
Department of Colorectal Surgery
Singapore General Hospital
Outram Road
Singapore 169608
E-mail: gcscpy@sgh.com.sg

Betty Cheng
BioInformatics Centre
National University of Singapore
Lower Kent Ridge Road
Singapore 119260
E-mail: betty@bic.nus.edu.sg

Eva Maria C. Cutiongco
Department of Paediatrics
University of the Philippines
College of Medicine
547 Pedro Gil Street
Ermita, Manila 1000
Philippines
E-mail: evamac@yahoo.com

Suthat Fucharoen
Thalassemia Research Center
Institute of Science and Technology for Research and Development
Mahidol University, Salaya Campus
25/25 M.3, Puttamonthon 4 Road
Puttamonthon, Nakornpathom 73170
Thailand
E-mail: grsfc@mahidol.ac.th

Jiro Fujimoto
First Department of Surgery
Hyogo College of Medicine
Nishinomiya 663-8501
Japan
E-mail: sfujimo@hyo-med.ac.jp

Leena A. Gole
Department of Obstetrics and Gynaecology
National University of Singapore
Lower Kent Ridge Road
Singapore 119074
E-mail: obggolel@nus.edu.sg

Akinobu Gotoh
Department of Urology
Kobe University School of Medicine
7-5-2 Kusunoki-cho
Chuo-ku, Kobe 650-0017
Japan
E-mail: gotoh@med.kobe-u.ac.jp

Kam Man Hui
Gene Vector Laboratory
Division of Cellular and Molecular Research
National Cancer Centre
11 Hospital Drive
Singapore 169610
E-mail: cmrhkm@nccs.com.sg

Mohd. Nizam Isa
Human Genetics Unit
School of Medicine
Universiti Sains Malaysia
16150 Kubang Kerian
Kelantan, Malaysia
E-mail: zam@kb.usm.my

Yasufumi Kaneda
Division of Gene Therapy Science
Graduate School of Medicine
Osaka University
Suita, Osaka 565-0871
Japan
E-mail: kaneday@gts.med.osaka-u.ac.jp

Pandjassarame Kangueane
BioInformatics Centre
#02-07 MD7 Medical Drive
National University of Singapore
Lower Kent Ridge Road
Singapore 119260
E-mail: kangs@bic.nus.edu.sg

P.N. Kantaputra
Department of Pediatric Dentistry
Faculty of Dentistry
Chiang Mai University
Chiang Mai 50200
Thailand
E-mail: dnpdi001@chiangmai.ac.th

Siok Im Koh
Chemical Process & Biotechnology Department
Singapore Polytechnic
500 Dover Road
Singapore 139651
E-mail: silk@sp.edu.sg

Oi Lian Kon
Division of Medical Sciences
National Cancer Centre
11 Hospital Drive
Singapore 169610
E-mail: dmskol@nccs.com.sg

Poh San Lai
Department of Paediatrics
National University of Singapore
Lower Kent Ridge Road
Singapore 119074
E-mail: paelaips@nus.edu.sg

Ann S.G. Lee
Division of Medical Sciences
National Cancer Centre
11 Hospital Drive
Singapore 169610
E-mail: dmslsg@nccs.com.sg

Joy Y. Lee
Department of Paediatrics
University of the Philippines
College of Medicine
547 Pedro Gil Street
Ermita, Manila 1000
Philippines
E-mail: joyyaplito@hotmail.com

Masafumi Matsuo
Division of Genetics
International Center for Medical Research
Kobe University School of Medicine
7-5-1 Kusunoki-cho
Chuo-ku, Kobe 650-0017
Japan
E-mail: matsuo@kobe-u.ac.jp

Vernon M.S. Oh
Division of Clinical Pharmacology & Therapeutics
Department of Medicine
National University Hospital
5 Lower Kent Ridge Road
Singapore 119074
E-mail: mdcohms@nus.edu.sg

Carmencita D. Padilla
Department of Paediatrics
University of the Philippines
College of Medicine
547 Pedro Gil Street
Ermita, Manila 1000
Philippines
E-mail: carpadil@fastmail.i-next.net

Purnomo Suryantoro
Pediatric Department
Faculty of Medicine
Gadjah Mada University
Sekip Utara
Yogyakarta 55281
Indonesia
E-mail: purnomo@yogya.wasantara.net.id

Stacey K.H. Tay
Department of Paediatrics
National University of Singapore
Lower Kent Ridge Road
Singapore 119074
E-mail: paetaykh@nus.edu.sg

Stephen D. Wilton
Australian Neuromuscular Research Institute
4th Floor "A" Block
QE II Medical Centre
Nedlands WA 6009
Australia
E-mail: swilton@cyllene.uwa.edu.au

Lim Soon Wong
Kent Ridge Digital Labs
21 Heng Mui Keng Terrace
Singapore 119613
E-mail: limsoon@krdl.org.sg

Eric P.H. Yap
Defence Medical Research Institute
Defence Science & Technology Agency
10 Medical Drive
Singapore 117597
E-mail: nmiv3@nus.edu.sg

FOREWORD

The most exciting development in human genetics in recent years has been our ability to study diseases at the molecular level. The precise DNA sequence alterations and protein defects are now known for many single-gene Mendelian disorders and some forms of cancers. As the Human Genome Project has just completed its draft sequencing phase, we can look forward to an explosion of molecular genetic information on human diseases, especially those with a multifactorial etiology.

Since the spectrum of gene mutations and DNA sequence variations often varies among different geographic regions and ethnic groups, it is important to develop efficient and affordable technologies for the collection of baseline data for individual communities.

In addition, while the utility of our current knowledge from human disease gene research has initially been limited to DNA diagnosis and carrier detection, we should look forward to novel treatments and rational therapies in the near future. Gene therapy represents a promising approach in the latter regard where some initial success has been reported, but much development is required before its impact can be fully recognized.

I congratulate Poh San and Eric for their tremendous effort in assembling this valuable volume of articles contributed by a group of young researchers in the Asia Pacific region. The book indeed captures a snapshot of the diverse approaches and solutions being developed at the frontiers of human genetics. It will be a valuable reference for researchers and students in molecular genetics and medicine as well as professionals in the biotechnology and pharmaceutical industries.

Lap-Chee Tsui
President, Human Genome Organisation (HUGO)
Geneticist-in-Chief and H.E. Sellers Chair in Cystic Fibrosis,
The Hospital for Sick Children,
Toronto, 2001

PREFACE

We are witnessing, at this transition between millennia, some exciting developments in human genetics research and discovery. While there has been unprecedented exposure in the public media of the human genome mapping project and its potential impact on human health and society, we have too, in the scientific communities, been awed by the rapid pace of progress as well as the new challenges and opportunities that this wealth in new biological information provides. This book was therefore born out of the need to document these discoveries in human genetics and molecular medicine, both as a practical help to those researching at the bench, as well as to capture a snapshot of the state-of-the-art in this rapidly evolving field. The title reflects the how's (Technologies) and why's (Diseases) that drive and motivate genetics research respectively. Three areas were picked to reflect major trends in genetics research, into which the contributions of this book have been arranged.

The first section, Emerging Technologies, highlights the importance of new techniques and tools of genetic analysis that have increased the throughput of generating biological information. Some of these have found their way into clinical practice (e.g. fluorescent *in situ* hybridization) while others are being used to hasten the gene/target discovery process (e.g. single nucleotide polymorphisms, discovery and genotyping). Research at the interface between biology and the physical/engineering sciences has resulted in advances in miniaturization (e.g. microelectromechanical systems and lab-on-chips), highly parallel processing (e.g. microarrays) and biological computation (e.g. data integration and prediction of macromolecular structures).

The second section, Genes & Diseases, deals with the discovery of disease genes and their significance. Some genes of known cellular function have yet to be implicated in disease (e.g. SNAREs), while on the other hand, there are syndromes in search of genes (e.g. mesomelic dysplasia). In yet others, the gene has been identified and established. For the latter, the challenges include developing efficient mutation screening methods for diagnosis (e.g. dystrophin and retinoblastoma); applying these tests in a clinical and public health setting particularly in developing economies (e.g. ambiguous genitalia, maple syrup urine disease, congenital adrenal hyperplasia); and determining the allelic diversity in different populations (e.g. thalassaemia, G6PD-deficiency, familial adenomatous polyposis). Establishing the genotype-phenotype correlation in these Mendelian diseases is important for genetic intervention in the form of screening, counseling and therapy. The search for the genetic aetiologies for human diseases has also gradually progressed from defining the causative genes for Mendelian syndromes to elucidating susceptibility genes of complex genetic traits. Affected relative pair study designs and non-parametric statistical tools have allowed the analysis of common multifactorial diseases such as myopia, hypertension and neoplasia, and infections.

The third section on Gene Therapy describes the nascent and rapidly evolving field of gene-based therapeutics for hitherto intractable genetic diseases. One of the basic challenges in this area is the development of effective and safe approaches for delivering genetic material to the specific cells of the patient. The various gene delivery methods and vehicles described here include naked DNA, liposomes, viral vectors and targeted delivery. Various approaches are also being attempted (e.g. antisense oligonucleotides, plasmid DNA, gene conversion) on several target tissues (e.g. muscle, liver, prostate).

Part of the content for this book is derived from an earlier conference: The International Symposium on Human Genetics and Gene Therapy, held in Singapore in February 1999. This meeting was organized by the Biomedical Research and Experimental Therapeutics Society of Singapore, the Singapore Society for Microbiology and Biotechnology and the Singapore Society for Biochemistry and Molecular Biology with the National University of Singapore and the International Center for Medical

Research, Kobe University School of Medicine, Japan, as co-hosts. Financial sponsorship for the meeting and the proceedings in this book by the Japan Society for the Promotion of Science is gratefully acknowledged. While the original abstracts of presented papers are appended, the full papers have been extensively updated and revised.

We wish to thank the scientists and clinicians who have kindly taken the time to contribute papers for this book. We are also grateful to Professor Lap-Chee Tsui, the President of the Human Genome Organisation (HUGO) for his support and encouragement of human genetics efforts in the region, and for authoring the Foreword. The artwork on the front cover was designed by Jimmy Low. Finally, Lim Sook Cheng and Alan Pui from World Scientific have eased this project from conception through the editorial hurdles to final print, and are therefore responsible to no small extent for the successful delivery of this publication.

The Editors

CONTENTS

II Genes & Diseases

III Gene Therapy

I EMERGING TECHNOLOGIES

FRONTIERS IN HUMAN GENETICS
Diseases and Technologies

© 2001 by World Scientific Publishing Co. Pte. Ltd.

FISH FOR THE OBSTETRICIAN AND GYNAECOLOGIST: A RAPID AND RELIABLE TOOL AIDING CLINICAL ANALYSIS

LEENA A GOLE*, C ANANDAKUMAR, YEE CHEE WONG and A BISWAS
Department of Obstetrics and Gynaecology
National University of Singapore
**E-mail : obggolel@nus.edu.sg*

Routine cytogenetic techniques of karyotyping and banding chromosomes have been the usual conventional testing procedures for detection of chromosomal anomalies in prenatal diagnostics. With the advent of molecular techniques of fluorescence *in situ* hybridization, the field of cytogenetics has been revolutionalized. Detection of some genes on chromosomes, which previously could not be detected by cytogenetics can now be visualized. Also, samples from which chromosomes in the metaphase plate maybe difficult to obtain, can be probed by FISH in all stages of the cell cycle. Hence applications in preimplantation genetics are increasing rapidly. The rapidity and efficiency of the technique makes it very attractive. This article reviews the applications as well as drawbacks of the FISH technique in different areas, mainly in prenatal and preimplantation diagnosis.
Keywords: fluorescence *in situ* hybridization, prenatal diagnosis, preimplantation genetic diagnosis

* Corresponding author

1 INTRODUCTION

Prenatal diagnosis is an important tool for the genetic evaluation of the unborn child. The conventional testing procedures have been a cytogenetic analysis of chorion villus biopsies starting from 6 to 10 weeks of gestation, amniotic fluid at 16 to 18 weeks with a reporting time of 2 to 3 weeks providing a final diagnosis at 20 to 21 weeks, or chromosome analysis on foetal blood or alternative fluids at 20 to 26 weeks of pregnancy. With the current trend of working towards non-invasive methods and rapid reporting time, so as to avoid undue anxiety to the patient, major advances are taking place in this field to provide a genetic diagnosis using early amniocentesis (EA) at 11 to 14 weeks. No doubt the existing banding techniques make possible the detection of minor chromosomal rearrangements, but the technique is still limited where localisation of genes is concerned, as the smallest band is yet a few thousand bases. With the advent of molecular biology techniques e.g. fluorescence *in situ* hybridization (FISH), in conjunction with conventional cytogenetics, there is immense additional knowledge pouring into the existing data. FISH[1 - 3] is becoming more and more relevant as an important future tool in prenatal and preimplantation genetic diagnosis. This technique can be applied to whole chromosome spreads as well as interphase cells. The potential use of this novel FISH technique in the diagnosis of numerical and structural chromosomal aberrations in routine karyotyping for prenatal and preimplantation diagnosis is immense.

2 PRENATAL DIAGNOSIS

Prenatal diagnosis with special reference to FISH, can be offered at differing gestational ages, the final aim being to analyse the chromosomes of the foetus. However, the target tissue differs. This maybe chorionic villus, amniotic fluid, foetal blood or alternative fluid samples e.g. fluids from cystic hygromas, ascites or pleural cavities based upon gestational age. Depending on which phase the cells are to be analysed, whether interphase or metaphase, specific probes can be utilised. By and large, interphase analysis is most commonly used for numerical anomalies, whilst metaphase analysis gives more definitive results for structural anomalies. Since **numerical anomalies**, e.g. trisomies and monosomies especially involving the chromosomes 13,18, 21, X and Y account for 90% of the abnormal foetuses, this is a very important application in prenatal diagnosis. Some sex-linked disorders require the necessity of prenatal diagnosis of foetal sex. Numerical anomalies include all the disorders caused due to presence or absence of extra chromosomes e.g. Down syndrome resulting from an extra chromosome 21, Patau syndrome due to trisomy of chromosome 13, Edwards syndrome due to trisomy 18 or Turner syndrome due to lack of one X chromosome. Repetitive sequences for the centromeric regions of 13, 18, 21, X and for the long arm of Y produce distinct signals in metaphase as well as interphase cells.[4–6]

Structural abnormalities are those due to translocations, deletions, amplifications or inversions of chromosomal fragments. With the use of painting probes, translocations in cultured leucocytes of foetal blood, alternative fluids and amniocytes can be easily detected, but more so in metaphase cells.[7–9] Used in conjunction with routine Giemsa banding, translocations involving a specific chromosome can be ascertained especially with the use of appropriate painting probes. Analysis of reciprocal translocations by chromosome painting has some limitations. As not all libraries have equal specificity and sensitivity in detecting different chromosomal regions, caution should be exercised during selection. However with the refinement of techniques, this problem will soon be minimised.

Deletions of the targeted region can be detected by loss of signals, whereas amplification can be detected by an increase in the area of the hybridization domain. Locus specific probes can detect presence or deletions of specific loci e.g. DiGeorge region (22q11.2), Prader-Willi/Angelman chromosome region probes (15q11-13)etc. Vysis (USA), Boehringer Mannheim (Germany) and Cytocell (U.K) are some of the commercial manufacturers for FISH probes.

3 APPLICATIONS IN PRENATAL DIAGNOSIS

3.1 Chorion Villus Biopsy

Sampling of chorion villi from the foetus is performed from 8 to 10 weeks of gestation onwards. The biopsy is usually taken under ultrasound guidance via transabdominal approach. The overnight short-term culture protocol is the most popular method to reduce reporting time, in spite of a slight increase in the rates of mosaicism as compared to the long-term cultures (1.26% v/s 0.66%). The number of metaphases and quality of chromosomes obtained is many times of very poor standards. Thus FISH can be useful for detection of suspected chromosomal anomalies even on the interphase cells of the villus sample. In fact FISH has been utilised to assess the effects of maternal cell contamination[10] on the sensitivity of prenatal diagnosis, with a conclusion that given a thorough dissection of villi, this issue does not pose any major problems in the results. However, the culture of chorionic villi still remains unpopular compared to the amniotic cell culture because of the artefacts and false positives and negatives generated.

3.2 Amniotic Fluid

Amniocentesis is usually performed at 16 to 18 weeks of gestation, however early amniocentesis from 11 to 14 weeks is being increasingly used. Routine culturing procedures for karyotyping take from 2–3 weeks. In comparison to this, FISH when carried out on uncultured amniocytes, results in a diagnosis within a few hours for the detection of a specific chromosomal abnormality. Use of multicolor FISH can facilitate diagnosis of upto 5 to 7 different chromosomes on the same cell.[11,12] This is a very significant application most appropriate for prenatal diagnosis. It overcomes the drawbacks of metaphase analysis, which requires cells to be cultured, which is labour intensive and time consuming. Using the probes for the most commonly occurring aneuploidies, e.g.13, 18, 21 and the sex chromosomes X and Y, FISH can be used as pre-screen for detection of aneuploidies and the sex of the foetus in uncultured amniocytes (Figs. 1a and b). The drawback of the technique is that all cells in the population may not show the trisomic signal due to chance overlap of the hybridization domains. Also mosaicism is not reliably detected. A comprehensive and all encompassing test is still the G-banded karyotype, which gives a detailed cytogenetic analysis of all the chromosomes. However, with the advent of M-FISH, all 46 chromosomes can be painted at the same time and all minor translocations deletions etc can be detected. Hence, though presently FISH is mainly used as complementation to routine cytogenetics, it will evolve into a powerful technique with the advance of sophisticated image analyzers and painting probes.

3.3 Foetal Blood Lymphocytes

Foetal blood analysis is routinely carried out in later gestational ages of 20 to 22 weeks as the viable cell populations in the amniotic fluid decrease with advancing gestational age. To minimise maternal blood contamination, blood is drawn from the cord or hepatic portal vein of the foetus. The lymphocytes are then cultured for routine cytogenetic analysis. FISH can be used to confirm any suspected chromosomal anomalies on the metaphase chromosomes (Fig. 1d) or it can also be used on uncultured lymphocytes.

3.4 Alternative Fluids

Alternative fluids e.g. serous fluids from cystic hygromas, the pleural filled cavities around the lungs or the abdominal region are a plentiful source of lymphocytes which can be utilised as an alternative to traditional amniotic fluid or foetal blood cultures, especially in pregnancies complicated by cystic hygromas or hydrops fetalis, where sometimes obtaining them can be difficult due to obstructions by large cysts or oligohydramnios. Along with the conventional culturing techniques, FISH can be carried out on these cells too, especially where metaphases maybe of poor quality as is the case most of the time[13 – 15] or on uncultured cells for interphase analysis.

3.5 Foetal Cells in Maternal Peripheral Blood

Compared to chorion villus biopsy, amniocentesis, foetal blood or alternative fluids, all of which are invasive techniques, prenatal detection of chromosomal anomalies in foetal cells in the maternal peripheral circulation is a relatively non-invasive technique. With the use of fluorescence activated cell sorters and antibodies specific for fetal cells, nucleated erythrocytes, which are foetal in origin, can be separated and probed with FISH for aneuploidies of chromosomes of interest, thereby eliminating a lot of laborious techniques previously used.[16,17]

The major limitation of the current technology available for prevention of genetic diseases is that selective abortion is the only choice after prenatal diagnosis. The development of new methods for diagnosis of genetic disease in the early stages of development of the human zygote and embryo (before implantation) is particularly needed for couples, who cannot accept the termination of a pregnancy and have a high recurrent risk for offspring with inherited diseases, e.g. X-linked diseases or translocation carriers. Preimplantation genetic diagnosis will make it possible to overcome the most sensitive problem in the management of genetic disease — the problem of abortion.

4 PREIMPLANTATION GENETIC DIAGNOSIS (PGD)

From diagnosing the post-implantation foetus for prenatal genetic anomalies by chorion villus biopsies and early amniotic fluids, a step ahead would be preimplantation diagnosis of human gametes and embryos. PGD could be, in principle, used the same as prenatal diagnosis, but at the moment is justified for high risk pregnancies. Rapid advances have recently taken place in this field to attempt to genetically diagnose the preimplantation embryo (day 3). With the advent of the assisted reproductive technologies (ART) e.g. *in vitro* fertilisation-embryo replacement (IVF-ER), gamete intrafallopian transfer (GIFT), tubal embryo transfer (TET) and intracytoplasmic sperm injection (ICSI), the problem of infertility in the majority of childless couples has been alleviated. However, a fair percentage of oocytes fail to fertilize in some patients and poor-quality embryos with abnormal cleavage and moderate to severe fragmentation are produced. These do not implant or are unable to sustain implantation after replacement. Chromosomal imbalance has been implicated for the high frequency of early embryonic loss and first trimester abortions. There are two main approaches for the genetic diagnosis before implantation. Either gametes i.e. sperms or oocytes can be analysed or secondly embryonic cells at various stages maybe analysed. However karyotypic analysis of chromosomes in all the above samples is very tedious and inconsistent. FISH has proved to be very useful in this aspect.

4.1 FISH on Sperm

Especially in cases scheduled for ICSI, oligozoospermia, asthenozoospermia, teratozoospermia or oligoasthenoteratozoospermia (OATS) are very common parameters. Many times there may be a high possibility that there are chromosomal abnormalities in the gonads, which may not be detected in the blood but may be present in the sperm. At such times it would be more advisable to carry out an analysis on the sperm before using them for ICSI. Though it is not feasible to use the sperm, which has been analysed, a gross percentage abnormality can be detected in the sample and the patient advised accordingly. A recent report of 12 ICSI pregnancies of which three sets were twins, found five cases of sex chromosomal aneuploidy,[18] two cases of 47,XXY, two of 45,X and one mosaic 45,X/46,X,dic Y. The parents of the cytogenetically abnormal foetuses had normal lymphocyte karyotypes. From this it was postulated that the fathers could have been mosaics with an aneuploid cell line confined to the germ tissue. It has long been noted that the frequency of chromosomal abnormality is increased in males selected for infertility and the trend is inversely proportional to the sperm concentration.[19] Karyotyping sperm, however, is very laborious and is usually done by allowing human sperm to fertilize zona-free hamster ova and then karyotyping decondensed sperm head preparations within the hamster ooplasmic matrix with approx. only 60% success. FISH can be carried out directly on sperm relatively easily. However, the nuclei of mature spermatozoa are highly condensed and protected with interprotamine disulfide bridges. The success of FISH relies on the partial decondensation of the sperm chromatin. The current procedures for sperm decondensation entail the use of reducing agents as well as detergents such as DTT (Fig. 1c) along with Triton X.

4.2 FISH on Oocytes

The majority of age-related non-disjunction occurs during maternal meoisis. Since direct analysis of the oocyte will result in destruction of the oocyte, polar body analysis and transfer of embryos derived from oocytes or pre-embryos with euploid polar bodies should drastically reduce the chances of an IVF couple giving birth to a child with a chromosomal aneuploidy. With the advanced techniques of micromanipulation, first and second polar bodies can be removed and subjected to FISH probes. A study on 45 IVF patients of advanced maternal age has been carried out by Verlinsky[20] 155 of 228 biopsied oocytes could be analysed, of which 23.2% were chromosomally abnormal. A similar correlation was found by Plachot[21] by routine cytogenetic analysis, in which the mean incidence of aneuploidy and structural anomalies was 25.6% and 2.8% respectively.

4.3 FISH on Embryos

Very few existing studies on the cytogenetic analysis of embryos[22] are documented in literature due to the low cell numbers, quality of metaphases and problems associated with spreading and banding. With the application of FISH and the commercial availability of probes for chromosomes – 13,18,21,X and Y, it is possible to use multiprobe single cell analysis on metaphases, even poor quality ones, for these embryos. A single 8 to 10 cell embryo at the cleavage stage can be biopsied for the removal of 1 or 2 blastomeres, which are subjected to FISH. This same embryo at the blastocyst stage can again be manipulated and some cells may-be removed for confirmation.[23] The only disadvantage being that these biopsied cells may not be representative of the entire embryo. Double target *in situ* hybridization has been performed with X & Y specific probes on embryos to check sex[24,25], as well as with autosomal probes to detect specific chromosomal aneuploidy and mosaicism.[26] The use of multicolor FISH has recently contributed to the confirmation of the incidence of numerical rearrangements in abnormal human embryos generated in IVF programmes and provided explanations to some of the causes of IVF failures in some patients.[27] Centromeric probes allow precise counting of chromosomes from blastomeres and as such give more reliable and accurate information on numerical rearrangements compared to the conventional method of fixation of embryos. With the introduction of FISH, preimplantation genetic diagnosis on single cells has been made feasible.

5 A BRIEF DESCRIPTION OF THE FISH TECHNIQUE

Fluorescence *in situ* hybridization entails the deposition of fluorescent molecules in the nucleus at the sites of specific DNA sequences. Specific DNA or RNA sequences of choice are labelled with reporter molecules. These "probes" and the target chromosomes are denatured. Complementary sequences in the probe and target are then allowed to reanneal. After washing and incubation with fluorescently labelled affinity reagents, a signal is made visible at the site of probe hybridization. Depending on how the probe is labelled, it can be detected directly or indirectly. For example, fluorochromes such as fluorescein can be directly coupled onto the probe and so visualised immediately, under the fluorescence microscope after hybridization to the target DNA. Indirect procedures require that the modified probe be detected by immunocytochemical means. For example, biotinylated d-UTP incorporated in the probe is detected by the fluorescein avidin anti-avidin system. The basic technique[1] comprises of preparation of the probe, preparation of the tissue, hybridization of probe to the tissue, stringent washes and visualisation of the probe.

The three main types of probes, which can be used, are satellite probes for repeat sequences, whole chromosome painting probes and locus probes.

5.1 Centromeric Probes

These are tandemly repeated sequences, several hundred to thousand times in the centromeric regions about 10^6 to 10^8 bp in size. On most human chromosomes, some part of the repeated sequence is sufficiently different, that FISH with a probe to the variant region produces a signal that is intense and chromosome–specific. Centromeric probes have now been isolated and cloned for human chromosomes and are commercially available for all the chromosomes.

5.2 Whole Chromosome Probes

These comprise of many different elements distributed more or less continuously over one chromosome so that the chromosome targeted by the probe appears continuously stained or painted. Translocations can be effectively detected, the only drawback being that not all libraries have equal specificity and sensitivity in detecting different chromosomal regions.

5.3 Locus Specific Probes

These are the most powerful approach to structural aberrations e.g. gene deletion or amplification in both metaphase and interphase cells. Once the important loci in a particular genetic disease have been identified, they can be studied using FISH with probes to this region. Ideally the probes should range from 15–50 kb.

6 CONCLUSION

Taking into account the powerful diagnostic capacity of the FISH technique, it could be used in prenatal diagnosis with a very practical and logical approach of carrying it out on the CVB (8–10 weeks) or early amniotic fluids (11–14 weeks) for a rapid diagnosis in emergency situations. In later weeks it could be done on routine amniotic or alternative fluids at 16–20 weeks. Confirmatory tests can be done on foetal blood at 22–26 weeks.

Conversely even before the prenatal period, FISH can be used on embryos for preimplantation genetic diagnosis. This could be the treatment of choice in the future, especially for translocation carriers or for the detection of sex in sex linked disorders. FISH on biopsied blastomeres before replacement of embryos would give hope for the birth of a normal child. As image analysis becomes more and more powerful, smaller regions on chromosomes will be detected. With the advent of commercial probes on the market, FISH will be easier to perform and can be widely used.

Acknowledgements

This study was funded by the National Medical Research Council Research projects NMRC RP 3960353 and RP 3690025.

References

1. Gole LA, Bongso A, Fluorescent *in situ* hybridization —Examples of its applications in clinical cytogenetics. *Singapore Medical Journal* 1997; **38**; 388–390.
2. Tkachuk DC, Pinkel D, Kuo WL, Weier HU, Gray JW. Clinical applications of fluorescence *in situ* hybridisation. *GATA* 1991; **8**(2): 67–74.
3. Trask BJ. Fluorescence in situ hybridisation: applications in cytogenetics and gene mapping. *TIG* 1991; **7**(5): 149–154.
4. Aviram-Goldring A, Daniely M, Chaki R, Lipitz S, Barkai G, Goldman B, Advanced FISH with directly labeled X, Y and 18 DNA probes as a tool for rapid prenatal diagnosis. *Reprod Med* 1999;**44**(6): 497–503.

5. Gersen SL, Carelli MP, Klinger KW, Ward BE, Rapid prenatal diagnosis of 14 cases of triploidy using FISH with multiple probes. *Prenatal Diagn* 1995; **15**: 1–5.
6. Verlinsky Y, Ginsberg N,Chmura M, Freidine M, White M, Strom C, Kuliev A, Cross hybridization of the chromosome 13/21 alpha satellite DNA probe to chromosome 22 in the prenatal screening of common chromosomal aneuploidies by FISH. *Prenatal Diagn* 1995; **15**: 831–834.
7. Rosenberg C, Blakemore KJ, Kearns WG, Giraldez RA et al. Analysis of reciprocal translocations by chromosome painting: Applications and limitations of the technique. *Am J Hum Genet* 1992; **50**: 700–705.
8. Speleman F, Van Roy N, Wiegant Joop et al. Detection of subtle translocations by fluorescence *in situ* hybridisation. *Clin Genet* 1992; **41**: 169–174.
9. Tucker JD, Morgan WF, Awa AA, Bauchinger M, Blakey D, A proposed system for scoring structural aberrations detected by chromosome painting. *Cytogenet Cell Genet* 1995; **68**: 211–221.
10. Bryndorf T, Christensen B, Xiang Y, Philip J.Prenatal diagnosis by FISH on chorionic villus cells; non-significance of maternal cell contamination. *Fetal Diagn Ther* 1994; **9**(2); 73–76.
11. Ried T, Landes G, Dackowski W, Klinger K, Ward DC, Multicolor fluorescence *in situ* hybridisation for the simultaneous detection of probe sets for chromosomes 13,18, 21, X and Y in uncultured amniotic fluid cells. *Hum Mol Genet* 1992; **1**(5): 307–313.
12. Ried T, Baldini A, Rand TC, Ward DC, Simultaneous visualisation of seven different DNA probes by *in-situ* hybridization using combinatorial fluorescence and digital imaging microscopy. *Proc Natl Acad Sci* 1992; **89**: 1388–1392.
13. Gole LA, Anandakumar C, Bongso A, Chua TM, Wong YC, Ratnam SS, Analysis of cystic hygroma, ascitic and pleural fluid by conventional lymphocyte culturing and fluorescent *in situ* hybridization. *Prenatal Diagn* 1997; **17**(12): 1151–1157.
14. Wax JR, Blakemore KJ, Soloski MJ, Gibson M, Stetten G. Fetal ascitic fluid: A new source of lymphocytes for rapid chromosomal analysis. *Obst Gynaecol* 1992: **80**; 533–535.
15. Ville Y, Borghi E, Pons JC, Lelorc'H M. Fetal karyotype from cystic

hygroma fluid. *Prenatal Diagn* 1992; **12**: 139–143.
16. Gaenshirt D, Garritsen HSP, Holzgreve W. Prenatal diagnosis using fetal cells in the maternal circulation. *Fetal and Maternal Medicine review* 1995; **7**: 77–85.
17. Bischoff FZ, Lewis DE, Nguyen DD, Muerrell S,Schober W, Scott J, Simpson JL, Elias S, Prenatal diagnosis with use of fetal cells isolated from maternal blood : Five color fluorescent in situ nybridization analysis on flow sorted cells for chromosomes X,Y 13,18 and 21. *Am J Obstet Gynec* 1998; **179**(1): 203–209.
18. In't Veld P, Brandenburg H, Verhoeff A, Sex chromosomal abnormalities and intracytoplasmic sperm injection. *Lancet* 1995; **346**(8977): 773.
19. Persson JW, Peters GB, Saunders DM, Genetic consequences of ICSI. *Human Reprod* 1996; **11**(5): 921–932.
20. Verlinsky Yu, Kuliev Anver M, *Preimplantation Diagnosis of Genetic Diseases: A New Technique in Assisted Reproduction*, Wiley–Liss, New York, 1993.
21. Plachot M, Cytogenetic analysis of oocytes and embryos. *Ann Acad Med* 1992; **21**(4): 538–544.
22. Bongso A., Fong C.Y., Ng SC., Ratnam S., Lim J, Preimplantation genetics: Chromosomes of fragmented human embryos. *Fertil Steril* 1991; **56**(1): 66–70.
23. Muggleton Harris AL, Glazier AM, Pickering S, Wall M, Genetic diagnosis using polymerase chain reaction and fluorescent *in situ* hybridization analysis of biopsied cells from both the cleavage and blastocyst stages of individual cultured human preimplantation embryos. *Human Reprod* 1995; **10**(1); 183–192.
24. Pellicer A, Rubio C, Vidal F, Minguez Y, Gimenez C, Egozcue J, Remohi J, Simon C, In vitro fertilization plus preimplantation genetic diagnosis in patients with recurrent miscarriage: an analysis of chromosome abnormalities in human preimplantation embryos. *Fertil Steril* 1998; **71**(6); 1033–1039.
25. Munne S, Magli C, Bahce M, Fung J, Legator M, Morrison L, Cohert J, Gianaroli L, Preimplantation diagnosis of the aneuploidies most commonly found in spontaneous abortions and live births: XY, 13, 14, 15, 16, 18, 21, 22. *Prenat Diagn* 1998; **18**(13): 1459–1466.

26. Coonen E, Harper JC, Ramaekers FCS, Delhanty JDA *et al.* Presence of chromosomal mosaicism in abnormal preimplantation embryos detected by fluorescence in situ hybridisation. *Hum Genet* 1994; **94**: 609–615.
27. Munne Santiago, Andrew L, Rosenwaks Z, Grifo J, Cohen J. Diagnosis of major chromosome aneuploidies in human preimplantation embryos. *Hum Reprod* 1993; **8**(12): 2185–2191.

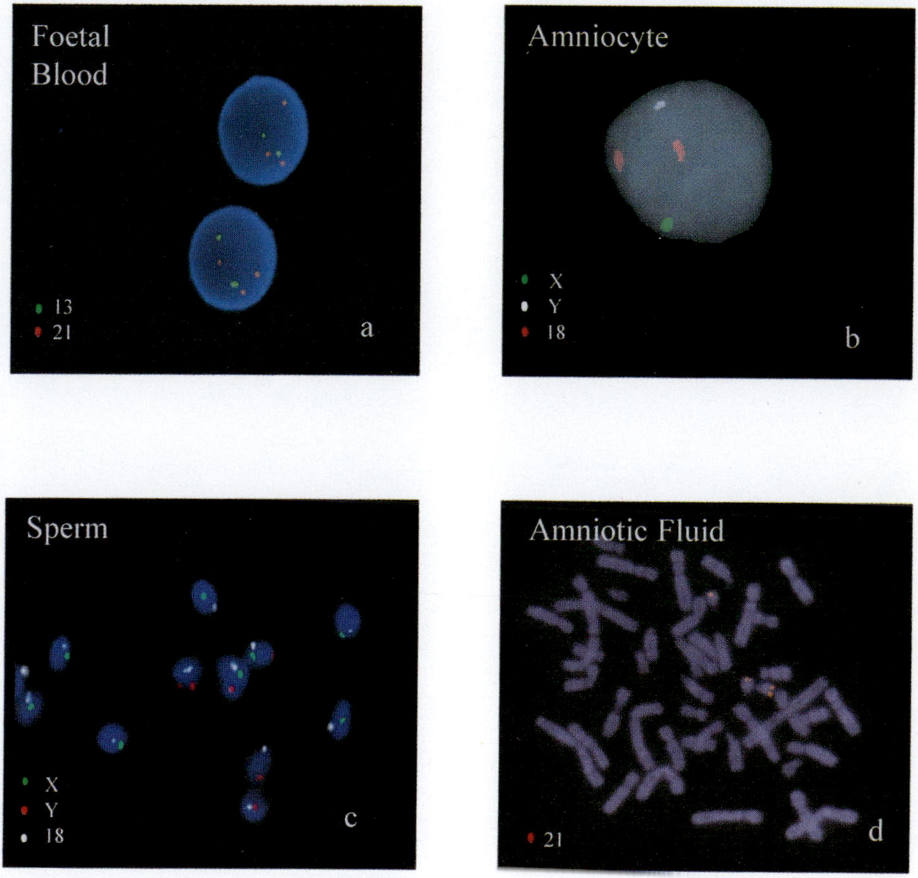

Fig 1: (a) Uncultured foetal blood showing trisomy 21 (b) Uncultured amniocytes with two signals for chromosome 18 and a single signal for X and Y (c) FISH on sperms showing abnormal disomic sperms in the center (d) Cultured foetal blood showing a metaphase with trisomy 21.

FRONTIERS IN HUMAN GENETICS
Diseases and Technologies

© 2001 by World Scientific Publishing Co. Pte. Ltd.

IN SILICO DETECTION AND PCR CONFIRMATION OF NOVEL SINGLE NUCLEOTIDE POLYMORPHISM (SNP) IN TISSUE INHIBITOR OF MATRIX METALLOPROTEINASE-2 (TIMP-2)

SI KOH[*,‡], CS LIM[*], GCH TAY[*], HM LIM[†], SGK SEAH[†], SSM TAN[†], RYY YONG[†], HM WU[†] and EPH YAP[†]

[*]*Chemical Process & Biotechnology Department, Singapore Polytechnic, 500 Dover Road, Singapore 139651.*

[†]*Defence Medical Research Institute, Defence Science & Technology Agency, 10 Medical Drive, Singapore 117597*

Single-nucleotide polymorphisms (SNP) are an important class of DNA polymorphisms as they are common, have relatively low mutation rates, and can be genotyped using rapid high throughput methods. Currently efforts are underway to discover and map SNP in the human genome. We report here a novel *in silico* approach to predict the presence of novel SNP in a specific gene, and demonstrate its utility in a candidate gene for myopia, the tissue inhibitor of metalloproteinase-2 (TIMP-2) gene. TIMP-2 proteins are inhibitors of metalloproteinases that are responsible for the degeneration of connective tissue including the sclera tissue of ocular globe. TheTIMP-2 gene is located on chromosome 17q25 and is encoded by five exons spanning 83 kb of genomic DNA. No intragenic or flanking DNA markers had previously been reported. Genomic, cDNA and EST (expressed sequence tags) sequences were obtained from DNA databases, and multiply aligned using sequence analysis software. Several putative sequence variants were found, and two were studied further by direct sequencing and PCR genotyping. New restriction sites were introduced by deliberate primer mismatches and PCR samples were genotyped by restriction digestion ("artificial RFLP"). The presence of one polymorphism in exon 3 was confirmed both by sequencing and by PCR genotyping of 187 unrelated individuals of Chinese descent. Allelic frequencies were 0.27 and 0.73, and observed heterozygosity was close to 0.40. This approach for *in silico* detection and *ex silico* confirmation of novel polymorphisms is extremely rapid and can be generalised to any gene, EST or genomic sequences. These polymorphisms may serve as useful markers for maps and for localising disease genes by linkage and association analysis.

Keywords: SNP polymorphism, sclera tissue degeneration, artificial RFLP, *ex silico* confirmation

---

‡ Corresponding author

1 INTRODUCTION

The study of DNA polymorphism (variation) provides information on our evolutionary history and on fundamental genetic mechanisms such as mutation, gene conversion and genetic drift. It also allows one to identify regions that may have functional important or related to development of rare genetic diseases. Study of human genetics has traditionally relied on classical (non-DNA) or protein polymorhisims, and important insights has been gained from direct analyses on DNA polymorphisms (viz. RFLP, dinucleotide polymorphisms, microsatellite DNA, VNTR studies) over the last ten years.

Single-nucleotide polymorphisms (SNP) are an important class of DNA polymorhpism. It has been estimated that random single base-substitutions occur once every 100 to 1000 nucleotides, approximately once in every 500 bases or so.[1] Many of these substitutions occur frequently within introns, non-coding regions (flanking regions) of genes, pseudogenes, and wobble bases of codons. Sequences of functionally importance such as promoters, enhancers, other regulatory regions and coding regions especially the first and second base of codons are much more conserved and stable unless there is a selective advantage. Current efforts are underway to discover the presence of novel SNP sites that are important in disease linkage and evolutionary studies. Because of the huge size of human genome, it is unwise to embark on the costly approach of sequencing hundreds of individuals in different loci. A more useful compromise is to adopt less expensive methods to a few chosen loci.

In this study, we attempt to analyse the single-nucleotide polymorphism (SNP) in one of the candidate genes for myopia — the tissue inhibitor of metalloproteinase-2 (TIMP-2) via a novel approach and confirm it by an ex silico approach using PCR technique. We predicted the presence of novel SNP sites in three different region of the TIMP-2 gene. TIMP-2 proteins are inhibitors of metalloproteinases that are responsible for the degeneration of connective tissue including the sclera tissue of ocular globe. TIMP-2 gene is located on chromosome 17q25 and is encoded by five exons spanning 83 kb of genomic DNA. No intragenic or flanking DNA markers had previously been reported.

2 MATERIALS AND METHODS

2.1 Genomic DNA Isolation

Blood (10 ml) was drawn into tubes with anticoagulants, and genomic DNA was isolated from several populations of Chinese descent (case controls include normal & myopes; paediatric and adult myopes; members from 42 families) using Qiagen blood kit. Concentration of DNA was assessed spectophotometrically. Working solution of DNA was prepared in nanopure water to a concentration of 100 μg/μl and stored at –20°C until needed.

2.2 Primers and PCR Condition

Confirmation of predicted SNP site at exon 3B were done by DNA sequencing as well as by PCR method using the designed primers described in Tables 1 and 2. PCR amplification was carried out on a Perkin Elmer GeneAmp PCR System 9600 thermal cycler. A 50-μl PCR reaction contained 10 mM Tris, pH 8.3, 50 mM KCl, 1.5 mM $MgCl_2$, 0.2 mM each of dATP, dTTP, dCTP, dGTP, 2.5 units of *AmpliTaq* DNA polymerase (Cetus), approximately 100 μg of target DNA and 10 μM primers. The cycling parameters were as described in Table 3.

2.3 Agarose Gel Electrophoresis of PCR Products

A volume of 8 μl of the amplified PCR products were digested by the respective restriction enzymes, *PstI* for exon 3A and *Fnu4H1* for exon 3B according to the manufacturer's instruction. The digested products were analysed by 4% (three parts Sigma typeIA agarose to one part of FMC Nusieve agarose) gels. Gels were stained in ethidium bromide and photographed under UV light. The expected genotypes of each locus are detailed in Table 2.

Table 1 Base sequence of designed primers and the estimated T_m

Primer	Sequences / Artificially Generated RE Sites	Length (nucleotide)	T_m °C
E1 (forward)	5' GCC CCC GAG ACA AAG AGG AG 3'	20	61
E2 (reverse)	5' GCA TTG CAA AAC GCC TGCTG 3' *Fnu4H1 site* 5'…GC↓AGC.. 3'	20	57
E3A1 (forward)	5' ACA CGG CCC CCT CCT CTG 3' *Pst* I site 5' ..CTGCA↓G.. 3'	18	57
E3A2 (reverse)	5' GGC TGA TGG CCC CAC TCA 3'	18	55
E3B1 (forward)	5' TGG GAA CGG AAT TCA CCA A 3'	19	51
E3B2 (reverse)	5' TTC TTT CCT CCA ACG TGCAG 3' *Fnu4H1* site 5' ..GC↓TGC.. 3'	20	55

Table 2 Primer characteristics and expected products after RE treatment.

Primer	Length (nt)	Predicted SNP Site / Position	Engineered RE site at the Primer	Amplicon	Possible Size of Alleles After Digestion
E1	20	None	None	Exon 1 299 bp	299 (allele 1) 280 + 19 (allele 2)
E2	20	2641th nt: A (non-cutter) G (cutter)	*Fnu4H1* GC↓ NGC		
E3A1	18	139th nt: T (non-cutter) C (cutter)	*PstI* CTGCA↓G	Exon 3A 97 bp	97 (allele 1) 20 + 77 (allele 2)
E3A2	18	None	None		
E3B1	19	None	None	Exon 3B 157 bp	157 (allele 1) 139 + 18 (allele 2)
E3B2	20	155th nt: A (non-cutter) G (cutter)	*Fnu4H1* GC↓ NGC		

Table 3 PCR amplification conditions for the three primer pairs

Steps	Exon 1 Amplification		E3A Amplification		E3B Amplification	
Primer pairs	E1 and E2		E3A1 and E3A2		E3B1 and E3B2	
Pre-heating	95°C	2	95°C	2	95°C	5
Denaturing	95°C	1	95°C	1	95°C	1
Annealing	72°C	1	60°C	1	58°C	1
Extension	72°C	1	72°C	1	72°C	1
No of cycle	35 cycles		35 cycles		35 cycles	
Final extension	72°C	5	72°C	5	72°C	5
Soaking	4°C	Infinite	4°C	Infinite	4°C	Infinite

2.4 Confirmation of Exon 3B SNP by Sequencing

DNA sequencing were carried out on three individuals of genotypes 11, 12 and 22 using the fluorescent ABI PRISM™ Dye Terminator Cycle Sequencing Kit. The gel was run in an ABI PRISM™ automated DNA sequencer.

3 RESULTS AND DISCUSSION

3.1 Predicted SNP Sites on Exons 1 and 3 of TIMP-2 Gene

We obtained four mRNA sequences for human TIMP-2 from the Genbank databases, (HUMANTIMP3, HUMMET, HUMTIMP2 and HSTIMP2M) and aligned these sequences to look for polymorphisms along the five exons of TIMP-2. By adopting biocomputing approach, we predicted three potential SNP loci located in exons 1 and 3 of TIMP-2 gene. Two of the G-

A SNP loci found in exons 1 and 3B are silent (i.e. no change in amino acid) whilst the other T-C SNP loci located in exon 3A results in an amino acid substitution (alanine ↔ valine; Fig. 1).

Exon 1 SNP: G-A polymorphism at position 2641th nucleotide is a silent SNP.

Primer E2

*Fnu4H*1 site

------- gtg cac cc**G** c↓ag cag g--------(cutter)

---------------- cc**A** -----------------(non-cutter)

Protein level: ---------------- Pro -------------------

No change in amino acid (silent SNP).

NB: We have not been able to establish the existence of this SNP as the PCR was not successful.

Reference: Genbank Accession No. U44381 (Exon 1)

Exon 3A SNP:
C-T polymorphism at position 139th nucleotide causes an amino acid substitution.

Primer 3A1

Pst I site

------gcc ccc tcc tcg tct g**C**a↓gtg tgt ggg------(cutter)
(Ala)

---------------------------- g**T**a ---------------(non-cutter)
(Val)

Protein level: Ala ↔ Val substitution

Exon 3B SNP: G-A polymorphism at position 155th nucleotide is also a silent SNP.

Primer 3B2

*Fnu4H*1 site

----------gtc tc**G** c↓tg gac---------------------- (cutter)

----------gtc tc**A** c tg gac------------------ (non-cutter)

Protein level: ------------- Ser -------------------
No change in amino acid (silent SNP).

Reference: Genbank Accession No. U44383 (Exon 3)

Fig. 1 Predicted three potential SNP sites at TIMP-2 gene based on Genbank data (HUMANTIMP3, HUMMET, HUMTIMP2)

3.2 Designing Primers to Create Artificial RFLP to Map SNP Loci

Restriction sites mapping and designing of modified primers were carried out using DNAStar Laser gene softwares. One of the bases at primers E2, E3A1 and E3B2 was modified in order to generate artificial *Fnu4H1*, *PstI* and *Fnu4H1* sites respectively near their 3' ends. The expected genotypes of the three SNP loci are detailed in Fig. 1.

3.3 Confirmation of Predicted SNP Loci by PCR

(i) Exon 1 SNP locus — unable to reconfirm using PCR

Several attempts were used to amplify the exon 1 SNP region using a combination of DMSO, etc., but we failed to amplify the expected PCR products. On analyses, we discovered that our exon 1 forward and reverse primers could assume a number of intra- and inter-secondary structures. A total of 4 hairpins and 12 dimers for forward primer whilst 8 hairpins and 9 dimers were found in the reversed primer (data not shown). The extensive secondary structures of the primer pairs may have hindered the amplification process.

(ii) Exon 3A SNP locus shows no significant difference in 23 individuals by PCR genotyping

We were unable to detect any significant differences amongst 23 unrelated individuals with respect to this SNP locus on exon 3A. All the PCR products from a small population screened were all cleaved by enzyme *Pst*I to give a 20 and 77 bp fragments (Fig. 2) implying that they all have a cytosine (C) residue at position 139 in both the alleles. As the SNP site in this region causes an amino acid substitution (Ala ↔Val), it may be a potential useful SNP marker. More works need to be done on screening a larger population including other ethnic groups before we exclude this SNP locus as potential useful marker for myopia linkage and association analyses.

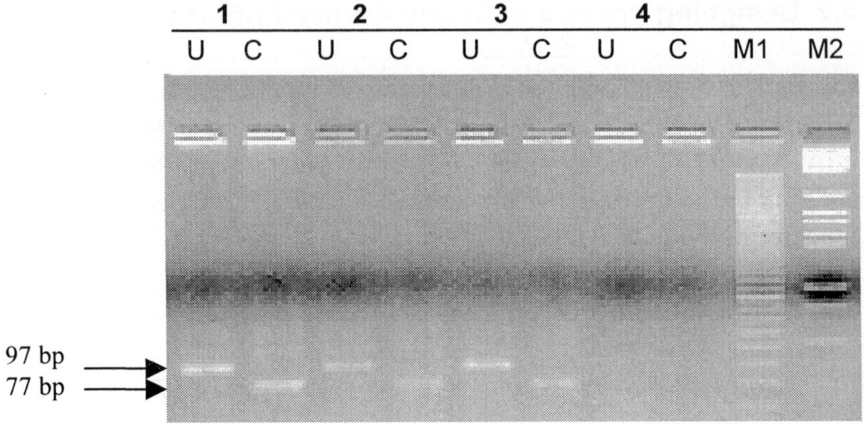

Fig. 2 A representative gel showing no significant difference between individuals of the population screened with respect to exon 3A *Pst*I SNP. Note that all the samples are cleaved by *Pst*I enzyme. 1-4 represent four unrelated individuals whilst M1, 20 bp marker and M2 is pGEM DNA marker.

(iii) Novel SNP site found at exon 3B

Table 4A shows the genotype data collected from 187 unrelated individuals of Chinese descent. Three different genotypes were evident (Fig. 3) and we use digital representation to denote the two alleles (allele 1 remains uncut whilst allele 2 is cut by the enzyme). Hence, homozygotes genotype 11 has both alleles remained uncut by enzyme *Fnu4H1* (expected size: 157 bp) whilst genotype 22, has both alleles being cut (shorter band, 139 bp). Heterozygotes, 12, has one of its alleles cut (139 bp) and the other remains uncut (157 bp).

3.3 Estimation of Allele Frequencies and Statistical Analyses

Since the above experiment unequivocally distinguished the alleles in this locus, allelic frequencies were estimated by the gene count method, which are also the maximum likelihood estimates. Agreement with the Hardy-Weinberg expectations of genotype frequencies was determined by the chi-

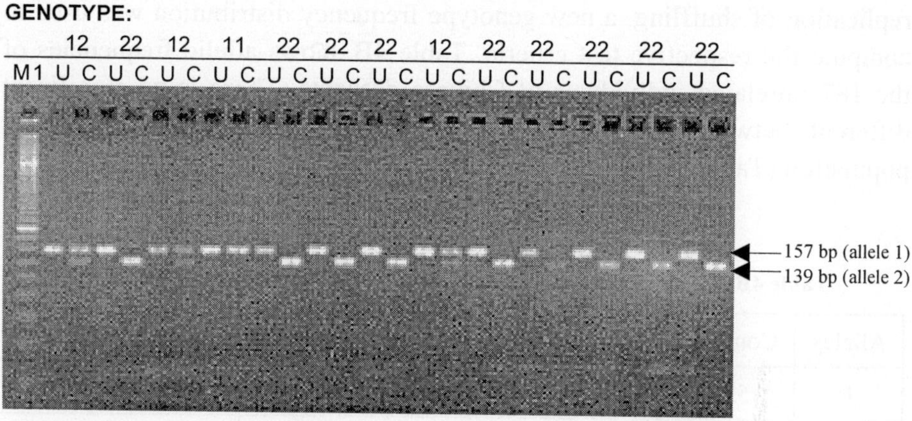

Fig. 3 A representative gel showing different genotypes of G-A polymorphisms in exon 3B SNP locus. All three genotypes (11, 12 and 22) are evidence in this gel. M1 represents 20 bp DNA size marker.

Table 4A Genotype distribution amongst 187 unrelated case normal and myope control groups

Genotypes	SNP (Exon 3B)	Control	Myopes	Total
11	A/A	6	7	13 (6.99%)
12	A/G	41	35	76 (40.6%)
22	G/G	45	53	98 (52.41%)
Total		92	95	187

Effective number of alleles in this sample population
= Reciprocal of the observed homozygosity = 1/ (0.5241 + 00699) = 1.78.

square test based on total heterozygosity at the locus. The levels of significance of the test procedure were determined empirically by a permutation-based simulation method. Briefly, it involved reconstruction of genotype frequencies by random shuffling of the two alleles. From every replication of shuffling, a new genotype frequency distribution was used to compute the respective test criteria. Table 4B shows allelic frequencies of the 187 unrelated individuals of Chinese descents. There is no significant different between the normal control and the myopic groups in this population (Table 4C).

Table 4B Allelic frequencies in the case studies of 187 unrelated individuals

Alleles	Control	Myopes	Total	Allele Frequencies	Genotype Frequency
1	53	49	102	0.273	0.07453
2	131	141	272	0.727	0.5285
Total	184	190	374	1.00	0.6031
					Heterozygosity = 0.3969

Table 4C Testing for significant difference between the normal and myope controls

Genotypes	Control % Freq.	Myopes % Freq.	Deviation	$\frac{\text{Deviation}^2}{\text{Control}}$
11	6.52	7.37	+ 0.85	0.1108
12	44.56	36.84	- 7.72	1.3375
22	48.91	55.79	+ 6.88	0.9678
Total	100	100		$\chi^2 = 2.4161$
Degree of freedom = 2; Probability = 20-30%				

Comments:

As this is not a significant χ^2 value, so there is no significant difference between the two groups of individuals studied. Thus, we conclude that there is no case association of this locus to the linkage of myopia.

52.41% (98 out of 187 individuals of Chinese descent) has a GC basepair at position 155 in both their alleles whilst 40.6% (76 out of 187) has a GC pair in one of the two alleles (Table 4A). Only about 7% of the population surveyed have an AT basepair at the equivalent position. The genotypes in this sample are conforming to the frequencies expected for a Hardy-Weinberg population within statistically acceptable limits (Table 4D).

Table 4D Testing for Hardy-Weinberg equilbrium of the population of 187 unrelated individuals

Genotypes	Observed cases	Expected cases (based on allele frequencies)	$\frac{(Obs - Exp)^2}{Exp}$
11	13	14	0.0714
12	76	74	0.05405
22	98	99	0.0101
Degree of freedom = 2; Probability = 90-95%			$\chi^2 = 0.1356$

Comment:
This is not a significant value and we may accept the hypothesis that the sample (hence presumably the population from which it was drawn) is conforming to the equilibrium distribution of genotype.

The allelic frequencies of the G-A SNP in exon 3B were calculated to be 0.73 (G) and 0.27 (A), respectively, with a heterozygosity of approx. 0.40 (Table 4D) based on the data observed in 187 unrelated individuals.

3.4 Other Populations Surveyed at This SNP Locus in Exon 3B

Data were also collected from another set of local Chinese population comprising 42 individuals from 21 families (fathers and mothers, presumably they are non-related) as well as 41 adults and 249 paediatric myopes. We computed the allele frequencies and determined whether the genotypes distribution in these different populations are in HWE equilibrium. Our results show that all the three groups produce non-significant chi-value and hence, the sampled population are all conforming to the HWE (Tables 5, 6 and 7). Table 8 summarises the allelic frequencies and the percentage heterozygosities in all the four different populations surveyed.

Table 5A Genotype distribution amongst 42 unrelated individuals (fathers and mothers of 21 families) with unknown phenotypes

Genotypes	SNP	No of individuals	Observed Genotype Frequency
11	A/A	2	4.76%
12	A/G	20	47.6%
22	G/G	20	47.6%
Total		42	100%

Effective number of alleles in this sample population
= reciprocal of the observed homozygosity = 1/ (0.476 + 0.0476) = 1.9.

Table 5B Allelic and genotype frequencies of the 42 unrelated individuals from 21 families (fathers and mothers)

Allele	No of Occurrence	Allele Frequency	Genotype Frequency	Expected Genotype number (Based on allelic Frequency)
1	24	0.2857	0.08162	Homozygotes 11: 3
2	60	0.7143	0.5102	Homozygotes 22: 21
	84	1.000	Ho = 0.5918	Heterozygotes 12:18
			Heterozygosity (He) = 0.4082	

Table 5C Testing for Hardy-Weinberg equilibrium of the above 42 unrelated individuals

Genotypes	Observed Cases	Expected Cases (based on allele frequencies)	Obs – Exp	$\frac{(Obs - Exp)^2}{Exp}$
11	2	3	-1	0.3333
12	20	21	-1	0.04762
22	20	18	+2	0.2222
Degree of freedom = 2; Probability = 70 – 80%			$\chi^2 = 0.6031$	

Comment:
We observed that the distribution of genotypes in the population studied conforms to Hardy-Weinberg equilibrium (HWE).

Table 6A Genotype distribution of paediatric myopes

Genotype	SM1 Plate	SM2 Plate	SM3 Plate	Total
11	2	4	4	10
12	37	32	27	96
22	55	58	57	170
Total	94	94	88	249

Table 6B Allelic frequency distribution of paediatric myopes

Allele	No of Occurrences	Allele Frequencies	Genotype Frequencies
1	116	0.2101	0.04414
2	436	0.7899	0.6239
Total	552	1.000	0.66804
			Heterozygosity = 0.33196

Table 6C Testing for Hardy-Weinberg equilibrium of the above 249 paediatric myopes

Genotypes	Observed Cases	Expected Cases (based on allele frequencies)	Obs – Exp	$\frac{(\text{Obs} - \text{Exp})^2}{\text{Exp}}$
11	10	11	–1	0.0909
12	96	83	+13	2.0361
22	170	155	+15	1.452
	Degree of freedom = 2; Probability = 10 – 20%		$\chi^2 = 3.5786$	

Comment:
Though the probability value is low, it is still within the acceptable statistical range. Hence, we conclude that the distribution of genotypes in the population studied conform to Hardy-Weinberg equilibrium (HWE).

Table 7A Alleles and genotypes frequencies distributions in adult myopes sample population

Genotype	MF Plate	Alleles	No. of Occurrences	Allele Frequencies	Genotype Frequencies
11	5	1	25	0.3049	0.09295
12	15	2	57	0.6951	0.48316
22	21	Total	82	1.0000	0.5761
Total	41			Heterozygosity = 0.4239	

Table 7B Testing for Hardy-Weinberg equilibrium of the above adult myope population

Genotypes	Observed Cases	Expected Cases (based on allele frequencies)	Obs – Exp	$\frac{(\text{Obs} - \text{Exp})^2}{\text{Exp}}$
11	5	4	+1	0.25
12	15	17	-2	0.2353
22	21	20	+1	0.05
	Degree of freedom = 2; Probability = 70 - 80%		$\chi^2 = 0.5353$	

Comment:
This population has a very high probability to be the same as the expected frequencies of genotypes. Hence, we conclude that the distributions of genotypes in this population are in accordance to Hardy-Weinberg equilibrium (HWE).

Table 8 Summary of the allele frequencies in different populations

Allele	Control	Myopes	Paediatric Myopes	Adult Myopes	FM Studies
1	53	49	116	25	24
2	131	141	436	57	60
No of individuals studied	92	95	276	42	42
Heterozygosity	0.4099	0.3828	0.33196	0.4239	0.3320
Average heterozygosity			0.3761		

3.5 Confirmation of SNP at Exon 3B by Sequencing

The three genotypes of exon 3B SNP comprises two homozygotes (11 and 22) and the heterozygote (12) were confirmed by fluorescence sequencing method using ABI PRISM™ Dye Terminator Cycle Sequencing Kit and a ABI PRISM™ automated DNA Sequencer. This confirms our prediction of the presence of the G-A polymorphism in exon 3B of TIMP-2 gene.

3.6 Myopic Families Studies

By the use of digits to denote alleles in the population, we can in principle analyse parentage based on simple Mendelian rules. For example, if the mother has genotype 11, and the child 12, then the true father must have contributed allele 2 and be of genotype 12 or 22, but not 11. With this principle, we have observed that the inheritance patterns of this SNP locus in exon 3B follows Mendelian inheritance for all our 90 individuals obtained from 22 different Chinese families. Figure 4 illustrates the transmission of the SNP exon 3B locus in a Chinese family with 3 children.

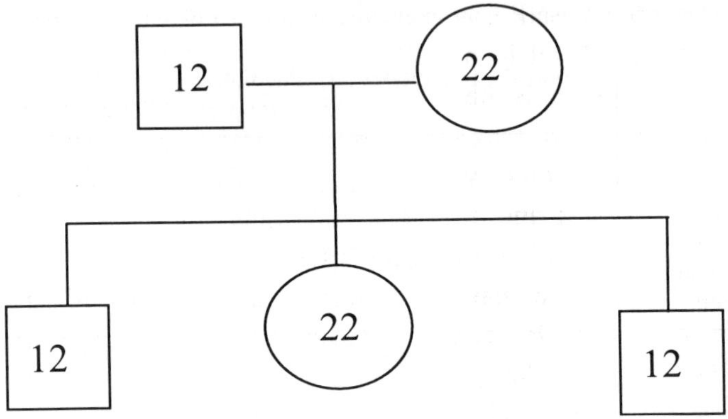

Fig. 4 Transmission of exon 3B SNP in a Chinese family with three children. Note that its inheritance pattern is typical to that of Mendelian genetics.

3.7 SNP Polymorphism and Its Implications

Single basepair substitutions occur much more frequently than VNTR loci, STRs, GC or CA dinucleotide repeats, etc. The rate has been estimated to be on an average of one nucleotide per every 500 bp.[1] It is believed that two unrelated individuals are likely to differ at several million nucleotide sites and that alternations at approximately 95% of the sites in the entire genome have very little effect on Darwinian fitness. SNP sites are common in human genome but their potential usefulness as genetic markers for case and linkage analysis are yet to be established. This study demonstrates the rapid method of screening for useful SNP, which can be generalized to any gene and any disease association analysis.

4 CONCLUSIONS

Of the three SNP loci predicted, we are unable to establish the exon 1 SNP G-A polymorphism at this moment as the PCR condition has not been worked out despite several attempts.

As for exon 3A SNP C-T polymorphism, it does not seem to exist at the 0.05 frequency level in our Chinese descent. However, since we only screened 20+ individuals here, the data may not truly reflect its non-existence. This SNP site may be potentially important, because it results in an Ala to Val substitution or vice versa. More works are therefore needed to screen a larger Chinese and other ethnic populations before we exclude this site to be a functional polymorphism marker.

In conclusion, we have discovered and confirmed it by DNA sequencing a new SNP at exon 3B in our local Chinese population. However, this SNP marker does not show any linkage or association to myopia, as there is no significant between normal and myopes population in terms of its genotypes distribution. The genotypes distribution at this locus does conform very well with the Hardy-Weinberg equilibrium suggesting that it could serve as a useful genetic marker for other genetic or evolutionary studies. It may be worthwhile to screen other ethnic populations at this SNP locus before we exclude exon 3B SNP to case association to myopia. Nevertheless, this study shows the potential application of biocomputing technique to speed up the discovery of genetic markers in linkage and association analysis. The approach adopted here to discover useful SNP sites is extremely rapid and could be generalised to any gene, EST or genomic sequences. With the same approach, we may generate more useful SNP sites at other regions of TIMP-2 gene for future myopic case association analysis.

Acknowledgements

The authors would like to acknowledge the financial supports provided by the Singapore Polytechnic, Singapore Totalisator Board (11-27801-45-2458) and the Defence Medical Research Institute (DMRI). In addition, we would like to thank the staff of the Singapore Eye Research Institute for helping us to collect the blood specimens used in this study. We thank the staff of DMRI for allowing us to work in their well-equipped laboratories. Their most cordial companion-ship, constructive discussion, and technical assistance during the course of the project were very much appreciated.

References

1. Kimura M,. Neutral evolution. *International Symposium on "Evolution of Life"* 1990, held in Kyoto, Japan. 26–28.

2. Bardien S, Ramesar R, Bhattacharya S, Greenberg J, Retinitis pigmentosa locus on 17q (RP17): fine localisation to 17q22 and exclusion of the PDGE and TIMP2 genes. *Hum Genet* **101** (1): 13–17.

3. Hammani K, Blakis A, Morsette D, Bowcock AM, Schmutte C, Henriet P, DeClerck YA, Structure and characterization of the human tissue inhibitor of metalloproteinase-2 gene. *J Biol Chem* 1996; **271**:25498–25505.

4. Rada JA, Brenza HL, Increased latent gelatinase activity in the sclera of visually deprived chicks. *Invest Ophthal Visual Sci*, 1995; **36** (8): 1555–1564.

5. NCBI database accession number J05593; U44381 and U44383.

6. Stetler-Stevenson WG, Brown PD, Onisto M, Levy AT, Liotta LA, Tissue inhibitor of metalloproteinase-2 (TIMP-2) mRNA expression in tumor cell lines and human tumor tissues. *J Biol Chem* 1990; **265** (23): 13933–13938.

FRONTIERS IN HUMAN GENETICS
Diseases and Technologies

© 2001 by World Scientific Publishing Co. Pte. Ltd.

A NOVEL METHOD OF GENOTYPING SINGLE NUCLEOTIDE POLYMORPHISMS (SNP) USING MELT CURVE ANALYSIS ON A CAPILLARY THERMOCYCLER

SOON-MENG ANG and ERIC P.H. YAP*

Defence Medical Research Institute, Defence Science & Technology Agency, 10 Medical Drive, Singapore 117597

**E-mail: nmiv3@nus.edu.sg*

We report the development of a homogenous assay for the genotyping of single-nucleotide polymorphisms (SNPs), utilizing a fluorescent dsDNA-binding dye. Termed T_M-shift genotyping, this method combines multiplex allele-specific PCR with sequence differentiation based on the melting temperatures of amplification products. Allele-specific primers differing in length were used with a common reverse primer in a single-tube assay. PCR amplification followed by melt curve analysis was performed with a fluorescent dsDNA-binding dye on a real-time capillary thermocycler. Genotyping was carried out in a single-tube homogeneous assay in 25 minutes. We compared the accuracy and efficiency of this T_M genotyping method with conventional restriction fragment genotyping of a novel single nucleotide polymorphism in the *Jagged1* (*JAG1*) gene. The flexibility, economy and accuracy of this new method for genotyping polymorphisms could make it useful for a variety of research and diagnostic applications.

Keywords: genotyping, SNP, allele-specific, multiplex, fluorescent

---

* Corresponding author

1 INTRODUCTION

Disease susceptibility and severity, in addition to many human traits such as drug handling ability, have been linked to single nucleotide polymorphisms (SNPs) at many distinct loci. Susceptibility genes have been identified in multigenic multifactorial diseases such as familial Creutzfeldt-Jakob disease, malaria, breast cancer, and familial hyperinsulinism. SNPs are also being used for association and linkage studies of candidate genes for polygenic diseases and in genome-wide scans[1] . Owing to their utility in disease gene mapping and their potential biological significance, SNPs in coding, intronic, promoter and intergenic sequences are being rapidly identified, either *in silico* by biocomputing prediction, or by mutation screening and sequencing. However, current limitations lie in the verification of these candidate SNPs, and in the rapid and economical genotyping of large numbers of samples. A biocomputing approach for novel SNPs discovery can facilitate the initial discovery process, and a standard PCR-RFLP method can be used for verification purposes. However, low-cost techniques for rapid genotyping are needed.

The *Jagged1* gene (*JAG1*) on chromosome 20p12 encodes a transmembrane receptor homologous to the Notch protein in *Drosophila melanogaster*, a protein with multiple roles in cell differentiation and cell fate decisions. The systemic abnormalities seen in patients with Alagille syndrome (AGS), an autosomal dominant disease with liver, heart and vertebral formation abnormalities have been linked to mutations in this gene[2]. A novel SNP, C3417T, was predicted in exon 26 of the *Jagged1* gene by alignment and filtering of EST and genomic sequences (Ang & Yap, unpublished), and confirmed by sequencing of genomic DNA samples from a local population. DNA samples had also been genotyped by PCR-RFLP which verified the polymorphisms at these positions[3].

We describe here the development of an allele-specific genotyping method combining multiplex PCR with melting temperature (T_M) analysis on a real-time fluorescent thermocycler. Using a fluorescent dsDNA-binding dye, SYBR® Green I, a method was developed which required minimal optimization, and which can be applied to other polymorphisms of interest. Amplification and melting curve analysis were carried out in the same capillary tube without manual intervention making this homogeneous assay ideal for applications requiring high throughput genotyping.

2 METHODS

2.1 Genomic DNA

Blood samples were drawn from 46 random individuals of Chinese ethnic backgrounds, and DNA samples extracted using the Qiagen Blood Midi Kit (Qiagen). DNA samples were quantitated on a UV-spectrophotometer at 260nm-280nm, and 5-100ng of DNA used in each 10 μl reaction.

2.2 Allele-Specific Primer Design

Allele-specific primers were designed in a fashion such as seen in Fig 1. The bases at the 3' end of the forward primers were designed to be either C allele-specific, or T allele-specific. The T-specific primer (15T) was 15nt long, while the C-specific primer was designed to be 20nt long. Two types of C-specific primers were tried, one with the entire sequence "complementary" to the target sequence (20C), and the second with an arbitrary "generic" 5' tail comprising G and C (20GC). A common 15-mer reverse primer (15R) was used with both forward primers in a multiplexed assay. To aid allele discrimination, the Stoffel fragment of *Taq* DNA polymerase was used to prevent 5' exonuclease activity interfering with the allele-specific amplification[4].

2.3 PCR Product Amplification and Melting Curve Analysis

All PCR reactions were performed in 10μl volumes with 10ng of genomic DNA in composite glass-plastic capillaries on the real-time fluorescent PCR machine, the LightCycler™ (Roche Diagnostics). Allele-specific PCR reactions were individually optimized with known controls to give specific products without primer-dimer or other non-specific amplification. Hotstart PCR reactions were set up as follows: 0.09 μM of either the generic-tailed (20GC) or the complementary (20C) C-specific forward primer, 0.36 μM of the shorter T-specific forward and the common reverse primer; 2 units of

Stoffel polymerase (Perkin Elmer) incubated with an equivalent amount of TaqStart™ antibody (Clontech); 1X Stoffel buffer (Perkin Elmer, 10 mM KCL, 10 mM Tris-HCl at pH 8.3); 2.5 mM $MgCl_2$; 250 μM each dATP, dCTP, dGTP and dTTP; 500 μg Bovine Serum Albumin (New England Biolabs); 0.05% Tween-20 (Roche Molecular Biochemicals); and 1X SYBR® Green I (FMC Bioproducts). The generic-tailed C-specific forward primer was GCCGGCCCATCAAGGATTAC (20GC), the complementary C-specific forward primer was CGGTCCCCATCAAGGATTAC (20C); the shorter T-specific forward primer CCCATCAAGGATTAT (15T); and the common reverse primer GGAGTTCTTGTTCTC (15R).

For the generic 20GC allele-specific amplification (20GC/15T-15R), an initial incubation of 90 sec at 95°C was followed by a PCR step of 0 sec denaturation at 95°C followed by 6 sec annealing/extension at 51°C for 5 initial cycles, after which the denaturation temperature was lowered to 85°C for 0 sec and 51°C for 6 sec in a run of 30 more cycles. Online analysis of fluorescence was performed during PCR. After the final denaturation, the DNA was then reannealed for 30 sec at 56°C, followed by gradual melting at 0.2°C/sec till 85°C. The PCR products were then cooled at 40°C for 30 sec. Fluorescent emission in each capillary was acquired by the software provided by the manufacturer and monitored onscreen during each annealing/extension step and continuously monitored during the melting program. SYBR Green I binds preferentially to dsDNA with a fluorescent emission wavelength which is monitored in the F1 channel of the LightCycler™. As the amount of dsDNA varies during PCR and melting curve analysis, this dsDNA dye can be used to monitor the quantity as well as sequence of the PCR products.

The second method using the complementary 20C primer (20GC/15T-15R) required a slightly different thermal protocol. This was because the optimal annealing temperatures for 20GC and 15T were different. After an initial denaturation of 120sec at 95°C, alternating cycles (which we termed '*flip-flop*') were used to amplify both the shorter T-products as well as the longer-tailed C-products under conditions optimal for each set of primers. This consisted of denaturation at 95°C for 0 sec followed by 7 sec annealing at 48°C, then denaturation at 95°C for 0 sec with a following step of 58°C for 4 sec. This was performed for 8 cycles

(8x2). This methodology had the effect of priming and amplifying a template base from which further amplification can take place without the preferential binding of the longer-tailed forward primer over the shorter one. 35 cycles of amplification were then carried out at a lower 85°C denaturation for 0 sec with a 51°C 7 sec annealing step. Subsequent steps were the same as above.

2.4 Data Analysis

Melting curves (Figs 2a and b) were generated automatically using the polynomial function of the software. The plots of raw fluorescence (F) data against cycle number, and -dF/dT against temperature allowed monitoring of product yield and melt profiles respectively. Unlike CCD-based sequence detectors, the capillary-to-capillary and cycle-to-cycle variation in the LightCycler™ was minimal, and raw fluorescence data was monitored above the initial baseline fluorescence in each capillary. Melting peaks (comparable to T_M) and peak heights were automatically calculated by the software.

Scatter graphs were constructed for the 46 samples genotyped with both the generic and the genomic methods. Two peaks, one at 71-72°C represented the shorter T-specific forward primer in combination with the reverse primer, the other 75–77°C representing either the GC-tailed 20GC or 20C forward primer with the reverse primer was observed. For plotting of the scatter graphs, the two peaks were compared relative to each other, and either peak was taken as a fraction of the larger peak. In both these cases the -dF/dT fluorescent peak value at 68°C was taken as the baseline and peaks above this value at 71, 77°C were then calculated. Therefore in Fig 2, homozygotes for the T-allele had only T-specific PCR products which consisted of 71–72°C peaks only, heterozygotes had both T-specific and C-specific peaks of 71–72°C and 75–77°C, while C-homozygotes had only C-specific 76–77°C products.

3. RESULTS

3.1 Assay Optimization

We used a multiplex format involving the use of two allele-specific primers and a common reverse primer so that genotyping could be performed in a single tube. A 15nt T-allele specific primer (15T) was combined with a 20nt C-allele specific primer (20C for the complementary tail, 20GC for the non-complementary tail). The polymorphic base was positioned at the 3' end of the primer for maximum allele-specificity (Fig 1). The entire amplicon was kept as small as feasible (about 35-50bp), so that the 5bp difference in amplicon sizes made a significant difference to melting temperature.

Optimization of both these assays differed due to the composition of the 5' tail. The 20C/15T-15R assay utilizing the 20C complementary primer with the 15T primer and the common reverse primer was optimized using a molar ratio of 1:4 of 20C:15T primers in order to reduce the preferential binding of the longer primer. However, the 2 primer-pairs had different optimal annealing temperatures, making it difficult to find a common protocol in which both 20C-15R and 15T-15R primer pairs were equally optimal. To overcome this problem, we developed a novel protocol which we have termed "*flip-flop*" PCR, where in the initial cycles of the PCR, alternating cycles had different annealing temperatures corresponding to the temperatures at which each primer pair worked best. Two alternate cycles of 95°C–58°C and 95°C–49°C were used for the first 8 cycles. At 58°C only the 20C-15R primer pair would anneal, and at 49°C although both 20C-15R and 15T-15R primer pairs would bind, the 4:1 molar ratio in favour of the 15T primer would bias the reaction towards the T allele. This would have the effect of amplifying each allele by about 2^8 fold.

Subsequent cycles then utilized a denaturation step of 85°C which was sufficient for denaturation of PCR products but not genomic DNA, therefore increasing the specificity of this assay[5]. The 51°C annealing temperature allowed both primer pairs to work at approximately equal efficiency when amplifying the PCR products formed during the initial cycles.

The optimization for the 20GC/15T-15R assay utilizing the 20GC generic tail was more straightforward. In theory both the 20GC and 15T primers would have more similar annealing temperatures, and amplification of the genomic DNA in the initial cycles could be carried out without having to use the "*flip-flop*" protocol to overcome the problem seen with the other assay. However, similar bias for the longer 20GC primer was seen in the amplification of PCR products as the 5' five nucleotides would be incorporated into products. Hence excess 15T primer was used at a molar ratio of 4:1. In both these assays it was also important that the 20GC and 20C primers were used at limiting concentrations.

Potential problems of primer-dimer products, which could be confused with the desired product, were avoided by the use of a hot-start protocol such as using TaqStart™ antibody.

These assays were tested for their effectiveness in genotyping samples at different DNA concentrations, and though both these assays were effective from 5–100 ng of genomic DNA, it was found that the optimal amount of genomic DNA required was 10 ng.

The allele-specificity of these primers was tested in combinations containing only one of each forward primers, such as 20GC-15R, 20C-15R, and 15T-15R sets. In separate experiments, these primer sets would only amplify the specific alleles. Non-specific amplification occurred after 1.5–2 times the number of PCR cycles required for the specific templates. A negative sample containing all the reagents but without DNA was used in each experiment to monitor the accumulation of primer-dimer products; with the use of TaqStart™ antibody, primer-dimers were not detectable.

3.2 SNP Genotyping

We tested the accuracy and efficiency of genotyping with both complementary-based and generic-based assays. The three genotypes for the *JAG1* C3417T polymorphism yielded distinct melting peaks in both assays, the 15T-15R product had a melting peak of 71–72°C, while the longer 20C-15R and 20GC-15R products had melting peaks of 76–77°C. These temperatures were reproducible in all experiments, with the 1°C variability due to pipetting errors affecting salt concentrations or fluorescence.

Using the 20GC/15T-15R assay, we genotyped 46 random DNA samples from ethnic Chinese which had been genotyped previously at the same SNP using PCR-RFLP. These samples were genotyped in a "single-blind" format where the results were compared to the previously obtained results. On the basis of melting curve analysis of PCR products post-amplification, all 8 homozygote CC, 21 heterozygote CT and 17 homozygote TT samples were genotyped correctly (Fig 2a). In the heterozygote, the mean height of the secondary peak as a fraction of the major peak was 0.514 (SD = 0.210). This is shown in the scatter plot of Figure 3a in which the 3 genotypes are well clustered.

Using the 20C/15T-15R assay, the same 46 DNA samples were also correctly genotyped (Fig 2b). In the heterozygote, the mean height of the secondary peak as a fraction of the major one was 0.432 (SD = 0.0753). (Fig 3b).

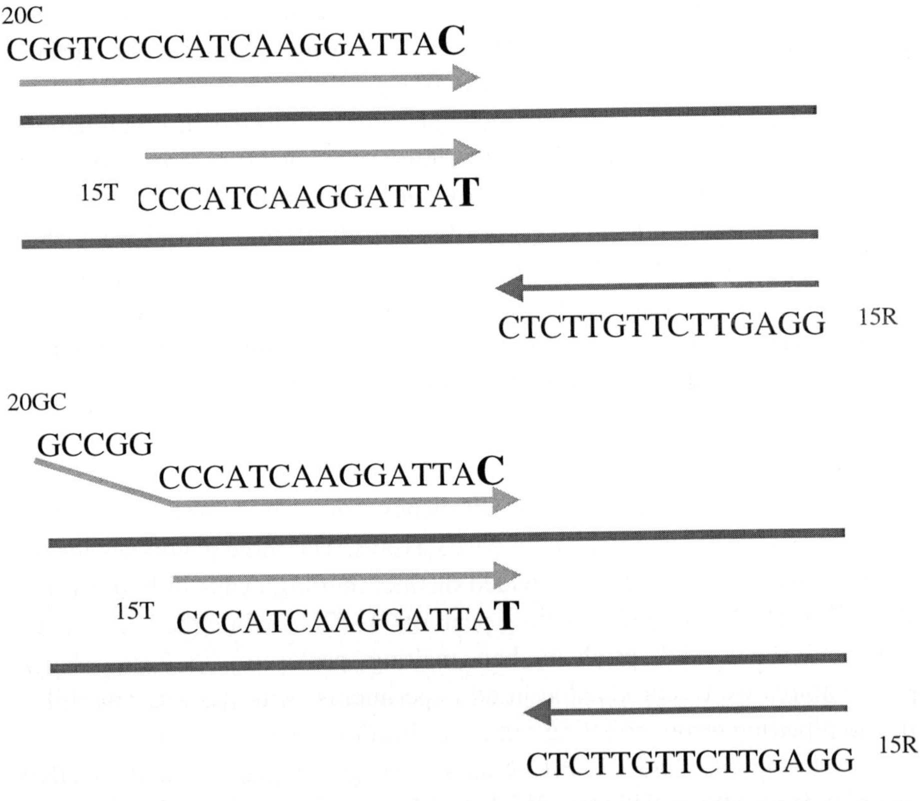

Fig 1 Design of allele-specific primers, the complementary 20C forward primer, the generic 20GC primer, the 15T primer and the 15R common reverse primer.

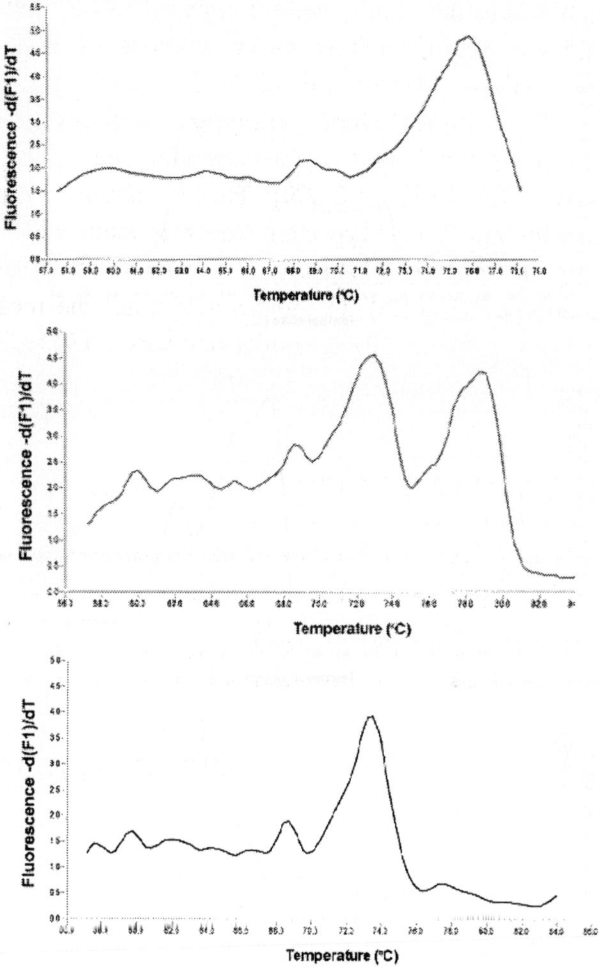

(a) Generic assay

Fig. 2 Melting peaks (-dF/dT) corresponding to the T_M of the PCR products. Amplification with the 20-mer generic or complementary C-specific primers in combination with the 15-mer reverse primer resulted in a melting peak at 76-77°C, while amplification of the 15-mer T-specific primer with the 15-mer reverse primer resulted in a melting peak at 71-72°C.

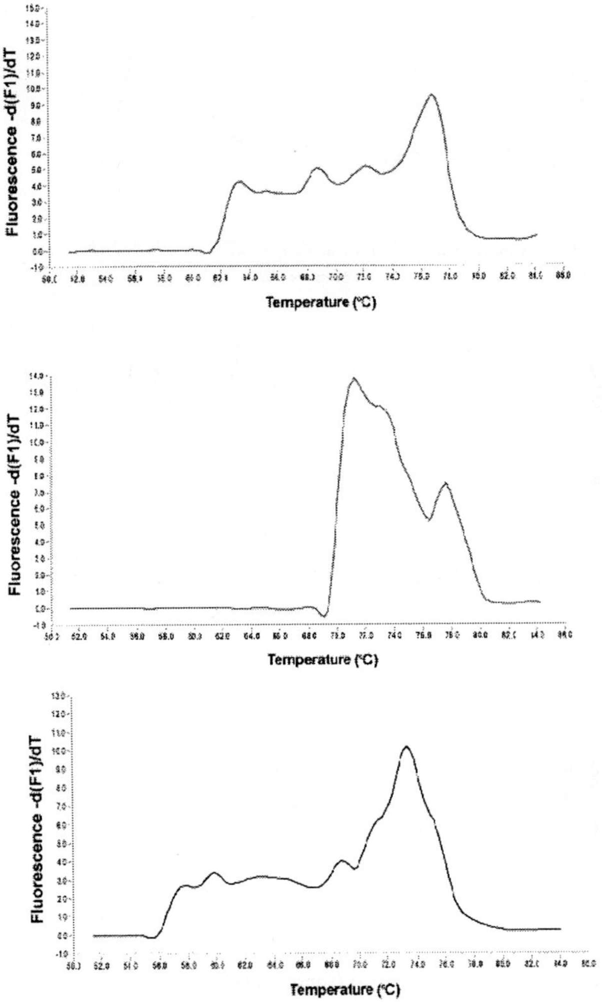

(b) Complementary assay

Fig. 2 (*Continued*)

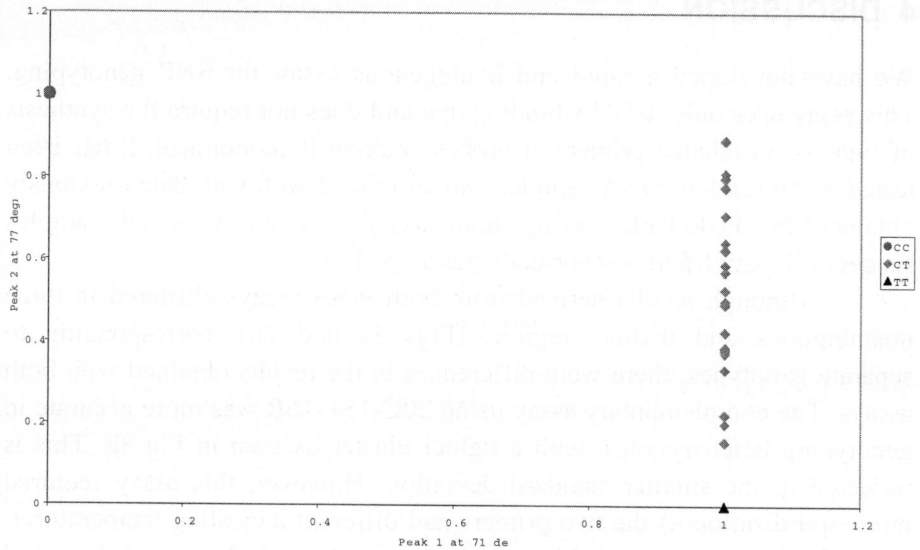

(a) Scatter plot of CC, CT and TT genotypes using the generic assay

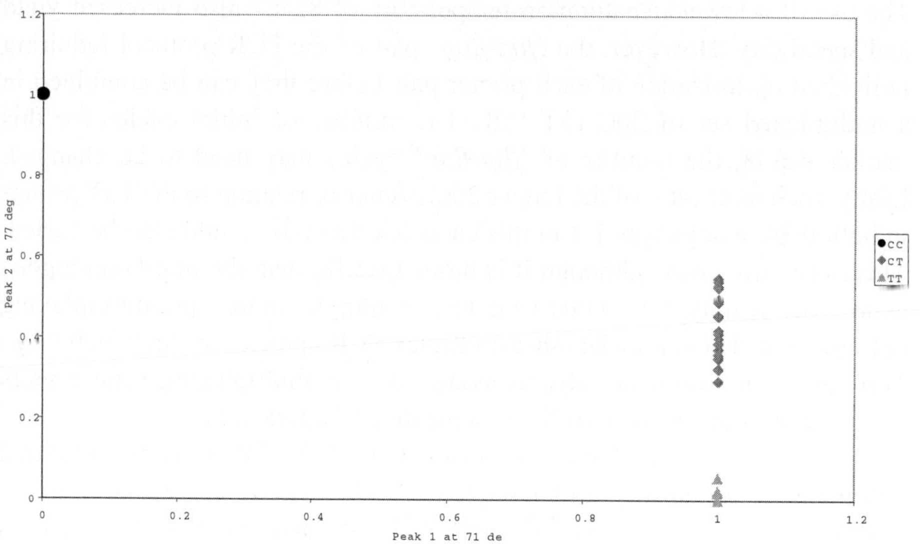

(b) Scatter plot of CC, CT and TT genotypes using the complementary assay

Fig. 3 Scatter plots of each of the three genotypes as analyzed with each assay format, generic and complementary. Samples were scored for the presence of melting peaks at 71°C and 77°C, and taken as a fraction of the largest peak.

4 DISCUSSION

We have developed a rapid and homogenous assay for SNP genotyping. This assay uses only dsDNA-binding dye and does not require the synthesis of fluorescent-labeled primers or probes, making it economical. It has been tested on 46 random DNA samples and correlated well with data previously obtained by PCR-RFLP using both assay formats, with all samples successfully amplified and correctly genotyped.

Although results derived from both these assays clustered in three unambiguous and distinct regions (Figs 3a and 3b), corresponding to separate genotypes, there were differences in the results obtained with both assays. The complementary assay using 20C-15T-15R was more accurate in genotyping heterozygotes, with a tighter cluster as seen in Fig 3b. This is reflected in the smaller standard deviation. However, this assay required more optimization as the two primers had different annealing temperatures. We had tried to overcome this to some extent through the use of the novel "*flip-flop*" PCR in the initial cycles to non-competitively amplify each allele. The use of a lower denaturation temperature of 85°C[4] also increased yield and specificity. However, the "*flip-flop*" part of the PCR protocol requiring individual optimization of each primer pair before they can be combined in a multiplexed set of 20C/15T-15R. The number of initial cycles for this marker was 8, the number of "*flip-flop*" cycles may need to be changed. Lastly, the molar ratio of the longer 20C primer in relation to the 15T primer in both these assays was 1:4 in this case, but this ratio could also be varied. Therefore, this assay, although it is more specific than the non-homologous assay in genotyping heterozygotes, requires more pre-multiplexing optimization. There may be other multiplex PCR applications in which "*flip-flop*" protocol would be advantageous, such as multiplexing primer pairs from different genes, with different annealing temperatures.

The non-homologous set of 20GC-15T-15R requires minimal optimization due to the similar sequences. The generic tail would 'flap' up during initial amplification of genomic DNA, and the bound sequences would have similar annealing temperatures in the initial cycles. A molar excess of 15T over 20GC primer is still required. However, in comparison, only a single optimization of either 20GC-15R or 15T-15R primer pairs is required before multiplexing is possible, and only the molar ratio would need to be optimized, minimizing upstream optimization time.

The stringency of the PCR amplification was also aided by the use of the Stoffel[3] fragment of *Taq* polymerase, an enzyme lacking the 3'-5' exonuclease activity required for proof-reading capability. This enzyme allows the mismatch at the terminal 3' nucleotide to discriminate between different alleles, and yet remains uncleaved during proof-reading.

The technique of T_M shift genotyping was first reported by Germer and Higuchi[6] who used allele-specific PCR with the addition of a GC-rich tail[7] to increase product differentiation. The T_M of PCR products are dependent on both GC composition as well as the length of the PCR product[5]. Therefore the influence of GC-tails depends on the total PCR product composition and length. This earlier work had the problem of primer-dimers with melting profiles similar to that obtained from template amplifications, complicating the interpretation of genotypes. With small PCR products such as this 30-35mer fragments, it was expected that we may have problems with primer-dimers. However, we also utilized TaqStart antibody, which through binding of the Stoffel fragment at low temperatures prevents amplification of primer-dimers but whose binding activity is inactivated once temperatures are raised >70°C. Due to the need to optimize the molar ratio of primers, we used limiting amounts of the longer primer, therefore overall primer concentrations were minimal. Low primer concentrations had the effect of minimizing amplification of primer-dimers, resulting in no amplification of primer-dimers in most cases. We were able to perform as many as 60–70 cycles without non-specific amplification of primer-dimers, increasing the confidence of genotyping results. This obviates the need for post-PCR analysis such as gel electrophoresis, with potential problems of carry over contamination. Short PCR products can now be distinguished by the gain in fluorescence of PCR products, which would otherwise not be possible by conventional means.

The interpretation of results used in these assays is based on the analysis -dF/dT, the change of fluorescence in relation to time as compared to the usual fluorescence with respect to time. We were able to look at melting peaks, which corresponds to the T_M of DNA. Due to the inter-assay variation of ± 0.5°C which is seen with the LightCycler™, we used a baseline fluorescence at 68°C, a temperature corresponding in most samples to the base of the melting peaks. This allowed us to compensate for intra-assay variation and capillary variation, while allowing us the ability to subtract baseline fluorescence.

In contrast to hybridization probe based assays, our assays relied on an economical detection method, the fluorescent dsDNA binding dye SYBR® Green I. The expense and optimization required of custom labeled probes makes it uneconomical for high-throughput genotyping. Compared with conventional PCR amplification, there are few additional reagents required. On the LightCycler™, 32 samples can be genotyped in 25 minutes for both amplification and melting curve analysis, per working day 12–13 runs can be carried out with preparation time included, a total of 384 samples per day. This technique can thus be applied for high-throughput genotyping. Even higher throughput would be possible on 96-well or 384-well format fluorometric thermocyclers. The speed for this assay allows rapid detection of mutations for rapid diagnosis of genetic disease.

It appears probable that these two assays can be readily applied to other genes and SNPs. While GC-rich target templates may not be as easily genotyped, keeping the PCR product small may help maintain the melt curve differences of the two alleles.

In conclusion, we have therefore developed two protocols, with different optimization requirements and accuracies, for the genotyping of DNA polymorphisms. Key features of these assays are the use of Stoffel fragment for fidelity, maintaining a small target size for the PCR product, the use of TaqStart antibody to minimize primer-dimers, the use of melting curve analysis and the lower denaturation temperature after five cycles. The value of SNPs in mapping and linkage studies, and in disease gene and pharmacogenetic analyses emphasize the importance of having a rapid, economical yet easily optimizable and flexible assay with high-throughput capabilities for SNP genotyping.

Acknowledgements

We wish to thank Shirley Seah for the initial PCR-RFLP based genotyping of the many samples and for the results of the population studies. We wish to thank Poh-San Lai for reviewing the manuscript and providing valuable advice and comments.

References

1. Kruglyak, L. The use of a genetic map of biallelic markers in linkage studies. *Nat. Genet.* 1997 **17:**21–24

2. Oda, T., Elkahloun, A.G., Pike, B.L., Okajima, K., Krantz, I.D., Genin, A., Piccoli, D.A., Meltzer, P.S., Spinner, N.B., Collins, F.S. and Chandrasekharappa, S.C. Mutations in the human Jagged1 gene are responsible for Alagille syndrome. *Nat. Genet.* 1997 **16:**235–242

3. Seah S.G.K., Lai P.S., and Yap E.P.H. Identification and characterization of single nucleotide polymorphisms in *Jagged1*, the candidate gene for Alagille syndrome and congenital heart defects. In abstracts of the combined annual Scientific meeting of the Biomedical Research and Experimental Therapeutics Society of Singapore, Singapore Society for Biochemistry and Molecular Biology and Singapore Society of Microbiology and Biotechnology, 13-14 August 1999, Singapore, pg 36

4. Lawyer, F.C., Stoffel, S., Saiki, R.K., Chang, S.Y., Landre, P.A., Abramson, R.D. and Gelfand D.H. High-level expression, purification, and enzymatic characterization of full-length Thermus Aquaticus DNA polymerase and a truncated form deficient in 5' to 3' exonuclease. *PCR Methods Applic* 1993 **2:**275–287

5. Yap, E.P.H. and McGee J.O'D. Short PCR product yields improved by lower denaturation temperatures. *Nucl. Acids Res.* 1991 **19:**1713

6. Germer, S. and Higuchi, R. Single-tube genotyping without oligonucleotide probes. *Genome Research* 1999 **9:**72–78

7. Sheffield, V.C., Cox, D.R., Lerman, L.S. and Myers R.M. Attachment of a 40-base pair G + C rich sequence (GC-clamp) to genomic DNA fragments by the polymerase chain reaction results in improved detection of single-base changes. *Proc. Natl. Acad. Sci*. 1989 **86:**232–236

FRONTIERS IN HUMAN GENETICS
Diseases and Technologies
© 2001 by World Scientific Publishing Co. Pte. Ltd.

BIOMEMS/"LAB-ON-A-CHIP": TOWARDS A CHEAPER, MORE RAPID, PORTABLE AND 'AUTOMATED' HIGH-THROUGHPUT GENOTYPING

TECK CHOON AYI[†] and ERIC PENG HUAT YAP*[,‡]
Defence Medical Research Institute, Defence Science & Technology Agency
10 Medical Drive, Singapore 117597
[†]*E-mail: nmiv30@nus.edu.sg*; [‡]*e-mail: nmiv3@nus.edu.sg*

With the completion of the human genome project, scientists will face the daunting task of analysing the function of all human genes obtained from the project. Global analysis of gene expression and DNA polymorphisms and mutations, however, are facilitated with the development of tools such as microarrays and "Lab-on-a-Chip" systems. This chapter gives an overview of the BioMEMs/ "Lab-on-a-Chip" systems, the components that constitute such systems and some commercial chip systems currently being developed or sold.
Keywords: BioMEMs, "lab-on-a-chip", microarray, chip components, commercial chip systems

1 INTRODUCTION

1.1 Mapping of Genes for Diseases and Traits

The sequencing of the entire human genome will provide the scientific community with the genetic blue print of mankind. With the advent of bioinformatics, genome sequences are already publicly available over the world-wide-web. We are now approaching the threshold of a new and exciting era. In this post-human genome project era, geneticists will be using data derived from the project to find genes that control human behaviour and physical attributes, or whose mutations cause diseases. Genetic markers for these phenotypes may exist as STRs (short tandem repeats), VNTRs (variable number tandem repeats), SNPs (single nucleotide polymorphisms),

* Corresponding author

chromosome deletions, rearrangements, etc., in both the extragenic regions, as well as in the genes themselves. Of these markers, the STRs, VNTRs, and SNPs are the most useful because they are highly polymorphic among individuals. SNPs in particular are estimated to occur at least once every 1kb interval of the human genome, and probably are the most informative, if all of them could be mapped. Indeed, such efforts are underway right now, with the NIH and a consortium of public labs and pharmaceutical companies among others, being involved (www.ncbi.nlm.nih.gov/SNP, and http://snp.cshl.org). The existence of markers associated with the phenotype of interest as determined from linkage analysis, indicate the presence of a gene or genes nearby. Having an integrated genetic and physical map and the location of all the genes in the human genome will enable identification of candidate genes rapidly. Thus, identifying disease genes by positional cloning will be made easier with the completion of the Human Genome Project.

Current methods of genotyping STRs and VNTRs, involve PCR and either polyacrylamide slab gel or capillary electrophoresis (CE), to separate the different DNA fragments according to their sizes. This approach is amenable to multiplexing. SNPs, on the other hand, are genotyped by PCR followed by DNA sequencing, RFLP, primer extension, or a variety of homogeneous single-tube assays or physical separation methods such as mass spectrometry or denaturing HPLC. Other novel approaches do not require prior DNA amplification e.g. the Invader Assay technology developed by Third Wave Technologies (www.twt.com, Madison, WI, USA). In order to find the markers for a phenotype that is caused by polymorphisms or mutations in multiple genes, such as the case in myopia, genome wide screening of thousands of genetic markers in large sample sets can be daunting and highly labour intensive. However, with the use of robots, e.g. the Biomek series from Beckman (www.beckman.com, Fullerton, CA, USA) and high-throughput 96 samples CAE (Capillary Array Electrophoresis) systems, e.g. MegaBace 1000 (Molecular Dynamics, www.moleculardynamics.com, Sunnyvale, CA, USA) and ABI PRISM® 3700 DNA Analyzer (PE Biosystems, www.pebio.com, Foster City, CA, USA), such analyses have become manageable. Alternatively, promising new technologies such as microarray and BioMEMS/Lab-on-a-Chip may be used to speed up gene or marker identification, reduce consumable and manpower costs, and increase throughput and accuracy.

1.2 Immunogenetics and Pharmacogenetics

Genetic polymorphism also plays a role in the ability of individuals to resist infection and to metabolise drugs. It also influences the virulence of a pathogen, and may be used for the identification of the variant pathogen. As an example of immunogenetics, a variant of the CCR-5 chemokine receptor with a 32 base pair deletion, promotes resistance towards HIV infection.[1,2] In contrast, a variant of the CX_3CR_1 receptor affecting only two amino acids, progresses towards full blown AIDS more rapidly than other haplotypes.[3] As for pharmacogenetics, polymorphisms in cytochrome p450 enzymes e.g. CYP2C19, CYP2D6, etc. affect the proteins' abilities to metabolise drugs. This lead to a scenario where drug dosages can be tailored to an individual based on one's P450 genotype, thus avoiding adverse drug reactions caused by an inability to detoxify the drug[4], or subtherapeutic dosing due to increase clearance.

It is obvious that both microarrays and BioMEMs/Lab-on-a-Chip will rival and surpass traditional technology in genotyping genes involved in immunity and drug metabolism. Briefly, a DNA microarray is constructed by spotting thousands of DNA fragments derived from PCRs or libraries, or even oligonucleotides onto silicone chips, glass slides or membranes, whose surfaces are chemically treated to bind the DNAs. A robot called a microarrayer is used to spot the DNA solution, down to a volume of less than one nanolitre per spot as a two dimensional (X, Y) grid on a surface. Each spot represents a distinct DNA entity e.g. one position (1,1) may be spotted with actin DNA, another (1,2) spotted with p53 DNA, etc. Simultaneous hybridisation of the microarray with cDNAs from test and control cells, each labelled with a different fluorophore, then facilitates the quantification of the different expression levels of each gene that had been spotted onto the microarray. Besides spotting, light directed chemical synthesis of oligonucleotides on a chip using photolithography, as pioneered by Affymetrix (www.affymetrix.com, Santa Clara, CA, USA), allows one to attach oligonucleotides to a chip at a density which surpasses the microarrays constructed by spotting. There are numerous literature and review articles written on microarrays due to the rapid development in this area.[5,6] In a point-of-care setting however, BioMEMs/Lab-on-a-Chip has an advantage over microarrays and other traditional technology because of its combination of low cost, speed, portability, low sample and reagent consumption, automation, and high-throughput parallel analysis of samples. In analysing samples, BioMEMs/Lab-on-a-Chip devices may be fabricated to incorporate a sample preparation component and a sample analysis component that uses either micro CE (capillary electrophoresis) for both DNA fragment length analysis and DNA sequencing or micro PCR (polymerase chain reaction) or electronic addressable arrays, e.g. Nanogen's array (www.nanogen.com, San Diego, CA, USA).

1.3 Other Applications

Besides the applications described above, BioMEMs and MEMs have also found applications in the field of *in-vivo* drug delivery,[7] immunoassays,[8,9] cell separation,[10] patterned cell attachment and growth,[11] patterned delivery of protein to a surface,[12] isoelectric focussing of proteins,[13] SDS gel electrophoresis of proteins,[14] and combinatorial chemical synthesis.[15] Since this chapter discusses the application of BioMEMs and "Lab-on-a-Chip" systems for genotyping, none of these other applications will be discussed in-depth.

This overview discusses in length each component that constitutes a BioMEMs/Lab-on-a-Chip system, and compares some commercial chip systems that are currently being developed or sold. The systems discussed will also include Nanogen's "Lab-on-a-Chip" microelectrode array, and Protogene's (www.protogene.com, Menlo Park, CA, USA) ink-jet printing (MEMs) technologies that can be use to create microarray on glass slides or membranes.

2. BIOMEMS/"LAB-ON-A-CHIP"

BioMEMs (Biomedical MicroElectroMechanical system) is a "Lab-on-a-Chip" micro-machine with the potential to analyse biological samples cheaply and rapidly in a parallel, high-throughput and automated manner. Due to their small size, BioMEMs has many advantages as compared to conventional devices for preparing and analysing biological samples (Table 1). MEMs (MicroElectroMechanical systems) microchip which encompasses BioMEMs, is already an established field and products that incorporate MEMs are found, for example, in the device that triggers inflation of airbags in car, the head of the ink-jet printer and the chemical sensor in a portable blood-analyser.[16] Instead of chemical analysis, BioMEMs involves the analysis of biological samples. The term "Lab-on-a-Chip" is generally used to define both MEMs for analysing chemicals and BioMEMs. Another term favoured by most authors to describe "Lab-on-a-Chip", is micro Total Analysis System or μTAS. "Lab-on-a-Chip" systems are produced using technologies derived from the microelectronic industries and incorporate microfluidics, microfilters, microreactors, microseperation columns, microfabricated pumps and other microcomponents, depending on the microchip's application. A fully integrated system with sample preparation, delivery, analysis and data output can be potentially built into BioMEMs. Hence BioMEMs are automated devices. However, all current commercially available BioMEMs have a 'desk-top' or benchtop unit besides the chip itself. This is for either sample preparation or data output for humans to read and access e.g. Cepheid's GeneXpert (www.cepheid.com, Sunnyvale, CA, USA) or Agilent's 2100 Bioanalyzer (www.chem.agilent.com, Palo Alto, CA, USA). Currently most BioMEMs do not have the ability to simultaneously analyse as many genes as microarrays.[17] In contrast to microarrays which analyse DNA or RNA samples based on their ability to hybridise with probes immobilised on the microarray, the stringency of which is controlled by temperature and laser, BioMEMs show more flexibility. BioMEMs can analyse DNA samples by selective amplification using micro-PCR in micro-wells or micro-CE separation, depending on the applications.

Table 1 Advantages of "lab-on-a-chip"

"Lab-on-a-Chip" has many advantages compared to conventional analysis system, due to its small size. Microchannels ranging between 5 and 200μM in width and depth have these properties:

(1) Rapid heat transfer that allows rapid cooling and heating time.
(2) Improved speed and efficiency of separation on micro CE.
(3) Reduced reagent consumption.
(4) Electroosmotic pumping which allows for valveless systems and less external pumping. This applies only to silica based chips.
(5) Highly portable and disposable.
(6) Amenable to parallelization with multiple separation channels, built-in chambers, etc.

Besides "Lab-on-a-Chip", there are also "Lab-on-a-Disc" and "Lab-on-a-Card" systems. An example of a "Lab-on-a-Disc" is the Gamera Bioscience/Tecan LabCD (www.tecan.com), which is still under development. This is a compact disc (CD) based μTAS that uses the centrifugal force of a spinning CD to move liquids. An example of a "Lab-on-a-Card" is PEBiosystem's RT-PCR card which is still under development.

2.1 The Components of a "Lab-on-a-Chip"

2.1.1 Chip materials

Many types of material were used for the fabrication of chips, from silicon and siloxane to different types of polymer, e.g. PDMS or poly(dimethylsiloxane), PMMA or poly(methyl methacrylate), polyamide, polycarbonate, polyethylene, etc. Each one of them has their own advantages and disadvantages. Early systems used silicon-based substrates because fabrication technologies were available from the microelectronic industries. Fused silica, quartz and glass are transparent, amenable to photolithographic and wet etching fabrication (Figure 1), dissipate Joule heat more rapidly than plastic materials and exhibit electroosmotic flow (EOF, refer section 2.1.3). These are popular materials for fabricating micro CE (micro Capillary Electrophoresis) based devices.[18-23] The silanol groups on silicon based device, however, will adsorb proteins and is suitable for PCR reaction without prior surface modifications. For silicon devices, the most reproducible results can be obtained by creating a layer of thermal oxide (silicon dioxide) on the surface.[24-26] Silanisation of the surface has also been attempted but PCR results tend to be less reproducible when compared to thermal oxide coating.[24]

For mass production of disposable chips, silicon, fused silica, quartz and glass based "Lab-on-a-Chip" are comparatively more expensive than polymer based chip. Polymers can be fabricated using different molding technologies such as hot embossing, injection molding and casting. The polymers are usually hydrophobic and do not support EOF, but electrophoresis in a microchannel filled with gel or polymer solution e.g. linear polyacrylamide (LPA) can still be carried out (refer section 2.1.3). A review on polymer based microchips and their method of fabrications has been reported by Becker H and Gartner C (2000).[27]

Another popular and cheaper material for chip construction is the use of elastomer, poly (dimethylsiloxane) or PDMS. PDMS is an excellent material for fabrication of "Lab-on-a-Chip". Using replica plating, features on the micron scale can be reproduced with high fidelity with PDMS. Multiple devices can be produced rapidly from a single master with only the minimal use of clean room facilities. PDMS cures at lower temperatures, exhibits reversible deformity and seals reversibly to itself and other materials on contact. It also seals irreversibly upon exposure to an air plasma, is optically transparent to wavelengths up to 280 nm, non-toxic to mammalian cells culture, and its hydrophobic surface upon plasma oxidation supports EOF. Fabrication uses the soft lithography techniques of rapid prototyping and replica molding. A review on fabrication of microfluidic systems using PDMS has been published by McDonald JC and co-workers (2000).[28]

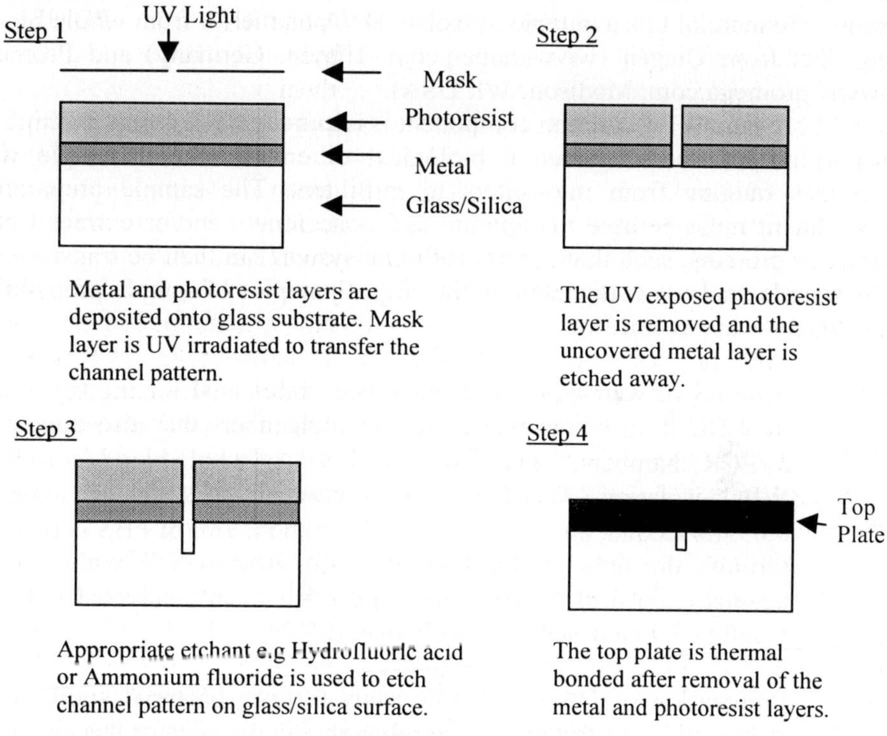

Fig. 1 Generic approach to chip fabrication using photolithography and wet etching. [Adapted from McCreedy T. (2000) *Trends in Analytical Chemistry* **19**(6).]

2.1.2 Sample preparation

To fabricate a truly integrated "Lab-on-a-Chip", it is not enough just to have sample separation and detection. Preparation of materials for analysis from biological fluids and the environment e.g. bacteria from soil, water and air, consumes time and usually involves multiple centrifugation steps. Most of the time the samples are diluted and need to be concentrated before analysis. Some of these materials may be biohazardous and it may thus be safer to automate the isolation process. As an example, preparation of DNA or RNA from HIV infected whole blood for PCR or RT-PCR analysis uses a gradient centrifugation step in Ficoll-Hypaque to separate blood plasma from white blood cells (WBCs) and red blood cells (RBCs) before concentrating the WBCs into a 'buffy coat'. Further processing of the plasma and 'buffy coat' is required to obtain RNA or DNA for RT-PCR or PCR reaction. There are many commercial kits available to isolate DNA and RNA from whole blood, e.g. kits from Qiagen (www.qiagen.com, Hilden, Germany) and Promega (www.promega.com, Madison, WI, USA).

A sample preparation component is required in a "Lab-on-a-Chip" to act as an interface between a biological or environmental sample with volumes ranging from microliters to millilitres. The sample preparation component must be able to separate and concentrate, and/or extract DNA, RNA or proteins, such that the microfluidic system can then be transfered to the sample analysis component of the chip. Examples of sample preparation components are as follows:

(i) A series of weir-type filters have been fabricated for the separation of WBC from whole blood on silicon chambers that also double up as PCR chambers.[29] Typically only 3.5μL of whole blood is used for WBC isolation. Whole blood is pumped through the chip at 0.035μl/second, and after the initial filtration, 1ml of PBS is pumped through the chip at 8ml/hour to wash. Micro PCR with primers against exon 6 of the dystrophin gene has been conducted with the result being analysed separately on CE.

(ii) Preparation of DNA for PCR analysis from bacterial spores on a microfluidic cartridge with an integrated minisonicator has also been reported.[30] The cartridge is programmed to deliver 5μl of lysate and 20μl of PCR reagent into a PCR tube, and then the PCR reaction subsequently analysed separately on a PCR machine.

(iii) Dielectrophoretic separation (refer to section 2.1.3) of *E. coli* from whole blood has also been performed.[31] The separated *E. coli* is subjected to electronic high-voltage pulses to lyse the bacteria, followed by proteolytic digestion on a single fabricated microelectrode array chip. The lysate, which contains DNA, RNA and proteins, is manually transferred and further examined by electronically enhanced hybridization on a separate microelectrode array chip (refer to section 2.1.3.4). No PCR is involved.

(iv) Separation of fluorescent and non-fluorescent *E. coli* on a microfabricated Fluorescence-Activated Cell Sorter (FACS).[32]

(v) A thin porous silicate micromembrane structure constructed between two microchannels for the concentration of DNA.[19] This membrane allows ionic current to pass but prevent larger DNA molecules from crossing. The entrapped DNA is injected into a micro CE channel, separated and then analysed.

2.1.3 Sample analysis: delivery and separation (microfluidics), and detection.

Microfluidics is the manipulation of liquids and gases in microchannels with cross-sectional dimensions in the order of 10-100μM. For biological analysis, microfluidics usually involve liquids.

Electrokinetic pumping or electrokinetically driven transport is the sum of the electrophoretic and electroosmotic forces acting on liquids in the microchannels (Figure 2). Electroosmotic flow or EOF can reach velocities of 5 cm/min or greater. The rate of electroosmotic flow is generally greater than electrophoretic migration of the individual ions and this effectively becomes the mobile phase pump of capillary zone electrophoresis (CZE). There are other pumping mechanisms available with low flow rate e.g. silicon or plastic diaphragm pumps with piezoelectric activators, but this technique of pumping may introduce backpressure leading to leakage. External pumps, excluding syringes, are bulky, expensive and not mobile. They often require much more engineering efforts. The MEMs community has evaluated and utilised an abundance of pumping principles to move liquids.[33]

Micro CE makes use of microfluidics to move liquids through bare capillaries or capillaries filled with sieving matrix e.g. hydroxyethyl cellulose (HEC) and LPA. Indeed, CZE makes use of electroosmosis to separate molecules based on their net charges in a bare glass or silica capillary (Figure 2). As an example, a microfabricated device uses a balance of capillary action and EOF to separate *Hind*III digested λ DNA into different sizes and detection can be achieved down to a single molecule level.[34] For applications such as DNA fragment length analysis and DNA sequencing, in order to better resolve DNA fragments, microcapillaries have been filled with relatively viscous sieving matrix such as HEC, LPA, poly(ethyleneoxide) (PEO) and poly(dimethylacrylamide) (PDMA). The presence of a sieving

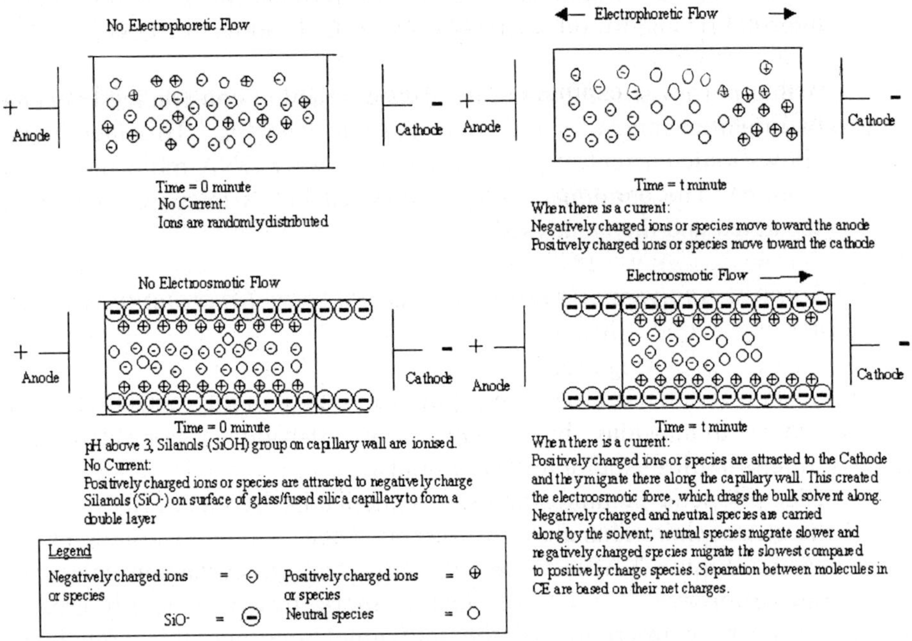

Figure2 Separation of charged molecules via electrophoresis and electroosmosis

matrix suppresses EOF by decreasing the zeta potential of silica and this reduces the probability of the viscous polymer being ejected out of the capillary by EOF. For silicone, fused silica or glass "Lab-on-a-Chip", a popular approach uses a modification of the Hjerten procedure[35] to covalently crosslink the capillaries with the sieving matrix LPA to reduce the effect of EOF. [36,23,21]

Separation of cell and bacteria in microcapillary is powered by another mechanism, the dielectrophoretic field-flow-fractionation (DEP-FFF). Dielectrophoretic field-flow fractionation has been used to separate cells based on their dielectric properties and sedimentation rates. Interdigitated electrodes are used to apply an alternating (AC) electrical field and the cells become levitated to different equilibrium height under the opposing influence of dielectrophoretic and sedimentation forces. When liquid that carries cells is moved through the interdigitated electrodes, the cells would levitate and travel at different velocities through the channel based on their degree of levitation, and separate. As an example, DEP-FFF have been used to enrich T-lymphocytes from a mixture of T-lymphocytes and MDA435, a human breast cancer cell line.[10] Separation of *E. coli* from whole blood using dielectrophoresis on a microelectrode array chip which do not contain microchannels for directing the flow of liquid is also possible (refer to section 2.1.3.4).[31]

Below are some examples of "Lab-on-a-Chip" systems used for the analysis of DNA.

2.1.3.1 Micro PCR for the amplification and detection of DNA

(i) Continuous-Flow PCR on a chip.[37] Three zones, made of thermostated copper blocks are kept at 95°C, 77°C and 60°C. A standard fused silica capillary, 40 μM deep and 90 μM wide, is fabricated, looping around the three zones. Sample and PCR reagents are introduced via three inlets and precision syringe pump forces the liquid through the three zones in a controlled loop, effectively thermal cycling and amplifying the samples 20 rounds in 4 minutes. The amplicon is then analysed and detected on agarose slab gel electrophoresis. This approach has been used to amplify and detect a 176 base pair DNA fragment from the DNA gyrase gene of *Neisseria gonorrhoeae*.

(ii) PCR and real-time quantitation of β-actin amplicons from human genomic DNA using Taqman chemistry has been done on silicon chips with a layer of thermal oxide.[26] Bacterial DNA have been amplified by PCR on surface treated silicon chips and the *C. jejuni* amplicons analysed by agarose slab gel electrophoresis.[24,25]

(iii) Infrared-Mediated Thermcocycling,[38] is a non-contact method for the rapid thermocycling of PCR mixtures in chip-like glass chambers using a tungsten lamp as an inexpensive infrared radiation source for heating and solenoid gated compressed air as coolant. Amplicons of T-cell receptor β-chain from IR-mediated PCR are analysed by both conventional CE and agarose slab gel electrophoresis.

2.1.3.2 Micro CE for the analysis of DNA fragment length and DNA sequence

(i) The separation and detection of multiplexed PCR samples containing the four loci *CSF1PO, TPOX, THO1* and *vWA* of STR, are done in less than 2 minutes on a micro CE filled with an LPA matrix operated under denaturing conditions at 50°C.[23] Laser induced fluorescence is used for detection of the fragments. Separation and detection on micro CE constitutes a 10 to 100 fold improvement in speed relative to normal capillary or slab gel system. Separation and detection of Herpes Simplex Virus PCR amplicons from clinical samples have also been achieved with micro CE in less than 110 seconds per sample per run.[39]

(ii) The use of ultrafast denaturing electrophoresis in short capillaries filled with the sieving matrix, agarose (BRE, No. 1503; FMC Bioproduct, ME, USA) and urea has been reported to separate and detect STR polymorphism in the endothelin 1 gene. Using this approach, the resolution of two fragments with a difference of 2 nucleotides is achieved in a capillary in 42 seconds at a temperature of 60°C implying that dinucleotide repeat polymorphism can be resolved on microcapillary. Compared to conventional slab gel electrophoresis, denaturing micro CE and laser-induced fluorescence detection resulted in a reduction of analysis time by a factor of 200.[40]

(iii) DNA sequencing of single stranded M13mp18 plasmid on a chip has been successfully attempted by Liu and colleagues.[21] Single-base resolution of DNA fragments extending over 500 bases was achieved with denaturing electrophoresis using microcapillaries filled with the sieving matrix LPA and urea, with temperatures ranging from 35°C to 40°C, and a run time of under 20 minutes. Four-colour fluorescence is detected using a laser confocal fluorescence detection system. In contrast to the rapidity of the micro CE separation and detection, conventional slab gel electrophoresis requires an overnight run time.

(iv) A 96 wells microplate with integrated micro CAE separation channels has been fabricated on a microchip by Simpson and colleagues.[20] The micro CAE chip is used to analyse the C282Y polymorphism which introduces a *RsaI* restriction site in the *HFE* gene, a gene correlated with hereditary hemochromatosis (HCC).[41,42] DNA isolated from peripheral blood and a segment of the *HFE* exon containing the variant is amplified by conventional PCR. Separation and detection of 96 samples is achieved in less than 8 minutes and this high throughput analysis is 50 to 100 times greater than conventional slab gels.

2.1.3.3 Micro PCR and micro CE on a chip for the amplification, detection and analysis of DNA fragments

(i) Deletions causing Duchenne/Becker muscular dystrophy, are detected using a silicon-glass chip for locus-specific, multiplex micro PCR of the dystrophin gene exons. The amplified DNA is manually transferred and analysed by micro CE on another chip.[36]

(ii) Waters LC and colleagues have constructed some integrated micro PCR and micro CE systems.[43,44] Micro PCR is performed on λ bacteriophages DNA, whole *E. coli* or *E. coli's* DNA on a glass chip and the DNA was stained with the dye, TO-PRO (Molecular Probes, Inc., www.probes.com, Eugene, OR, USA). The PCR product, while still on the chip, is transferred electrophoretically into the injection valve using the pinched sample injection technique.[45] The samples are resolved on micro CE with either HEC or PDMA sieving matrices in less than 3 minutes.

2.1.3.4 Electric field directed hybridization

Nanogen pioneered the use of electric field to concentrate and hybridise DNA to its complementary probe pre-immobilised at specific locations or addresses on a microelectrode array.[46,47] The microelectrode array is coated with a permeation layer, a layer of agarose with covalently attached streptavidin. The permeation layer acts as a physical barrier which separate DNA molecules in solution from the harmful reaction that often occur at the electrode's surface while allowing small ions to pass through to maintain conductance. A DNA molecule which has a net negative charge due to its phosphate backbone at neutral pH, is attracted and directed to a specific electrode by charging that electrode positive. Biotinylated single stranded DNA oligonucleotide probes are loaded sequentially and directed to specific locations or addresses of the microelectrode array where they are immobilised by binding to streptavidin. Fluorescence-labelled DNA samples are then introduced, targeted and concentrated at specific addresses of the microelectrode array pre-immobilised with the appropriate probes for hybridisation. Single mismatch mutation is differentiated from wild type DNA samples after hybridisation by reversing the polarity of the electrode from positive to negative and applying pulse electric currents. The interaction of mismatch DNA with the probe is considerably weaker compared to wild type DNA samples, and hence the mismatch DNA is easily repel by the current. Compared to passive hybridisation, electronic hybridisation shows greater sensitivity, speed and specificity.

Strand displacement amplification (SDA)[48] is an isothermal DNA amplification technology that is compatible with the microelectrode array chip format.[49] Decaplex anchored SDA has been used to amplify and detect human Factor V, HFE (hemochromatosis), TNF-α, Cyp19 (aromatase), Fas ligand and bacterial *parC, Pseudomonas's, Salmonella's*, and *E. coli's gyrA*, and *Chlamydia trachomatis*, from a mixtures of DNA templates. Similarly, RT-SDA has been used to amplify and detect Factor V, Fas ligand, TNF-α, and interleukin 1-β from *in-vitro* synthesised RNA templates.

2.1.3.5 Ink jet and bubble jet printing of microarray

The humble ink-jet or bubble-jet printer's printheads are MEMs, which use either rapid and extreme heating and cooling, or the vibration of a piezoelectric membrane to generate and eject ink droplets onto print media. They have been transformed into microarraying devices because the printheads could eject precise spots down to picolitres volume onto specific locations on a membrane or a glass slide. Oligonucleotides have been spotted using this printing approach as described by some authors.[50,51] *In-situ* synthesis of oligonucleotide microarray on chemically treated membrane using conventional phosphoramidite chemistry, which has a higher coupling yield compared to light directed chemistry, is currently being developed by Protogene and Rosetta inpharmatics (www.rii.com, Kirkland, WA, USA). This approach uses the printer to deliver small volume of either A, C, G or T phosphoramidites and other chemicals used by DNA synthesisers to make oligonucleotides onto precise locations or spots on a membrane or glass slide, where chemical coupling will take place to generate the microarray.

3 SOME COMMERCIAL "LAB-ON-A-CHIP" SYSTEM BEING DEVELOPED OR SOLD

3.1 Cepheid's GeneXpert (Prototype)

Cepheid's "Lab-on-a-Chip" cartridge is an automated and truly integrated system which uses microfluidics to do sample preparation on 5 ml of urine in less than 5 minutes, and includes filtration, cell lysis, DNA extraction, and addition of pre-loaded assay specific PCR reagents. The extracted DNA and PCR reaction mixture is delivered automatically to a closed, integrated reaction tube, where it undergoes rapid thermal cycling, PCR amplification and real-time optical detection using the Taqman chemistry in a Cepheid I-CORE® desktop module. This current system is designed for the detection of *Chlamydia* in urine and a prototype is available from Cepheid.

3.2 Agilent 2100 Bioanalyzer® (Benchtop Module) and Calipher LabChip® Kit

Agilent and Calipher technologies (www.calipertech.com, Mountain View, CA, USA) have jointly developed various LabChip® systems based on micro CE, for the analysis of DNA and RNA fragments. Agilent builds the benchtop modules and Calipher makes the chips, which will be available on the market soon.

3.3 Aclara Biosciences and Cellomics

Aclara Biosciences (www.aclara.com, Mountain View, CA, USA) and Cellomics (www.cellomics.com, Pittsburgh, PA, USA) are jointly developing the CellChip™ system. CellChip™ uses microfluidics to enable massive numbers of parallel live-cell assays to be performed on a chip, for a broad range of applications including the discovery of new drugs and stratification of patients for effective therapies and basic research. Aclara is also developing disposable plastic chips for microCE DNA fragment length analysis.

3.4 Micronics

Micronics (www.micronics.net, Redmond, WA, USA) is a medical diagnostic company employing microfluidics to develop a Micro Flow

Cytometer system and a micro separation and extraction device, the H-Filter™, for separating and isolating cells from mixed cells population such as blood. Beside cell isolation systems, the T-Sensor™ is being developed for clinical chemistry to identify metabolites, drugs, etc. in complex mixtures such as blood.

3.5 Nanogen

See section 2.1.3.4.

3.6 Protogene and Rosetta Inpharmatics

See section 2.1.3.5

4 Summary/Conclusions

BioMEMs/'Lab-on-a-Chip" devices potentially could have many applications for the analysis of different type of biomolecules and cell types, using different combinations of sample preparation and sample analysis modules. Currently, most systems are designed to analyse DNA or RNA via PCR or RT-PCR coupled with CE, or hybridisation to oligonucleotide probes laden electrodes. Sample preparation usually involves the separation of cells using fabricated microfilters and DEP-FFF in capillaries or electrodes. This is followed by whole cell PCR or DNA extraction from cells. Although BioMEMs/"Lab-on-a-Chip" can potentially be used for gene expression studies via quantitative RT-PCR, the future of these chips lies in genotyping applications. Major efforts are currently underway to find all human SNPs for use as genetic markers. Armed with this information on SNPs, BioMEMs/'Lab-on-a-Chip" devices can be fabricated for use by medical practitioners to diagnose multiple patients in a point-of-care setting for disease susceptibility. Alternatively, a patient may make a self diagnosis as Lab-on-Chips are automated devices. Indeed, major electronic companies such as Motorola and Hewlett Packard have invested substantially in BioMEMs/"Lab-on-a-Chip" research and development, hoping to reap the rewards of such an endeavour.

References

1. Liu R, Paxton WA, Choe S, Ceradini D, Martin SR, Horuk R, MacDonald ME, Stuhlmann H, Koup RA and Landau NR, Homozygous defect in HIV-1 coreceptor accounts for resistance of some multiply-exposed individuals to HIV-1 infection, *Cell* 1996; **86(3):** 367-377.
2. Dean M, Carrington M, Winkler C, Huttley GA, Smith MW, Allikmets R, Goedert JJ, Buchbinder SP, et. al., Genetic restriction of HIV-1 infection and progression to AIDS by a deletion allele of the CKR5 structural gene, *Science* 1996; **273:** 1856-1862.
3. Faure S, Meyer L, Costagliola D, Vaneensberghe C, Genin E, Autran B, et. al., Rapid progression to AIDS in HIV+ individuals with a structural variant of the chemokine receptor CX3CR1, *Science* 2000; **287:** 2274 – 2277.
4. Wolf CR and Smith G, Pharmacogenetics, *Br Med Bull* 1999; **55(2):** 366-386.
5. Schena M (Ed.), *Microarray Biochip Technology*, Eaton Publishing Company, Natick, MA, USA, 2000.
6. The Chipping Forest, *Nature Genetics* 1999; **21(Suppl):** 3-60.
7. Santini JT, Cima MJ and Langer R, A controlled-release microchip, *Nature* 1999; **397:** 335-338.
8. Chiem N and Harrison DJ, Microchip-based capillary electrophoresis for immunoassays: analysis of monoclonal antibodies and theophylline, *Analytical Chemistry* 1997; **69(3):** 373-378.
9. Colyer CL, Tang T, Chiem N and Harrison DJ, Clinical potential of microchip capillary electrophoresis systems, *Electrophoresis* 1997; **18:** 1733-1741.
10. Wang XB, Yang J, Huang Y, Vykoukal J, Becker FF and Gascoyne PRC, Cell separation by dielectrophoretic field-flow-fractionation, *Analytical Chemistry* 2000; **72(4):** 832-839.
11. Kane RS, Takayama S, Ostuni E, Ingber DE and Whitesides GM, Patterning proteins and cells using soft lithography, *Biomaterials* 1999; **20:** 2363-2376.
12. Delamarche E, Bernard A, Schmid H, Michel B and Biebuyck H, Patterned delivery of immunoglobulins to surfaces using microfluidic networks, *Science* 1997; **276:** 779-781
13. Hofmann O, Chi D, Cruickshank KA and Muller UR, Adaptation of capillary isoelectric focusing to microchannels on a glass chip, *Analytical Chemistry* 1999; **71(3):** 678-686.
14. Yao S, Anex DS, Caldwell WB, Arnold DW, Smith KB and Schultz PG, SDS capillary gel electrophoresis of proteins in microfabricated channels, *PNAS* 1999; **96:** 5372-5377.

15. Dewitt S, *Synthesis on a Chip in Strategic Use of Combinatorial Chemistry*, IBC, Frankfurt, June 1998.
16. Kopp MU, Crabtree HJ and Manz A, Developments in technology and applications of microsystems, *Current Opinion in Chemical Biology* 1997; **1:** 410-419.
17. Sanders GHW and Manz A, Chip-based microsystems for genomic and proteomic analysis, *Trends in Analytical Chemistry* 2000; **19(6):** 364-378.
18. McCreedy T, Fabrication techniques and materials commonly used for the production of microreactors and micro total analytical systems, *Trends in Analytical Chemistry* 2000; **19(6):** 396-401.
19. Khandurina J, Jacobson SC, Waters LC, Foote RS, and Ramsey JM, Microfabricated porous membrane structure for sample concentration and electrophoretic analysis, *Analytical Chemistry* 1999; **71(9):** 1815-1819.
20. Simpson PC, Roach D, Wooley AT, Thorsen T, Johnston R, Sensabaugh GF, and Mathies RA, High-throughput genetic analysis using microfabricated 96-sample capillary array electrophoresis microplates, *PNAS* 1998; **95:** 2256-2261.
21. Liu S, Shi Y, Ja WW, and Mathies RA, Optimization of high-speed DNA sequencing on microfabricated capillary electrophoresis channels, *Analytical Chemistry* 1999; **71(3):** 566-573.
22. Woolley AT, Hadley D, Landre P, DeMello AJ, Mathies RA and Northrup MA, Functional integration of PCR amplification and capillary electrophoresis in a microfabricated DNA analysis device, *Analytical Chemistry* 1996; **68:** 4081-4086.
23. Schmalzing D, Koutny L, Adourian A, Belgrader P, Matsudaira P, Ehrlich D, DNA typing in thirty seconds with a microfabricated device, *PNAS* 1997; **94:** 10273-10278.
24. Shoffner MA, Cheng J, Hvichia GE, Kricka LJ and Wilding P, Chip PCR. I. Surface passivation of microfabricated silicon-glass chips for PCR, *Nucleic Acids Research* 1996; **24(2):** 375-379.
25. Cheng J, Shoffner MA, Hvichia GE, Kricka LJ and Wilding P, Chip PCR. II. Investigation of different PCR amplification systems in microbabricated silicon-glass chips, *Nucleic Acids Research* 1996; **24(2):** 380-385.
26. Taylor TB, Winn-Deen ES, Picozza E, Woudenberg TM and Albin M, Optimization of the performance of the polymerase chain reaction in silicon-based microstructures, *Nucleic Acids Research* 1997; **25(15):** 3164-3168.
27. Beker H and Gartner C, Polymer microfabrication methods for microfluidic analytical applications, *Electrophoresis* 2000; **21(1):** 12-26.

28. McDonald JC, Duffy DC, Anderson JR, Chiu DT, Wu H, Schueller OJA and Whitesides GM, Fabrication of microfluidic systems in poly(dimethylsiloxane), *Electrophoresis* 2000; **21(1):** 27-40.
29. Wilding P, Kricka LJ, Cheng J, Hvichia G, Schoffner MA and Fortina P, Integrated cell isolation and polymerase chain reaction analysis using silicon microfilter chambers, *Analytical Biochemistry* 1998; **257:** 95-100.
30. Belgrader P, Okuzumi, M, Pourahmadi, F, Borkholder D, and Northrup MA, A microfluidic cartridge to prepare spores for PCR analysis, *Biosensors and Bioelectronics* 2000; **14:** 849-852.
31. Cheng J, Sheldon EL, Wu L, Uribe A, Gerrue LO, Carrino J, Heller MJ, and O'Connell JP, Preparation and hybridization analysis of DNA/RNA from E. coli on microfabricated bioelectronic chips, *Nature Biotechnology* 1998; **16:** 541-546.
32. Fu AY, Spence C, Scherer A, Arnold FH, and Quake SR, A microfabricated fluorescence-activated cell sorter, *Nature Biotechnology* 1999; **17:** 1109-1111.
33. Shoji S, in H. Becker and A. Manz (eds.), *Microsystem Technology in Chemistry and Life Science*, 1998: 164
34. Chou HP, Spence C, Scherer A and Quake S, A microfabricated device for sizing and sorting DNA molecules, *PNAS* 1999; **96:** 11-13.
35. Hjerten S, *J. Chromatogr* 1985; **347:** 191-198.
36. Cheng J, Waters LC, Fortina P, Hvichia G, Jacobson SC, Ramsey JM, Kricka LJ and Wilding P, Degenerate oligonucleotide primed-polymerase chain reaction and capillary electrophoretic analysis of human DNA on microchip-based devices, *Analytical Biochemistry* 1998; **257:** 101-106.
37. Kopp MU, De Mello AJ, Manz A, Chemical amplification: continuous-flow PCR on a chip, *Science* 1998; **280:** 1046-1048.
38. Oda RP, Strausbauch MA, Huhmer AFR, Borson N, Jurrens SR, Craighead J, Wettstein PJ, Eckloff B, Kline B and Landers JP, Infrared-mediated thermocycling for ultrafast polymerase chain reaction amplification of DNA, *Analytical Chemistry* 1998; **70:** 4361-4368.
39. Hofgartner WT, Huhmer AFR, Landers JP and Kant JA, Rapid diagnosis of herpes simplex encephalitis using microchip electrophoresis of PCR products, *Clinical Chemistry* 1999; **45(12):** 2120-2128.
40. Klepárník K, Malá Z, Pribyla L, Blazková M, Vasku A and Bocek P, Ultrafast detection of microsatellite repeat polymorphism in endothelin 1 gene by electrophoresis in short capillaries, *Electrophoresis* 2000; **21(1):** 238-246.
41. Calandro L, Thorsen T, Barcellos L, Griggs J, Baer D and Sensabaugh GF, *Blood Cells Mol. Dis*. 1996; **22**: 194a-194b.
42. Feder JN, Gnirke A, Thomas W, Tsuchihashi Z, Ruddy DA, Basava A, Dormishian F, Domingo R Jr, Ellis MC, Fullan A, et. al., A novel MHC class I-

like gene is mutated in patients with hereditary haemochromatosis, *Nature Genet.* 1996; **13(4):** 399-408.
43. Waters LC, Jacobson SC, Kroutchinina N, Khandurina J, Foote RS, and Ramsey JM, Multiple sample PCR amplification and electrophoretic analysis on a microchip, *Analytical Chemistry* 1998; **70(24):** 5172-5176.
44. Waters LC, Jacobson SC, Kroutchinina N, Khandurina J, Foote RS, and Ramsey JM, Microchip device for cell lysis, multiplex PCR amplification, and electrophoretic sizing, *Analytical Chemistry* 1998; **70(1):** 158-162.
45. Jacobson SC, Hergenroder R, Koutny LB, Warmack RJ and Ramsey JM, *Analytical Chemistry* 1994; **66:** 1107-1113.
46. Sosnowski RG, Tu E, Butler WF, O'Connell JPO and Heller MJ, Rapid determination of single base mismatch mutations in DNA hybrids by direct electric field control, *PNAS* 1997; **94:** 1119-1123.
47. Edman CF, Raymond DE, Wu DJ, Tu E, Sosnowski RG, Butler WF, Nerenberg M and Heller MJ, Electric field directed nucleic acid hybridization on microchips, *Nucleic Acids Research* 1997, **25:** 4907-4914.
48. Walker GT, Little MC, Nadeau JG and Shank DD, Isothermal in vitro amplification of DNA by a restriction enzyme/DNA polymerase system, *PNAS* 1992; **89:** 392-396.
49. Westin L, Xu X, Miller C, Wang L, Edman CF and Nerenberg M, Anchored multiplex amplification on a microelectronic chip array, *Nature Biotechnology* 2000; **18:** 199-204.
50. Stimpson DI, Cooley PW, Knepper SM and Wallace DB, Parallel production of oligonucleotide arrays using membranes and reagent jet printing, *Biotechniques* 1998, **25:** 886-890.
51. Okamoto T, Suzuki T and Yamamoto N, Microarray fabrication with covalent attachment of DNA using bubble jet technology, *Nature Biotechnology* 2000; **18:** 438-441.

FRONTIERS IN HUMAN GENETICS
Diseases and Technologies

© 2001 by World Scientific Publishing Co. Pte. Ltd.

SILICO BIOTECH

PANDJASSARAME KANGUEANE*,†,§ and MEENA K SAKHARKAR‡
BioInformatics Centre, #02-07, MD7, Medical Drive
National University of Singapore, Singapore 119260
**Department of Microbiology, MD4, Medical Drive*
National University of Singapore, Singapore 119260
E-mail: †kangs@bic.nus.edu.sg, ‡meena@bic.nus.edu.sg

The advancements in molecular biology have made biotechnology a billion-dollar business over the last two decades. Recent developments in instrumentation, nano-technology and information technology have provided the biomedical research community with enormous amounts of diverse information governing biological systems. Consequently, there is an urgent need for information storage, curation, analysis and retrieval (ISCAR) using bioinformatics tools. Though the very definition of bioinformatics is debatable, there is a general agreement about the importance of certain fundamental concepts. Broadly, bioinformatics is the marriage between modern biology and information technology to glean new knowledge from redundant databases. Bioinformatics helps researchers gather, standardize, combine and manipulate data to tease out the knowledge they contain. In future, it will guide in performing *in silico* biotechnological experiments to aid biomedical research and application. Hence "SILICO BIOTECH" highlights the simple relationships between different disciplines that govern the complex drug discovery process and its relevance in health care.
Keywords: health care, biomedicine, silico-biotech, bioinformatics, biotechnology, information technology

---

§ Corresponding author.

1 INTRODUCTION

Intellectual revolution creates new visions, new ideas and new hopes for a better social living. When new fields emerge from new ideas, old words are usually not adequate to describe these fields. "Silico Biotechnology" and "Bioinformatics" are best described with examples rather than with single words or short phrases. Unlike chemistry and physics, mathematical theories and quantitative methods (except statistics) have played a secondary role in the creation of knowledge in biomedicine. Most of the advancement in biomedicine has been due to improvements in experimental tools. Results are qualitative hence descriptive models are formulated and tested. Comparatively, biologists often have inadequate backgrounds in mathematics but are very strong with respect to laboratory tools and, more importantly, with respect to the interpretation of laboratory data from complex systems. Engineers usually possess a very good background in physical and mathematical sciences. Theories lead to mathematical formulation and comparing the predicted response to experimental data tests the validity of the theory. Biologists are usually better at the formation of testable hypothesis, experimental design and data interpretation from complex systems. Engineers are typically unfamiliar with the experimental techniques used by biologists. Hence, the skills of engineers and biomedical scientists are complementary. To convert the promises of genetic engineering for new processes to make new products requires the integration of these skills. Developments in genomics[1–4], proteomics[5], pharmacogenomics[6,7], nanotechnology[8,9] and bio-informatics[10,11] will pave the way for an information revolution in biology and medicine, which will result in personalized health care. A bioinformatics professional needs solid understanding of biotechnology, information technology and mathematics. An ideal bioinformatics professional can create a virtual skeleton for *in silico* biotechnological experimentation based on individual genetic constitution. Thus, "Silico Biotech" would serve as a platform for disease diagnosis, treatment and prevention.

2 EVOLUTION OF SILICO BIOTECH

Before the industrial production of penicillin, almost no chemical engineer sought specialized training in the life sciences. With the advent of modern antibiotics, the concept of biochemical engineering was established. The penicillin process also emanated a paradigm for bioprocess development and biochemical engineering. Biochemical engineering means the extension of chemical engineering principles to systems using biological catalyst to bring about desired chemical transformations. The use of recombinant DNA technology (rDNA) to improve a biochemical process lead to the creation of a new discipline called Biotechnology.[12] In otherwords, the use of rDNA technology to improve a *"process"* (chemical) is called biotechnology. This discipline draws professionals from electrical, mechanical, industrial, environmental, and chemical engineering.

The recent success of the international scientific community in decoding the genetic blueprint of the entire human genome[13] has lead to a post genomics era, where there is an overarching need for an intellectual fusion of biomedicine and information technology. The proposed marriage between biomedicine and information technology in a productive way is cardinal for knowledge discovery from information repositories. Bioinformatics plays a crucial role in data manipulation, data curation and knowledge extraction, thus bridging the gap between disparate information sources for subsequent improvements in biomedicine and health care. The knowledge base generated using bioinformatics tools will serve as input variables for in-silico biotechnological simulation specific for individual's genetic constitution. A schematic representation for silico biotech evolution is illustrated in Fig. 1. In Fig. 1, the integral components of silico biotech and the underlying disciplines fundamental to its development are clearly shown, showing its true inter-disciplinary nature.

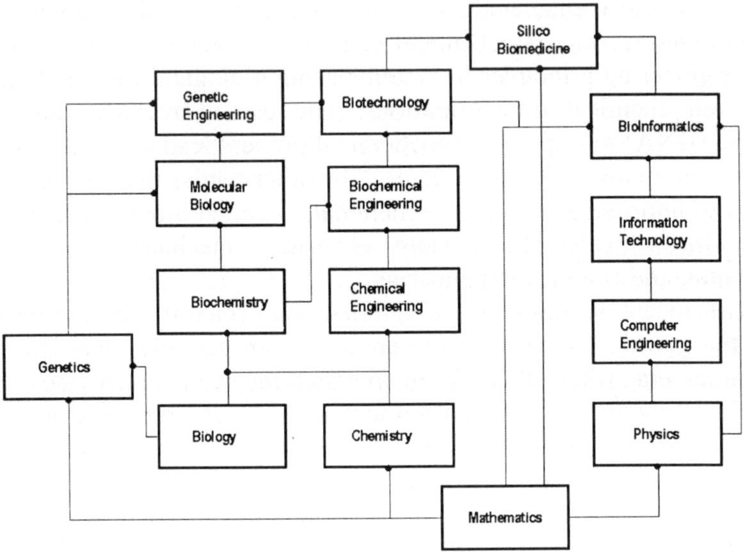

Fig. 1 Silico Biotech evolution

The convergence of genomic technology and computational advances are leading to innovative uses of existing data for rational drug design[14]. Information is the key because life at the molecular level can be understood as a process in which information is communicated between cellular compartments and adapted by a balanced process of selection. Mapping the connection between gene sequences and macroscopic life is fundamentally a problem of describing and modeling biological information processes. Simulation of molecular processes in cells using structured mathematical models and subsequent prediction of drug effects in humans will advance pharmaceutical research and speed up clinical trials. For an imaginary patient to benefit from the fruits of silico biotech it is imperative to tie rDNA technology with genome computation[15] and legal regulations[16].

3 CONCLUSION

From genomics to combinatorial chemistry, scientific advances are poised to revolutionize drug discovery and health care. New chemical and biological approaches are changing the way in which therapeutic agents are discovered, developed and administered. The complete mapping of human genes to their function in the context of disease and immunity using genomics, proteomics and pharmacogenomics techniques will ultimately result in the rational identification of individual specific drug targets. For example, SNPs, short for single nucleotide polymorphisms are places along the chromosomes where the genetic code tends to vary from one person to another by just a single base. SNPs in genes or control regions may influence susceptibility to common diseases. SNPs promised to pinpoint the genes involved in common diseases such as hypertension, cancer and diabetes. Also, molecular modeling of the minor histocompatibility antigen[17] HA-1 peptides binding to HLA alleles will be useful as an aid for defining a wider pool of HLA alleles in which HA-1 disparities among donor-recipient pairs can be investigated.[18]

Systematic quantification of the differences in function as a result of allelic variation within each of a protein family specific to a tissue or organ or system will lead to the development of a methodology for generation of a library of potential drug candidates *in silico.* The successful sampling of drug targets from a pool of chemical/biological entities using computational tools will result in faster and effective treatment of diseases.

Acknowledgements

We wish to express our sincere appreciation to all members of the research centre for discussions on the subject of this article, especially to A/P Tan Tin Wee, Director, BioInformatics Centre, National University of Singapore, Singapore and Dr Prasanna R. Kolatkar, Chairman, Research Council, BioInformatics Centre, Singapore.

References

1. Fleischmann RD, Adams MD, White O, Clayton RA, Kirkness EF, Kerlavage AR, Bult CJ, Tomb JF, Dougherty BA, Merrick JM, Whole-genome random sequencing and assembly of *Haemophilus influenzae Rd.*, *Science* 1995; **269:** 496–512.
2. U.S. Department of Health and Human Services and Department of Energy, Understanding Our Genetic Inheritance. The US Human Genome Project: The First Five Years (April 1990).

3. Collins, FS, Galas D, A New Five-Year Plan for the US Human Genome Project, *Science* 1993; **262:** 43–46.
4. Collins FS, Patrinos A, Jordan E, Chakravarti A, Gesteland R, Walters L, A New Five-Year Plan for the US Human Genome Project: 1998-2003, *Science* 1998; **282:** 682–689.
5. James P, Protein identification in the post-genome era: the rapid rise of proteomics, *Quarterly Reviews in Biophysics* 1997; **30:** 279–331.
6. Marshall A, Genset-Abbott deal heralds pharmacogenomics era, *Nature Biotechnology* 1997; **15:** 829–830.
7. Housman D, Ledley FD, Why pharmacogenomics? Why now?, *Nature Biotechnology* 1998; **16:** 492–493.
8. Freedman DH, Exploiting the nanotechnology of life, *Science* 1991; **254:** 1308–1310.
9. Bethell D, Schiffrin DJ, Supramolecular chemistry, Nanotechnology and nucleotides, *Nature* 1996; **382:** 581.
10. Altman RB, A curriculum for bioinformatics: the time is ripe, *Bioinformatics* 1998; **14:** 549–550.
11. Franklin J, Bioinformatics changing the face of information, *Annals of the New York Academy of Sciences* 1993; **700:** 145–152.
12. Kristapsons MZ, Iakobson IuO, Symposium on biotechnology and bioengineering, *Mikrobiologiia* 1978; **47:** 1129–1132.
13. http://www.ornl.gov/TechResources/Human_Genome/home.html
14. Drews J, Drug Discovery: A Historical Perspective, *Science* 2000; **287:** 1960–1964.
15. Sander C, Genomic Medicine and the future of Health care, *Science* 2000; **287:** 1977–1978.
16. Barton JH. Reforming the patent system, *Science* 2000; **287:** 1933–1934.
17. den Haan JM, Meadows LM, Wang W, Pool J, Blokland E, Bishop TL, Reinhardus C, Shabanowitz J, Offringa R, Hunt DF, Engelhard VH, Goulmy E, The minor histocompatibility antigen HA-1: a diallelic gene with a single amino acid polymorphism, *Science* 1998; **279:** 1054–1057.
18. Ren EC, Kangueane P, Kolatkar P, Lin MT, Tseng LH, Hansen JA, Molecular modeling of the minor histocompatibility antigen HA-1 peptides binding to HLA-A alleles, *Tissue Antigens* 2000; **55:** 24–30.

FRONTIERS IN HUMAN GENETICS
Diseases and Technologies

© 2001 by World Scientific Publishing Co. Pte. Ltd.

BIOINFORMATICS INTEGRATION SIMPLIFIED: THE KLEISLI WAY

LIMSOON WONG
Kent Ridge Digital Labs
21 Heng Mui Keng Terrace, Singapore 119613
E-mail: limsoon@krdl.org.sg

The power of the bioinformatics integration system called KRIS, better known as Kleisli, is reviewed through a series of examples.
Keywords: bioinformatics, data integration, scripting

1 INTRODUCTION

The Kleisli system was developed as a general solution to broad-scale data integration problems. Wong and his collaborators used the bioinformatics arena as the first testbed of his system [1] and succeeded in making an impression on the field [2,3,4] We provide here an account of the system.

Many problems in modern bioinformatics involve (a) accessing complex heterogeneous data sources that are geographically dispersed, (b) multiple sequential steps, and (c) passing information smoothly between these steps. Simple retrieval of data is not sufficient for modern bioinformatics. With the rapid growth of experimental data, in order to investigate a specific biological problem, the challenge is how to automate the process of manipulating and re-structuring of the information derived from various databases. This may require combining data derived from multiple public sources and local (private) sources and feeding the retrieved data into various application programs such as gene finding, protein structural prediction, functional domain or motif identification, phylogenetic tree construction, etc. All these procedures require specific input data sets and data formats. As observed by Baker and Brass, [4] many existing biology data retrieval systems [5,6] are not fully up to the demand of flexible and painless data integration. This is where the power of Kleisli comes into play.

Kleisli [1] is a powerful data integration system, implemented on top of a robust modern functional programming technology.[7] The system interfaces to a large number of data sources relevant to bioinformatics and uses a self-describing data model to allow data derived from different sources to be flexibly combined. There are more than two hundred biological databases and servers in the Internet.[8,9] Merely providing an interface to a collection of databases and analysis software is often not useful if it requires tedious programming to make use of the interface, as is the case with CORBA.[10] Kleisli goes one step further and provides a high-level query language called Collection Programming Language (CPL), based on elegant mathematical principles.[11,12] CPL offers a rich data model and many high-level operators to express complex queries and tranformations on these biology databases and analysis software in a straightforward manner that does not require extensive programming skill.

Many bioinformatics queries involving multiple databases and analysis software in multiple steps have a simple expression in CPL and are efficiently executed by Kleisli. In order to properly appreciate the virtue of the system, it is necessary to see some real-life examples. We describe below four queries taken from the many that were posed to us. These examples exercise many aspects of Kleisli and involves integration across Entrez,[5] scop,[13] HMMER,[14] WU-BLAST2,[15] Gapped BLAST,[16] patents, proteins, DNA sequences, feature tables, etc. We hope the succintness of these examples is sufficient illustration of the power, flexibility, and simplicity of the Kleisli system.

Section 2 is our first example, which asks what proportion of human mature peptides have prolines at their N-terminal. This simple example serves as a quick introduction to the basic syntax of CPL. Section 3 is our second example, which collects samples of TPR domains and uses these samples to contruct a model to recognize other TPR domains in Swissprot sequences. Section 4 is our third example, which asks what other protein sequences in the same superfamily of a given protein sequence have been patented. Section 5 is our last example, which selects from a file of pufferfish DNA fragments those that have no homology to vertebrate protein sequences but have some indirect relationship to some human or mouse ESTs. Section 6 rounds up the presentation by briefly discussing the architecture of the Kleisli system.

2 EXAMPLE: PROLINE AT N-TERMINAL

The first of our four examples is this query: *What proportion of mature peptides from human have prolines at their N-terminal?* Its implementation in Kleisli/CPL is given below.

```
1. { string-span (x.#sequence, f.#start, f.#end)
2.   | \x <- aa-get-seqfeat-general "human mature peptide",
3.     \f <- x.#feature, f.#name string-islike "%mat_peptide%"}
4. before (\mat-peptides =>
5.   {x | \x <- mat-peptides, x string-islike "P%" }
6. before (\prolines =>
7.   set-roughcount(prolines) / set-roughcount(mat-peptides)));
```

The syntax and meaning of CPL program is straightforward. e `before` $(\backslash x \Rightarrow e')$ means evaluate e and set the variable x to the result and then evaluate e' (which may use the value of x.) $\{e \mid \backslash x \leftarrow S, C\}$ means the set of e such that x is in the set S and the condition C is true; or in more prosaic words, iterate over each item x in the set S, if the condition C is true for this x, then evaluate e (which may depends on the value of this x) and include the result in the output set. $e.\#l$ means the value at the l field of the record e. Operators such as `aa-get-seqfeat-general` are provided in the Kleisli system to interface to various bioinformatics sources such as Entrez.

We now return to explaining the implementation of the proline query above. It basically consists of three steps. Firstly, Entrez is accessed to obtain GenPept reports of human amino acid sequences (line 2). For each report `x` of a sequence, its feature table is inspected to extract feature entries (line 3). For each feature entry `f`, we check if it annotates a mature peptide (line 3). If it is a mature peptide, the start and end positions are extracted from the annotation and are used to extract that segment of the sequence of `x` (line 1). The resulting sequence segments are assigned to the set `mat-peptides` (line 4).

Secondly, for each sequence segment `x` in the set `mat-peptides`, we check if it starts with the letter `P` (line 5). If so, `x` is a mature peptide that has a proline at its N-terminal. These resulting sequence segments are assigned to the set `prolines` (line 6).

Finally, we just count the number of items in the set `prolines` and the number of items in the set `mat-peptides`. Their ratio gives the desired answer (line 7).

3 EXAMPLE: TPR IN SWISSPROT SEQUENCES

It is not unusual for a biologist to "hunt" for a particular domain in large sequence databases. The second of our four examples is typical of domain hunting: *Build a hidden Markov model*[17] *of TPR domains*[18] *and use it to recognize TPR domains in Swissprot sequences.* Its implementation in Kleisli/CPL is given below.

```
 1. writefile
 2.  {(#uid: r.#uid, #title: r.#title, #accession: r.#accession,
 3.    #seq: string-span (r.#sequence, f.#start, f.#end))
 4.  | \r <- aa-get-seqfeat-general "TPR",
 5.    \f <- r.#feature,
 6.    (f.#end - f.#start) < 40, (f.#end - f.#start) > 25,
 7.    (f.#name string-islike "%mat_peptide%") orelse
       (f.#name string-islike "%region%"),
 8.    \d <--- f.#anno, d.#descr string-islike "%TPR%",
 9.    not (d.#descr string-islike "%putative%"),
       not (d.#descr string-islike "%hypothetical%")}
10. to "tpr.hmm" using hmm-localtraining;
11. hmm-localsearch (
12.  #name: "hmm-tpr", #model: "tpr.hmm", #threshold: 0,
13.  #level: 1, #dna: false, #window: 1000, #best: false);
14. readfile sprot from "sprot36.fa" using fasta-in;
15. {(#title: x.#title, #start: r.#seqstart, #end: r.#seqend)
16. | \x <- sprot,
17.   \h <- process <#hmmfs: x.#seq> using hmm-tpr,
18.   \r <- h.#hits, r.#score > SCORE};
```

This implementation has four steps. First, we extract known examples of TPR domains and construct a hidden Markov model (HMM) for it. This task is straightforward. We first obtain all GenPept reports that mention TPR from Entrez (line 4). For each such report `r`, we examine its feature table (line 5). For each feature `f`, we check if it has the expected length of a TPR domain (line 6) and if it is recorded as a mature peptide or as a region (line 7). If so, we check if there is an annotation `d` on that feature that contains the word `TPR` (line 8) and does not contain the word `putative` nor `hypothetical` (line 9). Such a feature can be safely regarded as a TPR domain. So we extract the sequence segment from the start and end positions of the feature `f` from the sequence `r` (line 3); other associated information of `r` are also extracted (line 2). The resulting set is then used to construct a HMM, which is written to the file `tpr.hmm`. This is the HMM of our TPR domains.

Next, a HMMER[14] connection `hmm-tpr` is established to the hidden Markov model `tpr.hmm` using standard parameters (lines 11–13).

In the third step, the set `sprot` is set to the contents of the file `sprot36.fa`, which is our copy of Swissprot (line 14).

In the last step, we iterate over every sequence in `sprot` (line 16). For each sequence `x`, we check it using our HMMER connection `hmm-tpr` (line 17). We examine the reply `h` to see if it contains a hit `r` with a score greater than the threshold `SCORE` (line 18). If so, we include the title of the sequence `x` and the start and end positions of the hit `r` in the output set as desired.

4 EXAMPLE: PATENTED PROTEINS IN SUPERFAMILY

Assessing the patent potential of a sequence is of practical importance. The third of our four examples is an interesting query on this subject: *Suppose we are given a protein sequence* `SEQ`. *List patented sequences that belong to the same superfamily as* `SEQ`. By a superfamily, we mean sequences that have the same tertiary structure and the same function, but they do not have to have similar sequences. The implementation in Kleisli/CPL is given below.

```
localblast-blastp (#name: "scop-blast", #db: "scopseq");
localblast-blastp (#name: "pat-blast", #db: "patseq");
setindex-access(#name:"sid2seq",#file:"scopseq",#key:"#sid");
scop-add "scop";
{(#sf:(#desc:i.#desc.#sf,#hit:x.#accession,#pscore:x.#pscore),
  #bridge: (#hit: s, #patent: p.#title, #pscore: p.#pscore))
| \x <- process SEQ using scop-blast, x.#pscore <= PSCORE,
  \i <- process <#sidinfo: x.#accession> using scop,
  \s <- process <#numsid: i.#type.#sf> using scop,
  \y <- process <#key: s> using sid2seq,
  \p <- process y.#seq using pat-blast, p.#pscore <= PSCORE};
```

The implementation consists of two steps. In the first step, we set up connections to our data sources. We need to use scop[13] to obtain superfamily classification. We also need access to patented sequences (we got them in the file `patseq`) and to representative sequences for the various superfamilies in scop (we got them in the file `scopseq`). So we establish a WU-BLAST2 [15] connection `scop-blast` to `scopseq` (line 1), a WU-BLAST2 connection `pat-blast` to `patseq` (line 2), and a connection `scop` to scop (line 4). To provide random access to representative sequences of scop superfamilies given their scop identifier (`sid`), we also establish the database index `sid2seq` (line 3). Having set all these connections up, we are ready to do the real work.

We now proceed to look for our patented sequences. We BLAST the given sequence `SEQ` against scop superfamily representative sequences to look for a hit `x` at a strong threshold `PSCORE` (line 7). `SEQ` is regarded to be in the same superfamily as `x`. We look up `x`'s record from scop (line 8). From this record, we ask scop for all the identifiers of representatives in the superfamily of `x` (line 9). Each representative `s` serves as a "bridge" to help us identify the desired patented sequences. For each such representative `s`, we use our index to look up its sequence `y` (line 10). We BLAST this sequence `y` against the patented sequences; those hits at the strong threshold `PSCORE` are the patented sequences we want (line 11). The relevant information (the superfamily information of `x`, the bridge `s`, and the patent information `p`) are returned in the output set (lines 5–6).

5 EXAMPLE: COMPACT GENOME OF PUFFERFISH

The small size of the pufferfish genome and the close correspondence of pufferfish genes exon-intron structures to homologous human genes make it an ideal shortcut to understanding the human genome. [19] Our last example is: *Given a file of pufferfish (expressed) DNA fragments, identify those that are "interesting."* A fragment is considered interesting if it is not known to be homologous to any known vertebrate proteins and if it has good homology to some human or mouse ESTs. Such a fragment is interesting because it is likely to be part of a pufferfish gene that is homologous to a human or mouse gene that is not yet sequenced. Here is the implementation in Kleisli/CPL.

```
1. writefile
     {(#uid: x) | \x <-aa-get-uid-general "vertebrata[Organism]"}
2. to "vert" using stdout;
3. writefile "vert" to "#uid" using setindex-create;
```

```
4. setindex-access(#name:"vert",#file:"vert",#key:"#uid");
5. webblast2-blastx-detail(#name:"nr-blast",#db:"nr",#level:1);
6. localblast-tblastn (#name:"est-blast", #db:"man-mouse-est");
7. readfile fugu from "fugu" using stdin;
8. writefile {(#fugu: f, #homologs: B)
9.    | \f <- fugu,
10.     \B == {(#title:x.#title,#uid:x.#uid,
                #pscore:x.#hits.list-head.#pscore)
11.             | \x <- process f.#seq using nr-blast,
                x.#hits.list-head.#pscore < NR-PSCORE}}
12. to "proteins" using stdout;
13. readfile proteins from "proteins" using stdin;
14. writefile {(#fugu: x.#fugu, #bridge: b)
15.    | \x <- proteins,
16.      set-isempty {z
17.        | \y <- x.#homologs,
18.          \z <- process <#key: y.#uid> using vert},
19.      \b <- x.#homologs}
20. to "bridges" using stdout;
21. readfile bridges from "bridges" using stdin;
22. writefile {(#bridge: b, #est: e)
23.    | \b <- set-unique {x.#bridge.#uid | \x <- bridges},
24.      \s <- aa-get-seq-by-uid (b.#uid),
25.      \e <- {(#uid:y.#uid, #title:y.#title, #pscore:y.#pscore)
26.             | \y <- process s.#sequence using est-blast,
                y.#pscore < EST-PSCORE}}
27. to "ests" using stdout;
28. writefile "ests" to "#bridge" using setindex-create;
29. setindex-access(#name:"ests", #file:"ests", #key:"#bridge");
30. {(#fugu: x.#fugu, #bridge: x.#bridge, #ests: H)
31. | \x <- bridges,
32.    \E == process x.#bridge using ests, not(set-isempty(E)),
33.    \H == {e.#est | \e <- E}};
```

The implementation consists of five steps. First, we need to set up connections to the data sources. To determine if a pufferfish DNA fragment is similar to any known proteins, we set up a Gapped BLAST2 connection `nr-blast` to the nonredundant protein sequence database at the National Centre of Biotechnology Information (line 5). To determine if a protein is similar to any human or mouse EST, we set up a local WU-BLAST2 connection `est-blast` to our private copy of human and mouse ESTs (line 6). To determine if a protein hit by `nr-blast` is vertebrate or not, we copy from Entrez all the identifiers of vertebrate amino acid sequences into a local file (lines 1–2). In order to speed up access to this file of vertebrate protein identifiers, we also create an index `vert` on this file (lines 3–4).

The pufferfish DNA fragments to be tested are given in the FASTA file `fugu`, which is now read in (line 7). We iterate over each DNA fragment `f` in the file (line 9). For each `f`, we BLAST it using `nr-blast`, keeping only those hits at a strong threshold `NR-PSCORE` (line 11) in the set `B` (line 10). This pufferfish DNA fragment and the corresponding set of protein homologs `B` are written to a file (lines 8, 12).

In the third step, we examine this file (line 13), as we wish only to keep those DNA fragments whose protein homologs are strictly nonvertebrate. For each DNA fragment `x` (line 15), we test each of its protein homologs against our vertebrate index to make sure they are not in the index (lines 16–18). If so, then each such protein homolog `b` is considered a "bridge" for the DNA fragment `f` (line 19) and is recorded in the file `bridges` (lines 14, 20).

In the fourth step, we examine this file of "bridges" (line 21), as we wish to find out those ESTs of human and mouse hit by them. For each "bridge" protein `b` (line 23), we get its sequence `s` from Entrez (line 24). Then we BLAST this sequence using est-blast and keep those hits at a strong threshold `EST-PSCORE` (lines 25–26). Each such hit `e` and the corresponding "bridge" `b` are written to a file (lines 22, 27). To allow very fast access to the EST hits of a "bridge", we also build an index on this file (lines 28–29).

Finally, all that remains to be done is to iterate over every pufferfish DNA fragment-protein "bridge" pair `x` in the file `bridges` (line 31), pull out the already computed EST hits of the "bridge" component of `x` and make sure it is nonempty (line 32), extract the relevant portion of these EST hits (line 33), and return the relevant information in the output set (line 30).

6 THE GUT OF KLEISLI

The examples show that the Kleisli system can handle a broad range of bioinformatics queries that require multiple data sources and analysis tools in a number of different specialized ways. The examples also show that queries that have a logically simple description also have a simple CPL description and thus a simple implementation in Kleisli. In fact, more often than not, the CPL program is considerably more succint and precise than the description in English! We hope we have succeeded in conveying the power of the Kleisli system.

We close this paper by briefly describing the architecture and the factors that made Kleisli a successful bioinformatics integration system.

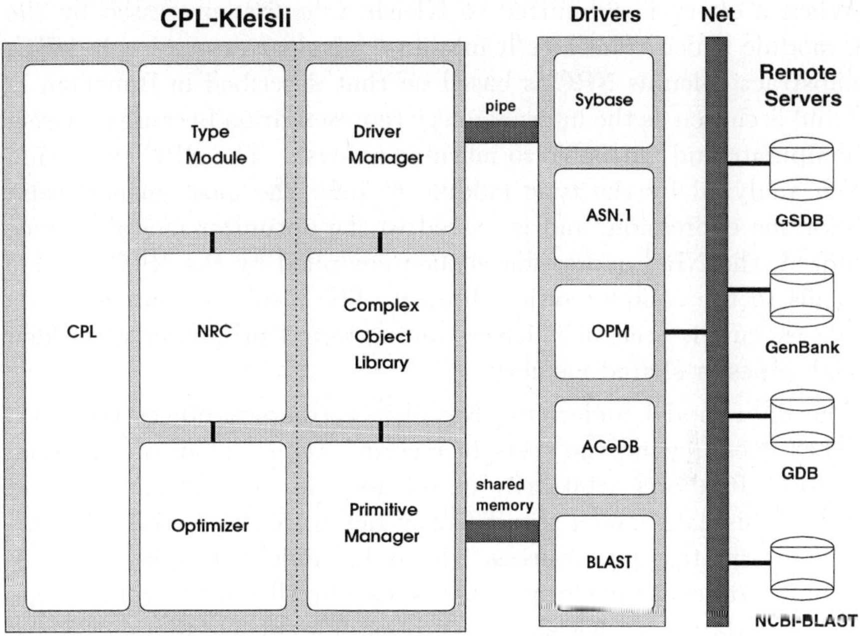

Fig. 1: Architecture of Kleisli.

The architecture of the system is depicted in Figure 1. Kleisli is extensible in many ways: It could be used to support many other high-level query languages by replacing the CPL module. Kleisli could also be used to support many different types of external data sources by adding new drivers. The installation at Kent Ridge Digital Labs contains drivers for all popular bioinformatics systems, including Sybase, Oracle, Entrez, WU-BLAST2.0, Gapped BLAST, PSI-BLAST, HM-

MER, ACEDB, Pfscan 2.0, NetStart, etc. Furthermore, the optimizer of Kleisli can be customized by adding different rules and strategies.

The core of the Kleisli system is divided into two main components, as shown by the dotted line in the figure. The first component provides high-level language support and consists of the CPL module, the type module, the Nested Relational Calculus (NRC) module, and the optimizer. The second component is the query engine and consists of the driver manager, the primitive manager, and the complex object library.

When a query is submitted to Kleisli, it is first processed by the CPL module which translates it into an equivalent expression in NRC. The abstract calculus NRC is based on that described in Buneman *et al.*,[12] and is chosen as the internal query representation because it is easy to manipulate and amenable to machine analysis. The NRC expression is then analyzed by the type module to infer the most general valid type for the expression, and is passed to the optimizer module. Once optimized, the NRC expression is then compiled by the NRC module into calls to the complex object library. The resulting compiled code is then executed, accessing drivers and external primitives as needed through pipes or shared memory.

The data model underlying Kleisli is a complex object type system that goes beyond the "sets of records" type system of relational databases. It allows arbitrarily nested records, sets, lists, bags, and variants. This data model is sufficiently rich to readily embed all common bioinformatics data sources. This data model is coupled to a self-describing data exchange format that is used by all external components to exchange data with Kleisli. All core modules of Kleisli are designed around or derived from this data model.

There are several key ideas behind the success of the system. The first is its use of complex object data model where sets, bags, lists, records, and variants can be flexibly combined. The second is its use of a high-level query language CPL which allows these objects to be manipulated easily. The third is its use a self-describing data exchange format, which serves as a simple conduit to external data sources. The fourth is its query optimizer, which is capable of many powerful optimizations.

Availability

The Kleisli system is available from KrisTech Inc. in California under the name KRIS.

References

1. Davidson S *et al.*, BioKleisli: A digital library for biomedical researchers. *International Journal of Digital Libraries* 1997; **1**:36–53.
2. Benton D, Bioinformatics—principles and potential of a new multidisciplinary tool. *Trends in Biotechnology* 1996, **14**:261–272.
3. Karp PD, Database links are a foundation for interoperability. *Trends in Biotechnology* 1996, **14**:273–279.
4. Baker PG, Brass A, Recent development in biological sequence databases. *Current Opinion in Biotechnology* 1998, **9**:54–58.
5. Schuler GD *et al.*, Entrez: Molecular biology database and retrieval system. *Methods in Enzymology* 1996, **266**:141–162.
6. Etzold T, Argos P, SRS: Information retrieval system for molecular biology data banks. *Methods in Enzymology* 1996, **266**:114–128.
7. Appel AW, *Compiling with Continuations.* Cambridge University Press, 1992.
8. Ashburner M, Goodman N, Informatics—genome and genetic databases. *Current Opinion in Genetics & Development* 1997, **7**:750–756.
9. Karp PD, Metabolic databases. *TIBS* 1998, **23**:114–116.
10. Siegel J, *CORBA: Fundamentals and Programming.* Wiley, New York, 1997.
11. Wadler P, Comprehending monads. *Mathematical Structures in Computer Science* 1992, **2**:461–493.
12. Buneman P *et al.*, Principles of programming with complex objects and collection types. *Theoretical Computer Science* 1995, **149**:3–48.
13. Murzin A *et al.*, SCOP: A structural classification of protein database for the investigation of sequences and structures. *Jour-*

nal of Molecular Biology 1995, **247**:536–540.

14. Sonnhammer EL *et al.*, Pfam: Multiple sequence alignments and HMM-profiles of protein domains. *Nucleic Acids Research* 1998, **26**:320–322.
15. Altschul SF, Gish W, Local alignment statistics. *Methods in Enzymology* 1996, **266**:460–480.
16. Altschul SF *et al.*, Gapped BLAST and PSI-BLAST: A new generation of protein database search programs. *Nucleic Acids Research* 1997, **25**:3389–3402.
17. Eddy SR, Hidden Markov models. *Current Opinion in Structural Biology* 1996, **6**:361–365.
18. Lamb JR *et al.*, Tetratrico peptide repeat interactions: to TPR or not to TPR? *Trends in Biochemical Sciences* 1995, **20**:257–259.
19. Elgar G, Quality not quantitiy: The pufferfish genome. *Human Molecular Genetics* 1996, **5**:1437–1442.

FRONTIERS IN HUMAN GENETICS
Diseases and Technologies
© 2001 by World Scientific Publishing Co. Pte. Ltd.

ISSUES IN SECONDARY STRUCTURE PREDICTION QUALITY

BETTY CHENG
BioInformatics Centre, National University of Singapore,
10 Kent Ridge Crescent, Singapore 119260
E-mail: betty@bic.nus.edu.sg

Secondary structure prediction is one of the most reliable of protein structure prediction methodologies, but the rapid developments require constant evaluation in order to select the best method. Key criteria such as selection of data for training sets, cross-validation, prediction accuracy, are reviewed and discussed in the context of judging secondary structure prediction methods. Where possible, a comparison with prediction methods is presented.
Keywords: secondary structure prediction, protein structure prediction.

1 INTRODUCTION

Development of methods for protein structure prediction has been extremely rapid in the last few years and several exciting tools have become available. Protein secondary structure prediction has seen especially rapid progress and the state-of-the-art predictive methods are reliable enough for general use provided the user is aware of the limitations of the prediction.

On-line protein structure prediction resources have proliferated and it is easy for the most casual user to access and obtain a prediction for their sequence. The reader is confronted with a bewildering variety of predictive methods. Some sites are poorly documented, making it difficult to evaluate the quality of the prediction.

Several criteria to evaluate predictive methods are reviewed here. Instead of reviewing the most accurate method of the moment, the focus will be on using these criteria to evaluate the quality of secondary structure prediction methods allowing the reader to select the most appropriate predictive method.

Protein structure may be divided into several categories, secondary structure, three-dimensional or tertiary structure and quaternary structure. Unlike, quaternary or tertiary structure, secondary structure gives one-dimensional information. Predictions may tell us that a particular portion of an amino acid sequence contains a helical segment but no information about how this helix packs against other portions of the sequence is contained within a secondary structure prediction.

Secondary structure refers to repeating structural motifs such as alpha helices, beta-sheets. Secondary structure was originally assigned by human judgment, as a result, there are several competing definitions, which do not always agree. Several automated methods use different definitions of secondary structure: DSSP is based on hydrogen bonding,[1] STRIDE uses dihedral angles and hydrogen bonds,[2] DEFINE uses interatomic distances,[3] the PRISM[4] method defines secondary structural *states*, defined as regions of the phi-psi space. Secondary structure definitions are often unsatisfactory because only small differences may separate residues, which are defined as helical from those that are not.

Commonly, secondary structure prediction methods attempt to predict three secondary structural states, helical (h), beta-sheet (s), or coil/loop (c or l) for each residue in a sequence (Table 1). Some methods will attempt to give more information by defining and predicting more than 3 states, an important point to remember when comparing performance for different methods.

Table 1 The first row shows an amino acid sequence, the second line is the secondary structure prediction. For each amino acid, there are three possible states helical (H), beta-sheet (or extended)(E), loop (L) and * = no prediction. A long stretch of helical states as shown below indicates the presence of a helix.

MKFIIAFFVATLAVMTVSGEDKKHDYQNEF...
***HHHHHHHHHHHHHHHHH**LLLLLLL**HHHHHH...**
Helix one Helix two

The latest generation of secondary structure prediction methods is reliable enough for general use but what is the utility of knowing where the helices, beta-sheets and loops are located without knowing how these elements pack against each other? In fact, secondary structure predictions do not give a tremendous amount of information except in the context of other structural studies, which produce complementary information such as X-ray crystallography, NMR, electron microscopy, protein structure prediction, or biochemical methods of extracting functional/structural information.

2 THE TRAINING SET

Most predictive models use experimental data and will be limited by the scope and accuracy of the data. Such methods work well on known structures but may have difficulty with sequences with completely novel folds. It is particularly important to be aware of the quality and type of data used in the training set. The ideal training set will be composed of high quality protein structures; these structures are usually obtained from X-ray crystal data. Currently, the set of known structures is large enough that training sets with resolution of 2.5 Angstrom or better are routinely used. Furthermore, the structures selected for inclusion in the training set should avoid redundancy. All homologous sequences should be removed. Typically, any sequence with 25% or more pairwise sequence similarity should be excluded from the training set in order to prevent bias. A training set composed of only sequences with the same fold, for instance, if the training set were to consist of only the globin family, will produce a method that can only predict globin secondary structure.

3 ACCURACY

The first criteria used to judge prediction is the prediction accuracy. This is simply the number of correctly predicted residues divided by the total number of residues. A reasonable way to estimate the accuracy and reliability of a predictive method is to submit a set of sequences with high-quality structures to the predictor to obtain the average accuracy for this set. However, because of the limited number of high-quality structures, all of these structures were probably used to train the predictive method giving abnormally high accuracy. This problem can be corrected using cross-validation.

4 CROSS-VALIDATION

For information-based predictive methods, some form of cross-validation is essential: Leave-One-Out cross-validation (the most accurate), cross-validation, or jack-knifing are all acceptable. For leave-one-out cross-validation, one data point (one protein structure) is withheld from the training set and the predictive method is trained. A prediction is made for the one data point, which is not part of the training set and the accuracy calculated. The process is repeated for all the remaining data points in the training set. This gives a very reliable estimate of prediction accuracy but the leave-one-out cross-validation is resource and labor-intensive. A reasonable compromise is to use cross-validation where instead of leaving out only one data point, a larger subset is left out of the training set, e.g. about 20%. This greatly reduces the labor while the estimate of the prediction accuracy remains reliable.

5 TYPES OF ERRORS

There are different types of prediction errors some more serious than others. Table 2 shows examples of different types of errors. The most obvious type of error is mis-prediction where the secondary structure for one residue is incorrectly predicted. For simplicity, I shall follow Rost and Sander's notation.[5] The accuracy matrix, **A,** is defined as A_{ij} = the number or residues predicted in secondary structure type i and observed to be in structure type j. The accuracy of a prediction is easily measured by the following equation:

$$Q = \sum_{i=1}^{3} A_{ii} / \sum_{i=1}^{3} \sum_{i=1}^{3} A_{ij}$$

where the numerator contains the number of correctly predicted residues and the denominator is simply the total number of residues in the data set.

However, there are other types of errors such as shift errors, which cannot be measured by accuracy alone.

Table 2 For a hypothetical amino acid sequence of nine residues, the correct secondary structure is shown in row A. Predictions with different types of errors are shown in rows below. (i) shows an incorrect prediction of a helical state for a coil (ii) shows a shift error (iii) incorrectly predicts two helices (iv) under-prediction of helix, (v) over-prediction of helix

	X_1	X_2	X_3	X_4	X_5	X_6	X_7	X_8	X_9
A	c	c	h	h	h	h	h	c	c
i	h	c	h	h	h	h	h	c	c
ii	c	h	h	h	h	h	h	c	c
iii	c	c	h	h	c	h	h	c	c
iv	c	c	l	l	l	l	l	c	c
v	h	h	h	h	h	h	h	h	h

The basic types of errors include shift errors, under or over prediction of helices/sheets/loops, incorrect prediction of the lengths of secondary structural elements, incorrect prediction of the number of secondary structural elements. Shift errors are errors in predicting the beginning or end of a secondary structural element such as a helix or a beta sheet. The endpoints of secondary structural elements are difficult to predict properly (due in part the definition of secondary structure).

Since secondary structure predictions are often used as starting points for 3-D structure predictions, certain types of errors are more serious than others. Many structural fold recognition algorithms rely on secondary structure prediction methods as a starting point. These fold recognition algorithms seek to match secondary structural elements to known fold profiles. Thus, a shift error may cause a helix to appear longer but this does not distort the fold profile whereas predicting an incorrect number of helices in a sequence will cause the fold recognition algorithm to match the sequence to a different profile. Thus, it is possible to have two predictions, with the same accuracy but one prediction has a much more serious error than the other. Examples (ii) and (iii) in Table 2 illustrates this situation.

Accuracy will not detect the different types of errors yet the differences are vital. It is imperative that additional measures to detect these different classes of errors be used. On this subject, Rost and Sander's article is most comprehensive.[5] For the convenience of the reader, the metrics for these more subtle errors are reproduced here; the interested reader is referred to the original articles for detailed discussions.[5-6]

$$I = 1 - \frac{\sum_{i=1}^{3} a_i \times \ln a_i - \sum_{i,j=1}^{3} A_{ij} \times \ln A_{ij}}{b \times \ln b - \sum_{i=1}^{3} b_i \times b_i}$$

$$b = \sum_{j=1}^{3} b_j = \sum_{j=1}^{3} a_j = \sum_{jj=1}^{3} A_{ij}$$

where I measures the information; when I=1, the prediction is perfect. I is useful for measuring under and over predictions. The average segment length, $\langle L_i \rangle$ where i= α,β, L is simply the lengths of all segments with structure, i, divided by the number of segments with structure i.

Some methods are capable of producing predictions with a probability for a sequence.[4-7] The highest probability prediction would be chosen as the final answer. However, the difference in probability between the most and second most probable predictions, the reliability index or RI, can be used as a simple measure of the confidence of the prediction. Two predictions with similar probabilities indicate that the different answers are almost equally likely.

In order to check for overlaps in segments, Sου, is calculated.

$$Sov = \frac{1}{N}\sum_{s}\frac{\min ov(s_1;s_2)+\delta}{\max ov(s_1;s_2)}\times len(s_1)$$

where N is the total # of residues, 1 (observed) or 2 (predicted) indicates the two compared sequences, the summation over S is over pairs of segments, each pair of segments must have at least one residue in the same secondary structure in common. *len* (s_1) is the length of the observed segment. Min ου is the number of residues for which both segments have the same state, i.e. **h**, while max ου is the number of residues where either S_1 or S_2 is in state **h** is the accepted variation allowing the comparison of segments of varying length.

The segment overlap measure is an attempt to quantify another human judgement. Secondary structure predictions are often used as starting points for protein fold predictions. This process often starts by matching secondary structural elements (segments with helices, sheets, loops) from the prediction to known proteins. How do we identify segments? Usually, three criteria are used: the number of segments, the length of each segment and the degree of similarity of each segment. Sου is one way to measure the degree of similarity.

Table 3 A comparison of prediction performance for a select group of prediction methodologies. Performance evaluations of the PRISM method should bear in mind that PRISM predicts four structural states not three

Measure	GORIII	PHD	PHD3	PRISM[4]	PSIPRED[6]
Q_3	58.9	72.5	72.5	62.9	78.3±7.8
Sov_3	50.2	75.6	69.4	NA	73.5 ±12.7
I	0.10	0.28	0.28	NA	NA

6 CONCLUSIONS

The development of these measures of different types of errors is a tremendous advance. The best way to measure these errors is during the cross-validation when the accuracy is also measured. Unfortunately, not all who work on improving prediction methods measure the performance of their methods and it is not easy for any others to obtain these measurements of errors (Table 3).

In addition, although the criteria discussed here do much to increase the reliability of secondary structural prediction methods, still there remain many other performance criteria. The methods referred to here are intended for use on globular proteins in aqueous solution. We would not expect these methods to produce accurate predictions for membrane or fibrous proteins, for example.

References

1. Kabsch W, and Sander C, *Biopolymers* 1983; **22:** 2577-2637.
2. Frishman D, Argos P, *Proteins* 1995; **23:** 566-579.
3. Richards FM, Kundrot, CE, *Proteins* 1988; **3:**71-84.
4. Lambert MH, Scheraga HA, *J Comput Chem* 1989; **10:** 770-797; Lambert MH, Scheraga HA, *J Comput Chem* 1989; **10:** 798-816; Lambert MH, Scheraga HA, *J Comput Chem* 1989; **10:** 817-831.
5. Rost B, Sander C, *J Mol Biol* 1993; **232**: 584-599.
6. Rost B, Sander C, Schneider R, *J Mol Biol*, 1994; **235**: 13-26.
7. Rost B, Sander C, *Proteins: Structure, Function, Genetics*, 1994; **19**: 55-72.
8. Jones D T, *J Mol Biol*, 1999; **292**: 195-202.

II GENES & DISEASES

FRONTIERS IN HUMAN GENETICS
Diseases and Technologies

© 2001 by World Scientific Publishing Co. Pte. Ltd.

AN INTEGRATIVE APPROACH TO THE IDENTIFICATION AND CHARACTERISATION OF HUMAN SNARE GS27

THUAN D. BUI* and WANJIN HONG
Membrane Biology Laboratory, Institute of Molecular and Cell Biology, 30 Medical Drive, Singapore 117609
**E-mail: dtbui@imcb.nus.edu.sg*

SNAREs participate in various transport events, in both the secretory and endocytotic pathways. Here we report the identification and characterisation of a new human SNARE, GS27. GS27 has two mRNA transcript isoforms as a result of alternative RNA splicing. GS27a has 6 exons and encodes a predicted protein of 212 amino acids, whereas GS27b has 7 exons and encodes a predicted protein of 213 amino acids. The gene spans 16 kb of genomic DNA and locates on the human chromosomal position 17q21. The antibodies raised against the cytoplasmic domain of GS27a protein showed specific staining in the Golgi apparatus. It could also inhibit transport *in vitro* from the endoplasmic reticulum (ER) to the *trans*-Golgi, suggesting a function in the ER-to-Golgi transport. It may be a coincidence but one of the loci implicated in human familial essential hypertension is lying near the GS27 gene locus making it a potential candidate for this disease.
Keywords: GS27, membrin, GOSR2, Golgi, SNARE

Intracellular protein trafficking is a process of transporting cargo proteins from one intracellular compartment to another within the cell. The precise delivery of cargo proteins is critical for the maintenance of physiological homeostasis. This involves the protein transport along the secretory and endocytotic pathways, which is primarily mediated by vesicles that bud from a donor compartment and fuse with a distal target compartment.[1-4] Due to the existence of diverse types of intracellular membrane compartments and the resulting transport vesicles, identifying proteins mediating the docking/fusion event and studying the mechanism governing the specificity of vesicle docking/fusion represent active areas of molecular cell biology.

The biochemical and genetic studies have established the role of N-ethylmaleimide-sensitive factor (NSF) and soluble NSF attachment protein (SNAP) in diverse trafficking events.[5] The participation of NSF and SNAP in vesicle transport is through a superfamily of proteins referred to as SNAP

* Corresponding author

receptors (SNAREs).[6-12] The SNARE hypothesis suggests that pairing of SNAREs associated with the transport vesicle (v-SNAREs) with those on the target membrane (t-SNAREs) represents a key event in the docking/fusion and in determining the specificity.[10,12]

Numerous v- and t-SNARE molecules have now been identified from a wide range of species. They all have a common motif of ~60 amino acids that are responsible for the v-t-SNARE interaction.[13] Based on the crystal structure data, SNAREs can alternatively be classified into Q- and R-SNAREs depending on whether a central residue in the SNARE motif is an arginine or a glutamine, respectively. On this basis, v-SNAREs namely synaptobrevins/VAMPs belong to R-SNAREs, and t-SNAREs namely syntaxins and SNAP-25 belong to Q-SNAREs.[14,15] The specific interaction between v- or R-SNAREs and t- or Q-SNAREs via the SNARE motif forms a SNARE core complex, which consists of four-helix bundles, formed from one R-SNARE and three Q-SNAREs (one helix from syntaxin and two helice from SNAP-25 in neurons or SNAP-23 in non-neurons). Although SNARE core complexes consisting of one Q-SNARE have also been reported.[16]

Specific SNARE molecules have been identified in all intracellular compartments and cytoplasm. Those locate in the ER and Golgi apparatus have attracted most attention because they represent the first step of vesicular transport along the exocytotic pathway. We reason that if any of the SNARE genes are involved in the genesis pathway leading to human diseases, it would most likely be Golgi and/or post-Golgi SNAREs.

Using yeast genetic approach, numerous intracellular proteins have been isolated to be essential for the ER-to-Golgi transport including rBET1, rSEC22b, syntaxin-5,[17,18] GS27 (also known as membrin or GOSR2)[19] and GS28 (also known as GOS28 or GOSR1).[20] This report will examine GS27 in details.

By searching the expressed sequence tag (EST) databases using a *Caenorhabditis elegans* protein sequence (P41941) as query that is weakly homologous to yeast protein *Bos1p*, a v-SNARE involved in the ER-to-Golgi transport, led to the identification of two sets of EST that encode two distinct isoforms of the human GS27, called GS27a (AF007548) and GS27b (AF229796) (Table 1).

Table 1 The human GS27 cDNA isoforms.

ESTs containing GS27a	Human tissues	ESTs containing GS27b	Human tissues
AW015395	Breast	AI871587	Lung
AW006073, AW006068, AI566820	Brain	AI144078	Ovary
AI969824	Prostate	AA888001	Colon
AI335781, AI129639, AA054473	Uterus	AI656283	B-cell
AI31125, W57584	Heart	AA449138	Total fetus 8-9 weeks
AI025286, AA860246	Testis		
AA643667, AI339895	Colon		
AA488981, AA237034	B-cell		
AA987968	Lung		
R78386	Placenta		

By screening the human BAC DNA library from Research Genetics Inc, six BAC clones were isolated to contain GS27 gene fragments: 9F11, 99N24, 101C24, 235L22, 484G10 and 504P18.[21] One clone 99N24 was subjected to partial sequence analysis and confirmed the authenticity of GS27 sequence, which is identical to the sequence recently released by the Human Genome Project available in the GenBank database (AC005670). By aligning human GS27 cDNA sequences against its DNA sequence revealed the complete gene organisation of human GS27 (Fig. 1).

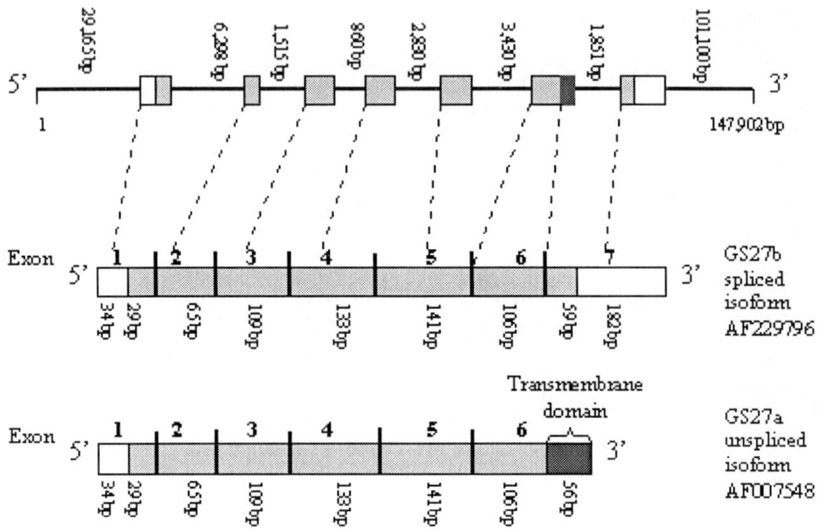

Fig. 1 The human GS27 gene organisation.

It can be seen that the sequence divergence of GS27 occurs at the end of exon 6 where the splicing of the intron between exons 6 and 7 is absent, which results in mRNA transcription of exon 6 into the intronic sequence. Without mRNA splicing, the unspliced GS27a isoform is produced. It has an 18 amino acid sequence at the most 3' terminus containing a motif for transmembrane domain. In the presence of mRNA splicing, the spliced GS27b isoform is produced. In this case, the most 3' terminal 18 amino acid sequence is replaced by a completely different 19 amino acid sequence that has no motif for transmembrane domain (Fig. 2). This may have a biological significance in term of protein localisation.

```
ggagggtcctgcgaccggaagccggaaaccggaagggggctgtgaggacgtgttccgag 29,160
gaagcCAGAGCCGGAGCCGTGGCCTGCGGGGCCGGCGACATGGATCCCCTGTTCCAGCAA 29,220
                                        M  D  P  L  F  Q  Q
ACGCACAAgtgagggccggtcggggagc>>>intron 1 (6,298 bp)<<<cctccct 35,513
 T  H  K
cttcctttgatagGCAGGTCCACGAGATCCAGTCTTGCATGGGACGCCTGGAGACGGCAG 35,573
               Q  V  H  E  I  Q  S  C  M  G  R  L  E  T  A
```

```
ACAAGCAGTCTGTGCACAgtgagtaattaactgtggag>>>intron 2 (1,515 bp)
D  K  Q  S  V  H
<<<ctttttttctttttgtacagTAGTAGAAAACGAAATCCAAGCAAGCATAGACCAGAT 37,142
                      I  V  E  N  E  I  Q  A  S  I  D  Q  I
ATTCAGCCGTCTAGAACGTCTGGAGATTTTGTCCAGCAAGGAGCCCCCTAACAAAAGGCA 37,202
 F  S  R  L  E  R  L  E  I  L  S  S  K  E  P  P  N  K  R  Q
AAATGCCAGACTgtaagtgctgacctctgaca>>>intron 3 (860 bp)<<<tctct 38,058
 N  A  R  L
ctccatcaattccagTCGGGTTGACCAGTTAAAGTATGATGTCCAGCACCTGCAGACTGC 38,118
                 R  V  D  Q  L  K  Y  D  V  Q  H  L  Q  T  A
                  GS27BF →                     GS27BF1 →
GCTCAGAAACTTCCAGCATCGGCGCCATGCAAGGGAGCAGCAGGAGAGACAGCGAGAAGA 38,178
 L  R  N  F  Q  H  R  R  H  A  R  E  Q  Q  E  R  Q  R  E  E
GCTTCTGTCTCGAACCTTCACCACTAACgtaagccaggcccgtggtgag>>>intron 4
 L  L  C  R  T  F  T  T  N
(2,830 bp)<<<tgcctgtgtttctttcacagGACTCTGACACCACCATACCAATGGAC 41,062
                                   G  S  D  T  T  I  P  M  D
                                                  GS27AF →
GAATCACTGCAGTTTAACTCCTCCCTCCAGAAAGTTCACAACGGCATGGATGACCTCATT 41,122
 E  S  L  Q  F  N  S  S  L  Q  K  V  H  N  G  M  D  D  L  I
                                   GS27AF1 →
TTAGATGGGCACAATATTTTAGATGGACTGAGGACCCAGAGACTGACCTTGAAGgtgggg 41,182
 L  D  G  H  N  I  L  D  G  L  R  T  Q  R  L  T  L  K
tccctgctggggga>>>intron 5 (3,430 bp)<<<tctgttctcttctgccccagG 44,606
GGACTCAGAAGAAGATCCTTGACATTGCCAACATGCTGGGCTTGTCCAACACAGTGATGC 44,666
G  T  Q  K  K  I  L  D  I  A  N  M  L  G  L  S  N  T  V  M
← GS27AR1
GGCTCATCGAGAAGCGGGCTTTCCAGGACAAGTACTTTATGATAGgtgggatgctgctga 44,726
R  L  I  E  K  R  A  F  Q  D  K  Y  F  M  I  G  G  M  L  L
              ← GS27AR
cctgtgtggtcatgttcctcgtggtgcagtacctgacatgagccagccacgctcagtggc 44,786
T  C  V  V  M  F  L  V  V  Q  Y  L  T  *
                                                     ← GS27BR1
t>>>intron 6 (1,851 bp)<<<aaggctgtcctgtcccccagGCACCCAAGGATCC 46,575
                                               G  T  Q  G  S
                ← GS27BR
TGCCAGACAGCACACTTTGGAGGAAGGTCTGCAGGGAGCAGCTGAGCCATTTGTTCTTGA 46,635
 C  Q  T  A  H  F  G  G  R  S  A  Q  S  S  *
ACTCTGGGAGGCAGAAGTCCCCGCACCCATCATGCGTGGACTGATAGGACATCTTTTCGT 46,695
GGTGTGCACCAGTGCTTTCCACACTTGACAGTGGTTGGCTTTGATGAACCCTCATGCTGC 46,755
ACCTTCAGAGCCAGTCCTCTAGTTTGGAATAAAAATTGCAGAGGTGGttttgggtcttt 46,815
accacctgcggctggtggacagcagccagtgtgtctggacacccaggggcattgagactg 46,875
```

Fig. 2 The nucleotide and amino acid sequences of human GS27. Intron (lower case) and exon (upper case, underlined), the predicted amino acid sequence (upper case, bold), and start and stop codons (dark grey shading, bold) are highlighted. The C-terminal amino acid sequences distinguishing two GS27 isoforms are highlighted (light grey shading, box). Numbers on the right indicate the nucleotide position as reported in the GenBank™. AC005670 contains the complete human GS27 gene and is derived from a human BAC clone hRPK.63_A_1 by the Whitehead Institute/MIT Center for Genome Research.

The endogenous mRNA expression of GS27a and GS27b has been confirmed by nested-PCR analysis in the human brain, colon, kidney and placenta (Fig. 3). It is still yet to shown that both GS27a and GS27b proteins are endogenously present in the human tissues due to a lack of specific antibodies.

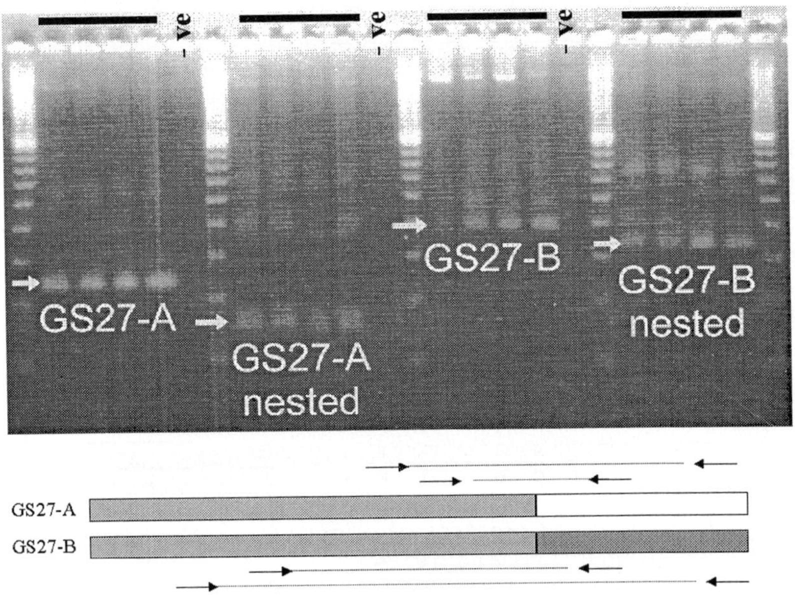

Fig. 3 Nested PCR analysis of human GS27a and GS27b mRNA in 4 different human tissue types.

The chromosomal localisation of human GS27 gene was determined by PCR analysis on Standford G3 radiation hybrid panels using a pair of sequence specific primers at the 3'-UTR of human GS27 (5'-CGC TCA GTG GCT GAA CAG CA-3' and 5'-AGC ACC GCC AAT GGT GGT GG-3'), and by fluorescence in situ hybridisation (FISH) using GS27 BAC clones 101C24 and 99N24.[21] Both experiments strongly support the mapping of human GS27 gene on chromosomal position 17q21 (Fig. 4).

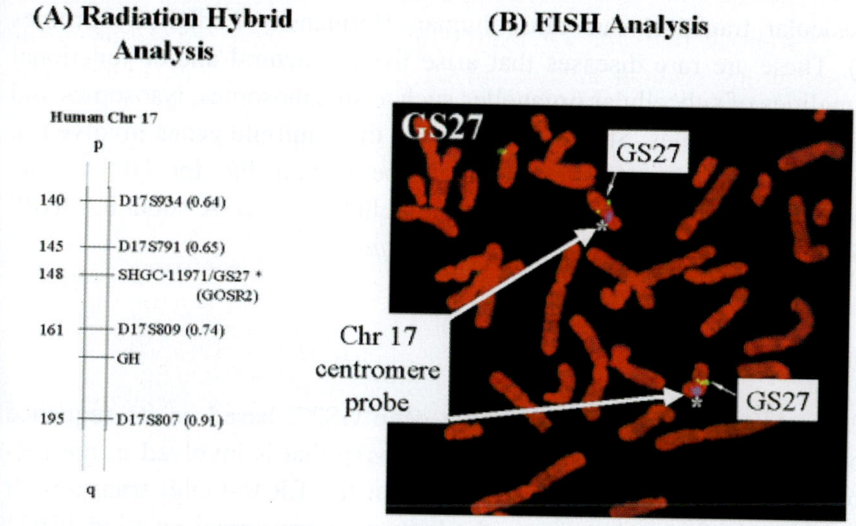

Fig. 4 Radiation hybrid mapping (A) and FISH analysis of the human GS27 gene. (A) The number on the left-hand side indicates the SHGC map position. The number in parentheses on the right-hand side indicates the chromosome position based on FISH analysis. Asterisk indicates linkage between two markers. GH = growth hormone.

Using fusion protein containing the cytoplasmic domain of GS27a, rabbit polyclonal and mouse monoclonal antibodies were generated and showed to stain specifically the Golgi apparatus and its surrounding vesicular structures. The excessive amount of GS27a fusion protein or GS27 antibodies have also been shown to inhibit transport *in vitro* from the ER to the *trans*-Golgi. This implicates a role of GS27 in the ER-to-Golgi transport.[19]

Within the chromosome 17q21, there is a putative gene(s) implicated in human familial essential hypertension lying between D17S934 and GH (Fig. 4).[22] Due to the known functional nature of GS27 protein and its chromosomal location between D17S934-GH, it is possible to envisage that mutations leading to a reduced or complete abolition of GS27 activity, can affect the delivery of target cargo proteins and result in hypertension.

So far, there is no SNARE gene known to either associate and/or link to a human disease. However, recent evidence suggests that defects in the vesicular transport can cause human Hermansky Pudlak Syndromes (HPS). These are rare diseases that arise from structural and/or functional abnormalities of subcellular organelles such as melanosomes, lysosomes and platelet dense granules. Studies have shown that multiple genes involved in the post-Golgi protein trafficking might be responsible for HPS.[23] The putative mutated genes identified so far include AP3 β3A subunit,[24] HPS protein[25] and syntaxin 13 interacting protein *pallid*.[26]

Summary

Our laboratory has identified a human gene GS27, based on its sequence homology to the yeast protein v-SNARE Bos1p that is involved in the ER-to-Golgi transport. GS27 also participates in the ER-to-Golgi transport. It has 2 mRNA isoforms, locates on the human chromosomal position 17q21 and the gene organisation has been completely determined. Hopefully, these information will allow future studies to investigate whether GS27 is associated and/or linked with any human disease.

References

1. Mellman I, Simons K, The Golgi complex: *in vitro* veritas, *Cell* 1992; **68:** 829-840.
2. Palade G, Intracellular aspects of the process of protein synthesis, *Science* 1975; **189:** 347-358.
3. Rothman JE, Wieland FT, Protein sorting by transport vesicles, *Science* 1996; **272:** 227-234.
4. Schekman R, Orci L, Coat proteins and vesicle budding, *Science* 1996; **271:** 1526-1532.
5. Whiteheart SW, Kubalek EW, SNAPs and NSF: general members of the fusion apparatus, *Trends Cell Biol.* 1995; **5:** 64-69.
6. Ferro-Novick S, Jahn R, Vesicle fusion from yeast to man *Nature* 1994; **370:** 191-193.
7. Pfeffer SR, Transport vesicle docking: SNAREs and associates, *Annu. Rev. Cell Biol. Dev. Biol.* 1996; **12:** 441-461.
8. Rothman JE, Warren G, Implications of the SNARE hypothesis for intracellular membrane topology and dynamics, *Curr. Biol.* 1994; **4:** 220-233.

9. Scheller RH, Membrane trafficking in the presynaptic nerve terminal, *Neuron* 1995; **14:** 893-898.
10. Sollner T, Whiteheart SW, Brunner M, Erdjument-Bromage H, Geromanos S, Tempst P, Rothman JE, SNAP receptors implicated in vesicle targeting and fusion, *Nature* 1993; **362:** 318-324.
11. Sudhof TC, The synaptic vesicle cycle: a cascade of protein-protein interactions, *Nature* 1995; **375:** 645-653.
12. Weber T, Zemelman BV, McNew JA, Westermann B, Gmachi M, Parlati F, Sollner TH, Rothman JE, SNAREpins: minimal machinery for membrane fusion. *Cell* 1998; **92:** 759-772.
13. Weimbs T, Low SH, Chapin SJ, Mostov KE, Bucher P, Hofmann K, A conserved domain is present in different families of vesicular fusion proteins: a new superfamily, *Proc. Natl. Acad. Sci. USA* 1997; **94:** 3046-3051.
14. Fasshauer, D, Sutton RB, Brunger AT, Jahn R, Conserved structural features of the synaptic fusion complex: SNARE proteins reclassified as Q- and R-SNAREs, *Proc. Natl. Acad. Sci. USA.* 1998; **95:** 15781-15786.
15. Jahn R, Sudhof TC, Membrane fusion and exocytosis, *Annu. Rev. Biochem.* 1999; **68:** 863-911.
16. Sutton RB, Fasshauer D, Jahn R, Brunger AT, Crystal structure of a SNARE complex involved in synaptic exocytosis at 2.4 A resolution, *Nature* 1998; **395:** 347-53.
17. Bennett MK, Garcia-Arraras JE, Elferink LA, Peterson K, Fleming AM, Hazuka CD, Scheller RH, The syntaxin family of vesicular transport receptors, *Cell* 1993; **74:** 863-873.
18. Rowe T, Dascher C, Bannykh S, Plutner H, Balch WE, Role of vesicle-associated syntaxin 5 in the assembly of pre-Golgi intermediates, *Science* 1998; **279:** 696-700.
19. Lowe SL, Peter F, Subramaniam VN, Wong SH, Hong W, A SNARE involved in protein transport through the golgi apparatus, *Nature* 1997; **389:** 881-884.
20. Subramaniam VN, Peter F, Philp R, Wong SH, Hong W, GS28, a 28-kilodalton golgi SNARE that participates in ER-golgi transport, *Science* 1996; **272:** 1161-1163.
21. Bui TD, Levy ER, Subramaniam VN, Lowe SL, Hong W, cDNA characterization and chromosomal mapping of human golgi SNARE GS27 and GS28 to chromosome 17, *Genomics* 1999; **57:** 285-288.
22. Julier C, Delepine M, Keavney B, Terwilliger J, Davis S, Weeks DE, Bui T, Jeunermaitre X, Velho G, Froguel P, Ratcliffe P, Corvol P, Soubrier F, Lathrop GM, Genetic susceptibility for human familial essential hypertension in a region of homology with blood pressure linkage on rat chromosome 10, *Hum. Mol. Genet.* 1997; **6:** 2077-2085.
23. Spritz RA, Multi-organellar disorders of pigmentation: intracellular traffic jams in mammals, flies and yeast, *Trends Genet.* 1999; **15:** 337-340.

24. Dell'Angelica EC, Shotelersuk V, Aguilar RC, Gahl WA, Bonifacino JS, Altered trafficking of lysosomal proteins in Hermansky-Pudlak syndrome due to mutations in the beta 3A subunit of the AP-3 adaptor, *Mol. Cell* 1999; **3:** 11-21.
25. Oh J, Bailin T, Fukai K, Feng GH, Ho L, Mao J, Frenk E, Tamura N, Spritz RA, Positional cloning of a gene for Hermansky-Pudlak syndrome, a disorder of cytoplasmic organelles, *Nature Genet.* 1996; **14:** 300-306.
26. Huang L, Kuo YM, Gitschier J, The pallid gene encodes a novel, syntaxin 13-interacting protein involved in platelet storage pool deficiency, *Nature Genet.* 1999; **23:** 329-332.

FRONTIERS IN HUMAN GENETICS
Diseases and Technologies

© 2001 by World Scientific Publishing Co. Pte. Ltd.

THE IDENTIFICATION OF *SRY* GENE AS A CLINICAL INVESTIGATIVE TOOL FOR SEX AMBIGUITY: AN EXPERIENCE IN H.U.S.M.

MN ISA* and MZ FUZIAH†

**Human Genetics Unit, †Department of Paediatrics, School of Medicine, Universiti Sains Malaysia, 16150 Kubang Kerian, Kelantan, Malaysia.*

Nine newborns and infants with sex ambiguity and congenital adrenal hyperplasia (CAH) were analysed for the presence of the testis-determining factor (TDF). They required an urgent genetic evidence for a sex assignment both for clinical management and parental request. The sex-determining region on Y chromosome (*SRY* gene) was used as a molecular marker in combination with the Fluorescence *in situ* hybridization (FISH) to identify the chromosome X and Y material on the interphase cells to complement the conventional karyotyping. The molecular analysis of *SRY* gene to define the association with the testicular-determining factor (TDF) supports the presence of the Y chromosome material. Our results showed that the sex genetics of the newborns were assigned then as male and female respectively and one newborn was reassigned as female. In the nine samples that we carried out the *SRY* analysis, seven showed the presence of *SRY* gene and in two samples it was absent. Sex assignment earlier not determined by the clinicians in three patients showed the presence of *SRY* gene in two patients while one was absent. All these sex reassignment and confirmation were supported by the molecular-cytogenetics and karyotyping studies. This approach provided a fast sex genetic information when the hormonal assay and karyotyping were not available and/or delayed. It is very useful to complement an immediate information for the clinicians to rule out cases of ambiguous genitalia due to CAH and subsequently clinical investigations that affects the long term management.

Keywords: testis-determining factor (TDF), *SRY* gene, ambiguous genitalia, fluorescence *in situ* hybridization (FISH)

---

* Corresponding author

1 INTRODUCTION

The development of male and female characteristic in an individual is a subject of interest since ancient times. The specific role of Y chromosome as a direct controller for the formation of testis and gonads was presumed due to the specific assignment of the testis-determining factor (TDF), specifically the sex-determining region on Y chromosome (*SRY* gene) on the short arm of the Y chromosome [1]. This gene which is located proximal to the pseudoautosomal region of the Y chromosome demonstrated was recently associated with the clinical manifestation of XX male and XY female syndrome [2]. Recent compelling evidence of SRY as an elusive testis-determining factor has then revolutionized the determination of maleness. A mutation in the HMG box of the *SRY* gene with XY then lead to sex reversal were also reported [3]. This has allowed the application of *SRY* gene as a unique molecular marker and the clinical investigative tool for precise sex determination. It is a useful tool for sex assignment in an individual with sex ambiguity. Further application of the *SRY* gene as molecular marker include in the sex typing in forensic samples, assessment of bone marrow transplantation and pathological investigation of gonadal dysgenesis, testicular feminization and other aberrations in sex reversal in human [3-10].

In this paper we present our experience of using the *SRY* gene as a marker for the clinical tool to assign sex in newborns and also in the management of patients with ambiguous genitalia. We have applied this investigation in combination of fluorescence *in situ* hybridization (FISH) using specific X and Y probes to support and complement the conventional chromosomal karyotyping.

Our aim is to provide an urgent genetic evidence for sex assignment of the newborns and in those who needed sex reassignment. It supports both the anatomical and hormonal investigations in individuals with testicular feminization.

2 MATERIALS AND METHODS

2.1 Patients

Nine newborns and infants with ambiguous genitalia mainly presenting with enlarged phallus, labial fusion and impalpable gonads were studied. A volume of 2–5 ml of blood were collected in lithium heparin.

2.2 Cytogenetics Analysis and Fluorescence *In Situ* Hybridization (FISH)

Metaphase spreads were obtained from synchronised cell cultures following standard procedure. The spread were G-banded for routine cytogenetic analysis.

Fluorescence *in situ* hybridization (FISH) using both chromosomes X and Y specific probes were performed on interphase and metaphase cells as recommended by the supplier (Boehringer Mannheim GmbH). The signals were analysed using the Zeiss microscope filter.

2.3 Molecular Analysis of *SRY* Gene and Chromosomes X and Y Specific Sequence

Immediately after the cell culturing, the DNA was extracted from the leftover blood samples using the commercial DNA extraction kits (BIO–101, USA). Polymerase chain reaction (PCR) of the *SRY* gene and chromosomes X and Y specific sequences using the specific primers previously described [4,7,8] and analysed by subsequent agarose gel electrophoresis.

3 RESULTS

A total of nine infants requested for genetic information for an urgent sex assignment clinically presented with ambiguity of the external genitalia. Generally, their physical examination showed impalpable gonads, clitoromegaly, scrotalization of labioscrotal folds and single urogenital sinus or with suspicion of congenital adrenal hyperplasia (Fig. 1). Majority of the patients presented with these features. No immediate hormonal assay was available for the patients. We performed the *SRY* gene analysis by immediately extracting the DNA following the cell culture for karyotyping from the remaining blood samples.

In the nine samples that we carried out the SRY analyses, seven showed the presence of *SRY* gene and two were absent (Table 1). Sex assignment, earlier not determined in three patients (patients 1,4,6) by the clinicians, showed the presence of *SRY* gene in two patients and absence of this gene in one patient. An infant earlier assigned as female (patient 7) was shown to have the *SRY* gene and was karyotyped as 46,XY. Another male infant (patient 9) who was admitted for failure to thrive and adrenal crisis did not carry the *SRY* gene and no Y material was detected by FISH and the karyotype was 46,XX. Patients 2 and 3 who were initially assigned as males were confirmed to have SRY gene both have karyotype 46,XY. Patients 5 and 8 who were not definitely sex assigned (? male) were found to have *SRY* gene.

Patient 7 was born with ambiguous genitalia. A genitogram revealed the presence of vagina and uterus. Baby was brought up as a girl and treated as CAH with cortisone acetate and fludrocortisone but was later stopped after the genetic results were available. The karyotype was 46,XY with *SRY* gene present. These findings lead to further imaging studies. Magnetic resonance image (MRI) of the pelvis and abdomen revealed an elongated structure measuring 3 cm posterior to the bladder and was most probably a vagina or rudimentary uterus (Fig. 2). Ovaries or testes were not visualised in abdomen or in the inguinal region. No suprarenal mass lesion seen. In view of the *SRY* gene result coupled with the MRI findings, a review of the diagnosis and gender assignment was mandatory. The baby underwent a laparotomy whereby a mullerian cyst was identified (confirmed histologically) and cystectomy done. Testes were identified in upper abdomen and were brought down to the inguinal region. However, no biopsy was performed. The parents then decided to change the sex name.

4 DISCUSSION

The accuracy of the clinical assignment of the sex as compared to the sex genetics of these newborns and infants is 50 %. The presence or absence of *SRY* gene in these patients helped the clinicians to urgently assign their sex. The identification of *SRY* gene and chromosomes X and Y material using FISH on interphase cells provide urgent diagnostic tool in these newborns with sexual ambiguity. This approach also complements the karyotyping in cases of poor metaphase and culture failure. It provides an additional clinical management for the patients and parental counseling. Our experience showed that with the hormonal assay and karyotyping either not available urgently and/or delayed, the presence or absence of *SRY* gene as a marker is very informative in situation with lack of up-to-date facilities.

Incorrect sex assignment sometimes lead to family embarrassment as it is impossible for the parents to accept the uncertainty of the sex of their newborn. This causes more embarrassment to reassign the sex of their children in the later age. Similarly, delayed naming and assigning the sex posed an extended family embarrassment in the community [11]. The naming occasion in this community is normally conducted on the 4th to 14th day of birth.

When methods of identifying the *SRY* gene are available, it is essential in deciding a proper and long term management of patients with ambiguous genitalia. Newborns with congenital andrenal hyperplasia (CAH) is the most common cause of ambiguous genitalia and always pose uncertainty for the clinician to assign the sex. The presence or absence of *SRY* gene thus can evoke other genetic studies including analysis of the androgen receptor and *SRY* gene mutations [3,4]. It is a necessary information for the planning of surgical reconstruction in infants with sex ambiguity as in patient 7, even the MRI was non conclusive in identifying the gonads. However, the presence of the *SRY* gene lead us to perform the laparotomy in searching for the testicular tissues. This is to avoid future complication at the later age as in the development of gonadal tumours. These are very important aspects of management to prevent future refusal of surgical recorrection and social and emotional embarrassment during adolescence.

From our experience, we noticed that the incidence of sex ambiguity in the newborns required an urgent decision to assign their sex both for clinical and family counseling. It is also an urgent decision because of the extended family and religious background. To further enhance the decision required, we are now including the mutational analysis of the androgen receptor and *SRY* gene in our battery of genetic studies both to determine an urgent sex assigment of the newborns and in those with follow-up of sex ambiguity.

Acknowledgments

We are grateful to the JSPS for the fellowship award and the Department of Human Genetics, University of Tokyo. This work is supported partly by the USM IRPA Grant USM 323/0501/401.

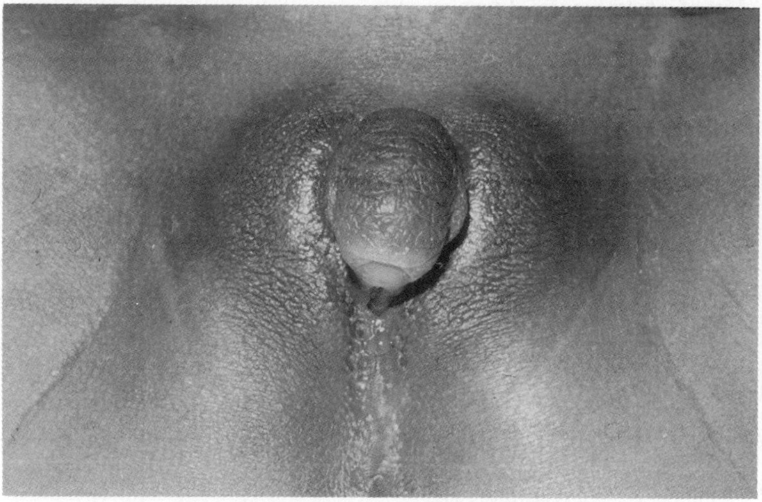

ig 1. External genitalia of a newborn showing an enlarged phallus and hyperpigmented .bioscrotal folds. The gonads were not palpable.

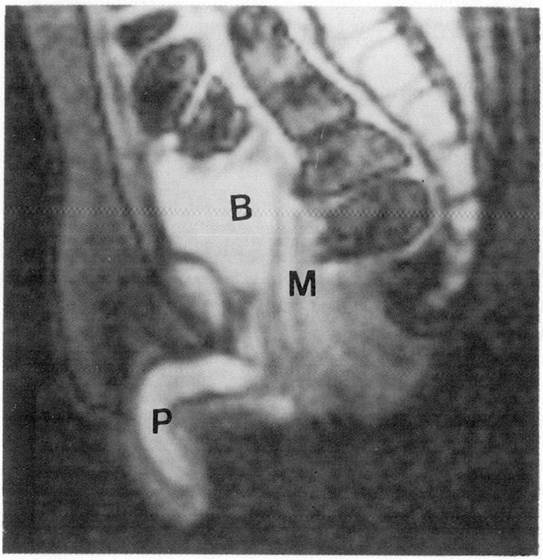

; 2. MRI abdomen of patient 7. **P** – penis, **B** – bladder, **M** – Mullerian cyst that was removed at ·arotomy.

Table 1. List of infants requested for sex genetic assignment

Patient	Age of Presentation	Sex	Clinical data	SRY	X	Y	Karyotype
1	10 days	ND	Phallus 1.3 cm Bifid scrotum ? Blind vagina Gonads palpable	+	+	+	46,XY
2	60 days	M	Ambiguous genitalia	+	+	+	46,XY
3	4 months	M	Bifid scrotum Hypospadias	+	+	+	46,XY
4	2 months	ND	Enlarged phallus Clitoromegaly Labial fusion Impalpable gonad	-	+	-	46,XX
5.	13 days	?M	Hypospadias	+	+	+	46,XY
6	22 days	ND	Phallus 1.5 cm A palpable gonad	+	+	+	46,XY, (isoYq)
7	3 months	? F	Enlarged phallus Scrotalisation of labial fold No gonad palpable	+	+	+	46,XY
8	11 days	? M	Incomplete fusion of labioscrotal folds Urogenital sinus	+	+	+	46,XY
9	2 months	M	Enlarged phallus Impalpable gonad Hyperpigmented labioscrotal folds	-	+	-	46,XX

(ND, sex not determined; M, male; F, female; SRY, sex determining region on Y chromosome).

The sexes of these patients were either assigned clinically or not determined (ND). The PCR amplification for the SRY gene and the X, and Y-specific primers were performed immediately after the cell culture. Total time required was approximately 12 hours after receiving the blood samples. Karyotyping was available only after 7–8 hours working day.

References

1. Sinclair AH, Berta P, Palmer MS, Hawkins JR, Griffiths AM, Smith MJ, Foster JW, Frischauf AM, Lowell-Badge R, Goodfellow PN. A gene from the human sex-determining region encodes a protein with a homology to a conserved DNA-binding motif. *Nature* 1990;**346**:240-243.

2. Hawkins JR. Sex determination. *Human Molecular Genetics* 1994;**3**:1463-1467

3. Domenice S, Nishi MY, Billerbeck AEC, Latronico AC, Medeiros MA, Russell AJ, Vass K, Carvalho FM, Frade EMC, Arnhold IJP, Mendonca BB. A novel missense mutation (S18N) in the 5' non-HMG box region of the SRY gene in a patient with partial gonadal dysgenesis and his normal male relatives. *Human Genetics* 1998;**102**:213-215.

4. Iida T, Nakahori Y, Komaki R, Mori E, Hayashi N, Tsutsumi O, Taketani Y, Nakagome Y. A novel missense mutation in the HMG box of the SRY gene in a patient with the XY sex reversal. *Human Molecular Genetics* 1994;**3**:1437-1438.

5. Berta P, Morin D, Poulat F, Taviaux S, Lobaccaro JM, Sultan C, Dumas R. Molecular analysis of the sex determining region from the Y chromosome in two patients with Frasier syndrome. *Horm Res* 1992;**37**:103-106.

6. Hamdorf B, Gregg K. Rapid and precise gender determination from DNA samples using polymerase chain reaction. *Asia Pacific J Mol Biol Biotech* 1994;**2**:136-142.

7. Nagafuchi S, Seki S, Nakahori Y, Tamura T, Numabe H, Nakagome Y. PCR detection of structurally abnormal Y chromosomes. *Jpn J*

Human Genetics 1992;**37**:187-193.

8. Van der Auwera B, Van Roy N, De Paepe A, Hawkins JR, Liebaers I, Castedo S, Dumon J, Speleman F. Molecular cytogenetics analysis of XX males using Y-specific DNA sequences, including SRY. *Human Genetics* 1992;**89**:23-28.

9. Zenteno JC, Lopez M, Vera C, Mendez JP, Kofman-Alfaro S. Two SRY-negative XX male brothers without genital ambiguity. *Human Genetics* 1997;**100**:606-610.

10. Louis CK, Low, Cheung PT. Ambiguous genitalia. *J Paed. Obst Gyn* 1996; **Jan/Feb**:5–10.

11. Abdullah MA, Katugumpola M, Al-Habib S, Al-Jurayyan N, Al-Samarrai A, Al-Nuaim A, Patel PJ, Niazi M. Ambiguous genitalia: Medical, socio-cultural and religious factors affecting management in Saudi Arabia. *Ann Tropical Paed* 1991;**11**:343-348.

FRONTIERS IN HUMAN GENETICS
Diseases and Technologies

© 2001 by World Scientific Publishing Co. Pte. Ltd.

G6PD DEFICIENCY AND APPLICATION OF THE MPTP TECHNIQUE

PS LAI*,¶, T SHIRAKAWA†, K NISHIYAMA†, M MATSUO‡, R JOSEPH§ and SH QUAK*

**Department of Pediatrics, National University of Singapore, Lower Kent Ridge Road, Singapore*

†Faculty of Health Science, Kobe University School of Medicine, Tomogaoka, Suma-ku, Kobe 654-01, Japan

‡Division of Genetics, International Center for Medical Research, Kobe University School of Medicine, Kusunoki-cho, Chuo-ku, Kobe 650, Japan

§Department of Neonatology, National University Hospital, Lower Kent Ridge Road, Singapore

Glucose-6-phosphate dehydrogenase (G6PD) deficiency is an X-linked recessive enzyme disorder that affects an estimated 200-400 million people. More than four hundred biochemical variants have been identified reflecting the vast heterogeneity of this disorder. At the molecular level, less than a hundred mutations have been characterized indicating a need for a rapid and efficient approach for mutational screening of G6PD deficient individuals in different populations. A technique called multiplex PCR using multiple tandem forward primers and a common reverse primer (MPTP technique) was recently developed. It was initially tested in Singapore population where the incidence of this deficiency occurs at approximately 3% of newborn males. The MPTP technique allows quick and comprehensive screening of point mutations and can be adopted to detect common variants such as 1376G→T (G6PD Canton) and 1388G→A (G6PD Kaiping) in mutation hot spots for specific populations. This paper presents a brief review on G6PD deficiency and the application of the MPTP technique for screening this disorder.

Keywords: glucose-6-phosphate dehydrogenase, G6PD deficiency, multiple tandem forward primers with common reverse primer (MPTP), G6PD Canton, G6PD Kaiping, neonatal jaundice

¶Corresponding author

1 INTRODUCTION

Glucose-6-phosphate dehydrogenase (G6PD) deficiency is a common genetically determined enzyme defect.[1] This X-linked disorder causes hemolytic anemia in the presence of environmental triggers such as infections, or exposure to substances such as broad beans and mothballs and to certain drugs like sulphonamides and nitrofurantoin.

An anecdotal report from the school of Pythagoras attributes the recognition of G6PD deficiency as far as back the sixth century B.C.[2] Apparently, the great philosopher, Pythagoras was said to have admonished his disciples that they should not retreat into a field of fava beans even if their lives were at risk. One century later, this notion was enshrined into one of Hippocrates' aphorisms, "kuamon apecesqe" ("avoid fava beans"). Favism became recognized as a distinct clinical entity in Italian medical literature as early as in the 19th century. The compounds, vicine and isouramil, abundant in fava beans, are hypothesized to precipitate hemolytic episodes.[3] However, it was not until the 1950s before G6PD deficiency was discovered as a result of investigations of hemolytic anemia arising from the administration of 8-amino-6-methoxyquinoline antimalarial drugs such as primaquine.[3,4] During this period, a large-scale project founded by the United States Army was undertaken to determine the cause of drug related sensitivity to primaquine. It was found that 15% of black soldiers were sensitive and that red blood cells from these individuals were deficient in G6PD. In these individuals, contact with the drug precipitated severe oxidative stress and induced hemolysis. Most hemolytic events are usually self-limiting and management of G6PD deficiency usually involves the avoidance of precipitating factors (Table 1).[5,6] It has since become apparent that clinical manifestations of G6PD deficiency can vary from asymptomatic to neonatal jaundice, drug or infection induced acute red cell hemolysis, favism and severe chronic non-sphercytic hemolytic anemia.[5]

Table 1 A list of drugs that may be environmental triggers in G6PD deficiency

	Sulphonamides and sulphones	Other bacterial compounds	Analgesics	Anthelmintics	Miscellaneous
Primaquine	Sulfanilamide	Nitrofurans - nitrofurantoin, furazolidone, nitrofurazone	Acetylsalicylic acid (aspirin)	β-Naphthol	Vitamin K analogues (eg menaphthone)
Pamaquine (Chloroquine)	Sulfapyridine			Stibophan	Naphthalene (moth balls)
	Sulfadimidine	Nalidixic acid	Acetophenetidin (Phenacetin)	Niridazole	
	Sulfacetamide (Albucid)	Chloramphenicol	Safe alternative: Paracetamol		Probenecid
	Sulfafurazone (Gantrisin)	*p*-Aminosalicylic acid			Dimercaprol (BAL)
					Methylene blue
	Sallcylazosulfa-pyridine (Salazopyrin)				Arsine
					Phenylhydrazine
	Dapsone				Acetylphenyl-hydrazine
	Sulfoxone				Toluidine blue
	Glucosulfone sodium (Promin)				Mepacrine
	Septrin				

2 BIOCHEMISTRY

Glucose-6-phosphate dehydrogenase (G6PD) oxidizes glucose-6-phosphate (G6P) to 6-phosphoglucono-δ-lactone (6PGL) and reduces nicotinamide adenine dinucleotide phosphate (NADP+) to NADPH (Fig. 1). This redox reaction is the primary step in the pentose phosphate pathway and maintains normal cellular levels of NADPH that is required for biosynthetic processes and for protection against oxidative damage in red blood cells.[7,8] It is believed that variants arising from *G6PD* gene mutations affect the enzymatic activity in either of two ways. One, variants that map at the dimer interface may affect the cohesive strength of the molecule. Active G6PD exists either as a dimer or tetramer of identical subunits and the dimer-tetramer equilibrium is affected by ionic strength and pH. Although the physiological significance of the dimer-tetramer interconversion is unknown, it is believed that the formation of this oligomeric structure is essential for enzyme activity and for NADPH formation. Secondly, variants may also harbor mutations that affect the enzyme active sites thereby influencing the reaction rate.

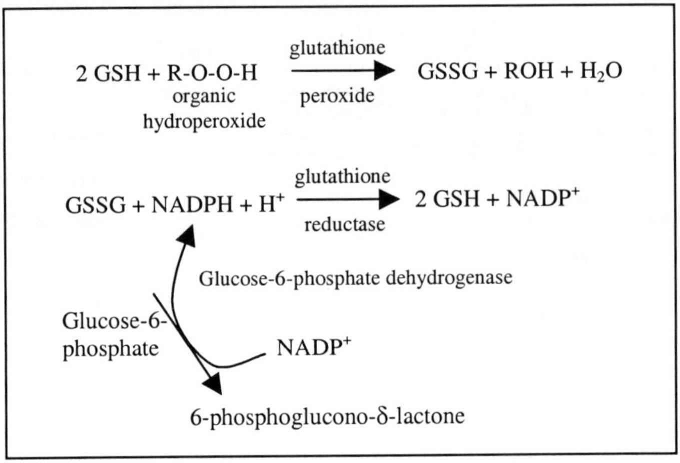

Fig. 1 Rate of NADPH production in the pentose phosphate pathway is controlled by the rate of glucose-6-phosphate dehydrogenase reaction

3 DIAGNOSIS

A wide spectrum of symptoms may be manifested such as acute hemolysis, jaundice, fatigue, paleness, shortness of breath, dark or red urine, fever, enlarged spleen, abdominal pain, etc.[1,3,9] In newborns, neonatal jaundice is common and treatment of this condition, if persistent, is through phototherapy or by exchange transfusion to alleviate jaundice. Clinical diagnosis can be made using a number of tests (Table 2). To confirm diagnosis of G6PD deficiency, laboratory tests measuring enzyme activity is usually carried out.[8] Based on the residual G6PD activity, the deficiency classes can be typed according to standards set by the World Health Organization (WHO).[6,11] A commonly used method to determine red blood cell G6PD activity has been described by Beutler which expresses G6PD activity as international units per gram hemoglobin at 37°C and differs from the WHO method carried out at 25°C.[10] Another popular test is to estimate enzyme activity using fluorescent ultraviolet light by measuring increases in absorbance at 340nm wavelength at 30°C.[11,12]

Table 2 Tests used in clinical diagnosis of glucose-6-phosphate dehydrogenase deficiency

Tests for G6PD
Elevated bilirubin levels
Low serum haptoglobin
Haemoglobinuria
Elevated absolute reticulocyte count
Low red blood cell count and haemoglobinuria
Heinz bodies present on examination of the peripheral blood smear using stains
Methylene blue test
Methemoglobin reduction test

4 GENETIC HETEROGENEITY

G6PD variants have been characterized on the basis of parameters such as electrophoretic mobility, specific activity in red cells (IU/G Hb), affinity constants (K_m) for G6P and NADP, and for NADPH (K_i), activity on substrates analogues, pH dependence of enzyme activity, thermostability, etc.[6,8,11] Using such parameters, over 400 different variants have been identified.[13,14] These results indicate G6PD to be a genetic locus with a high level of heterogeneity and the molecular identification of the *G6PD* gene in 1984 has further allowed molecular characterization of variants.[14,15,16] The latter allows a more definite identification of the DNA mutations and permits correlation between previous biochemical data with the molecular information. For instance, samples that had been previously identified as a particular biochemical/population variant has been found to yield further heterogeneity at the molecular level.[1,5] Examples of these are cases of G6PA$^-$ and G6PD Mediterranean. Conversely, some variants that have been assigned different names have been found to be the same at the molecular level, for instance, in cases of G6PDA$^-$, G6PD Mediterranean, G6PD Walter Reed, G6PD Beverly Hills, G6PD Geneva, G6PD Kerala-Kalyan, etc. To date less than a hundred different gene mutations have been identified.[13,14,15,17] As more laboratories switch to sequencing the actual gene alterations in G6PD deficient samples, an increasing number of molecular variants are expected to be identified.

5 MOLECULAR GENETICS

The *G6PD* gene maps to Chromosome Xq28.[18] It is 18 kb in size and spans thirteen exons. The protein-coding region is divided into twelve segments, ranging in size from 12 to 236 bp, with an intron present in the 5' untranslated region.[19,20] The mRNA consists of 2269 nucleotides and codes for a 59.2 kDal protein consisting of 515 amino acids. Sequencing of G6PD variants have so far shown only the presence of missense or silent mutations arising from nucleotide base substitutions. With the exception of rare cases of a triplet base deletion which resulted in formation of G6PD polypeptide chain without the single amino acid ("in-frame" mutation), no gross deletions or nonsense mutations have been found.[5,8] This supports the concept that G6PD enzyme is necessary for life-sustaining metabolic reaction in the body and that mutations resulting in total loss of G6PD activity are lethal so that a fetus with such severe mutation cannot survive to post-uterine life.

6 POPULATION GENETICS AND EPIDEMIOLOGY

It is estimated that G6PD deficiency affects over 200 to 400 million human males worldwide.[5] Originally thought to be prevalent only among males of African or Mediterranean ancestry, its epidemiology extends from Africa, Southern Europe, Middle East and South East Asia to Oceania.[3,5,8] Its recent prevalence in America is thought to be due to migration patterns. The geographic distribution of this enzyme defect in the mushy and malarial areas of the Mediterranean initially prompted the "malaria hypothesis" in the 1960s linking G6PD deficiency with resistance against *Plasmodium falciparum*.[4,5] It has since been found that G6PD deficiency occurs at polymorphic frequencies only in populations living, or have been living in generations where malaria is endemic. In addition, experimental evidence has proved that *Plasmodium falciparum* grows more slowly in G6PD deficient erythrocytes than in normal cells, and that female heterozygotes are protected from the lethal effects of the parasites.[21,22] Hence, selective pressure of malaria has influenced the geographical distribution of G6PD deficiency. In Southern Europe, G6PD deficiency is still common although malaria is not prevalent due to its recent eradication only over the last two to three generations. In native America where malaria is not endemic, G6PD deficiency is not found in polymorphic frequencies and the incidence of G6PD can be accounted for by migration.

Gene frequencies in a number of populations have been extensively reported. Gene frequencies are between 0–7% in parts of Italy, 26% in parts of Africa and as high as 70% among Kurdish Jews.[5,14,23] In Chinese groups in Southern China, Taiwan, Hawaii and South East Asia, the most common variants are R459L (1376G→T) and R463H (1388G→A) in exon 12, known also as the Canton and the Kaiping variants respectively.[9,13,24,25,26] The incidence of G6PD deficiency in these groups were reported to be between 1.9% to 16%. In Filipinos, G6PD Union and G6PD Vanua Lava (1360C→T and 383T→C) predominate while G6PD Taiwan-Hakka (1376G→T) is the most common variant in Taiwan.[13,27,28] In Japanese populations, variants such as G6PD Musashino (185C→T), G6PD Asahikawa (695G→A), G6PD Kamiube (1387C→T), G6PD Fukushima (1246G→A), G6PD Morioka (1339G→A), G6PD Iwate, G6PD Niigata and G6PD Yamaguchi (1160G→A) have also been reported.[15] In Thailand, the prevalence of G6PD deficiency is high in various parts of the country: 6–13.6% in Central, 10–15% in the North, 8–24% in the Northeast and 2.8–9.3% in the South.[29,30] Some of the common variants are G6PD Mahidol (487G→A), G6PD Viangchan (871G→A), G6PD Coimbra (592C→T), G6PD Union (1360C→T) and G6PD Canton (1376G→T) and G6PD Gaozhou (95A→G). In Singapore, the incidence of G6PD is approximately 3% among the male population consisting of Chinese, Malays and Indians.[31,32] The common mutations are G6PD Canton (1376G→T), Kaiping (1388G→A).

The molecular basis of G6PD heterogeneity has yet to be fully elucidated. If G6PD deficiency had spread geographically by human migration and genetic drift, it would be expected that a common mutant would exist.[1,5] Molecular analysis however shows that different variants are polymorphic in different parts of the world. This suggests that multiple point mutations must have arisen independently and then spread to establish themselves at polymorphic frequencies. Some variants thus appear to have arisen recently while others may be ancient and widespread. More studies are still required to correlate molecular variants with biochemical characteristics and clinical manifestations.

7 G6PD IN SINGAPORE

G6PD deficiency in Singapore was first described in 1963 and was the major cause of kernicterus leading to deaths in neonates until a mass screening programme for G6PD deficiency was introduced.[25,31,33] Recent data reveals an incidence of 1.62% in all newborns, occurring at 3.15% in males and 0.11% in females. Racial variations were observed among males from the three major ethnic groups in the population : Chinese (3.94%), Malays (2.95%) and Indians (0.66%). G6PD deficiency is diagnosed from the modified Bernstein's technique to measure enzymatic activity. Intermediate deficiency has also been identified at a frequency of 1.83% among Chinese female newborns. Some 69 different biochemical variants have been identified and it is believed that four common molecular variants predominate (1376G→T , 1388G→A , 563C→T , 95A→G).

In the molecular characterization of variants, a number of approaches have been used, namely by dideoxy fingerprinting assay, single-strand conformation analysis, restriction analysis, allele-specific oligonucleotide hybridization, etc.[32] Most of the mutations in G6PD are point mutations. To detect new nucleotide changes, direct sequencing technique is usually required due to limitations in current screening methods. The large number of G6PD biochemical variants has yet to be fully confirmed at the molecular level. To facilitate studies on the molecular pathogenesis of G6PD deficiency, rapid DNA screening methods are thus required.

8 MPTP SCREENING TECHNIQUE

A novel method for scanning point mutations in G6PD gene called the multiplex PCR using multiple tandem forward primers and a common reverse primer (abbreviated as MPTP from underlined letters) technique was first described by Shirakawa and co-researchers in 1997.[34] In this approach, mutation scanning can be performed for gene sequences by amplifying the targeted region using flanking primers. Subsequently using the amplified product as template, a second PCR is carried out using multiple short-length forward primers arrayed in tandem together with a common reverse primer. These forward primers consist of between thirteen to fifteen bases and overlapped each other by three to four nucleotides. The absence of amplified product from the second multiplex PCR then defines the site of a mutation within a narrow region of the primer recognition site. Sequence analysis is then carried out to confirm the presence of a nucleotide change in the region identified by the initial scanning.

9 EXPERIMENTAL PROCEDURES

9.1 Materials and Methods

Blood samples from G6PD deficient patients were obtained from male newborns born at the National University Hospital in Singapore. The samples were identified as G6PD deficient from the assay of G6PD enzyme activity using an in-house modification of the Bernstein's technique.[10] Genomic DNA from these samples was extracted by standard phenol chloroform method. To evaluate the MPTP technique, exon 12 was selected for molecular analysis as two mutation hot spots are located here, namely G6PD Canton (1376G→T) and G6PD Kaiping (1388G→A). Amplification by polymerase chain reaction (PCR) was carried out in a Perkin Elmer thermocycler using primers and conditions previously described.[32,34] The full length of exon 12 was first amplified. The second multiplex PCR involved twelve different fragments in two sets of amplification PCR reactions (Fig. 2).

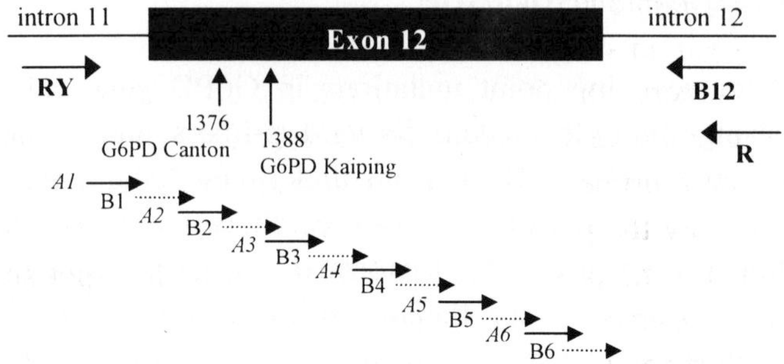

Fig. 2 MPTP Technique to scan exon 12.
The exon 12 region of the G6PD gene was amplified by a first PCR using primers RY and B12.[32,34] Using the first PCR product as template, a second PCR is then carried out by multiplex amplification of two sets of six alternative forward primers together with a common reverse primer (R). Set A: Forward primers A1 to A6, Set B: Forward primers B1 to B6. All the forward primers range from 13 to 15 mers and are designed to be in tandem and to cover the entire exon 12 region with overlaps of three to four nucleotides. The location of the two common mutations G6PD Canton and G6PD Kaiping are indicated by vertical arrow at nucleotides 1376 and 1388, respectively.

A common reverse primer was used in each of the two sets of PCR reactions containing six primers. The forward primers were designed such that they covered the entire exon 12 sequence with three to four nucleotide overlaps, thus covering nucleotide numbers 1376 and 1388 where the sequence alterations of G6PD Canton and Kaiping variants are found. The PCR products were analyzed on 4% agarose gel containing 0.5μg/μl ethidium bromide. Mutations were confirmed by direct sequencing using a dye terminator kit (PE Applied Biosystems) on the ABI Prism 310 DNA sequencer (Perkin Elmer).

9.2 Results and Discussion

The use of alternative (tandem) primers in two separate PCR multiplex reactions resolved the problem of separation of PCR products that would otherwise differ by ten nucleotides if sequential primers were used. The absence of amplification product in either set of reaction indicated a mutation at the respective primer annealing sites.

Using the MPTP technique, it was clear that the presence of the G6PD Canton and Kaiping variants could be detected by failure of amplification using the forward primers corresponding to the mutation sites. No other mutations were detected in the thirty-four G6PD deficient patients screened. The use of such tandem primers allows all mutations to be detected along the sequences walked by the primers. In addition, a specific PCR reaction test could be designed using primers selected for mutation hot-spot sites to screen for absence or presence of these targeted variants.[35]

Our study on thirty-four G6PD deficient patients using the MPTP technique has shown that the most common G6PD gene mutations are 1376G→T (Canton variant) and 1388G→A (Kaiping variant) at 44% and 32% incidence respectively (Table 3). Both mutations are located in exon 12 (Fig. 3). It has been previously reported that the Canton mutation is most common among Orientals and results in a codon change from arginine to leucine.[5] The high incidence of these two missense mutations in the Singapore population reflects the high proportion of individuals with Southern Chinese origins. The Kaiping mutation has similarly been reported in Thailand, Taiwan and China.[9,13] In contrast, the most common mutation in Southern Thailand is the G6PD Mahidol variant (487G→A) while G6PD Union (1360C→T) and G6PD Viangchan (871G→A) predominate in the Philippines.[27,29,30] In Malaysia, the common variants are G6PD Viangchan (871G→A), G6PD Mediterranean (563C→T), G6PD Gaozhou (95A→G), Canton (1376G→T), G6PD Kaiping (1388G→A) and G6PD Banks-Torres (383T→C).[16,36] In Indonesia, the common variants reported are G6PD Mediterranean (563C→T), G6PD Canton (1376G→T), G6PD Mahidol (487G→A) and Kaiping (1388G→A).[37,38,39] These comparative observations on South East Asian populations seem to indicate that some mutations may be population specific and reflective of the ethnic origins of the samples.[32,35] However, it should be noted that such observations are dependent on the number of samples analyzed and a large population-based screening is required to reflect the mutation spectrum more accurately.

Table 3 Distribution of eight major G6PD gene mutations in Asian population analyzed by MPTP method*

Variants	Mutations	Exon	Singapore	Indonesia	Malaysia	Philippines
A (+)	376A→G	5	0	0	0	0
Bank-Torres	383T→C	5	3 (9%)	0	6	12
Mahidol	487G→A	6	4 (12%)	2	3	0
Taiwan-Hakka	1376G→T	6	0	0	0	2
Mediterranean	563C→T	6	1 (3%)	5	9	0
Union	1360C→T	11	0	0	0	33
Canton	1376G→T	12	15 (44%)	3	2	0
Kaiping	1388G→A	12	11 (32%)	0	3	0
Total number of mutations			34	10	23	47

*Molecular characterization was carried out for other exons besides exon 12 by a similar approach using MPTP technique.[32,34]

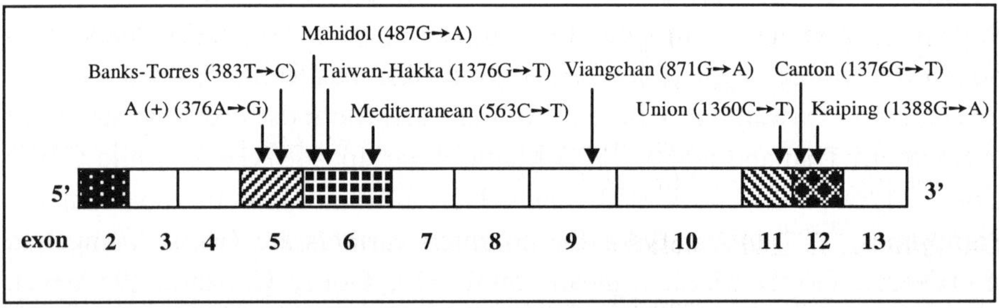

Fig. 3 Common G6PD variants and mutation sites in Asian populations (Adapted from Lai PS, 1999)[32]

10 SUMMARY

The prevalence of G6PD deficiency varies greatly throughout the world.[1,5] G6PD variants have been well characterized based on biochemical characteristics in contrast to molecular mutation analysis. It is clear that the G6PD gene displays vast genetic heterogeneity with more than 400 biochemical variants being documented to date.[14,26] However, only a small portion of this heterogeneity has been confirmed at the molecular level with less than one hundred different gene mutations being reported so far. The MPTP technique is currently being used to screen for G6PD deficiency in a number of other Asian population.[27,32,35,36,39] This method appears to be a fast and efficient approach for scanning of point mutations as well as for screening specific known mutations. This technique also allows analysis of large number of samples from patients. It can thus be applied for fast characterization of G6PD deficient patients at the molecular level.

Acknowledgements

We wish to thank Ms MH Liew for assistance in the preparation of this manuscript and Ms Dorothy SH Teo for the methodology on the modified Bernstein's enzyme assay. We acknowledge partial funding from the MINDEF–NUS/06/00 grant.

References

1. Beutler E, Study of glucose-6 phosphate dehydrogenase: History and molecular biology, *A J Hematol* 1993; **42**: 53–58.

2. Scriver CR, Disease, war and biology: Languages for medicine and pediatrics, *Pediatr Res* 1995; **38**(6): 819–29.

3. Beutler E, G6PD deficiency, *Blood* 1994; **84**: 3613–3636.

4. Carson PE, Flanagan CL, Ickes CE, Alving AS, Enzymatic deficiency in primaquine-sensitive erythrocytes, *Science* 1956; **124**: 484–485.

5. Beutler E, G6PD: Population genetics and clinical manifestations, *Blood Reviews* 1996; **10**: 45–52.

6. WHO Working Group, Glucose-6-phosphate dehydrogenase deficiency, *Bull WHO* 1989; **67**: 601.

7. Beutler E, The red cell: A tiny dynamo, in MM Wintrobe, (eds.), *Blood Pure and Eloquent*, McGraw-Hill, New York, 1980: 141–168.

8. Luzzatto L, Mehta A, Glucose-6-phosphate dehydrogenase deficiency, in CR Scriver, AL Beaudet, WS Sly, D Valle, (eds.), *The Metabolic and Molecular Bases of Inherited Disease*, McGraw-Hill, New York, 1995: 3367–3398.

9. Li PN, Thompson JN, Wang XY, Song L, Analysis of common mutations and associated haplotypes in Chinese patients with glucose-6-phosphate dehydrogenase deficiency, *Biochem Mol Biol Int* 1998; **46**(6): 1135–1143.

10. Bernstein RE, A rapid screening dye test for the detection of glucose-6-phosphate dehydrogenase deficiency in red cells, *Nature* 1962; **194**: 192.

11. World Health Organization. Standardization of Procedures for the Study of Glucose-6-phosphate dehydrogenase, WHO Technical Report Series No.366. World Health Organization, Geneva, 1967.

12. Normah J, Choo KE, Oppenheimer SJ, Selamah G, Glucose-6-phosphate dehydrogenase enzyme activity in normal, hemizygote and heterozygote Kelantanese Malays, *J Paediatr Child Health* 1991 Dec; **27**(6): 376–379.

13. Tang TK, Huang CS, Huang MJ, Tam KB, Yeh CH, Tang CJC, Diverse point mutations result in glucose-6-phosphate dehydrogenase (G6PD) polymorphism in Taiwan, *Blood* 1992; **79**(8) : 2135–2140.

14. Cappellini MD, Montemuros FM, Dotti C, Tavazzi D, Fiorelli G, Molecular characterization of the glucose-6-phosphate dehydrogenase (G6PD) Ferrara II variant, *Hum Genet* 1995; **95**: 440–442.

15. Hirono A, Fujii H, Takano T, Chiba Y, Azuno Y, Miwa S, Molecular analysis of eight biochemically unique glucose-6-phosphate dehydrogenase variants found In Japan, *Blood* 1997; **89**(12): 4624–4627.

16. Ainoon O, Joyce J, Boo NY, Cheong SK, Hamidah NH, Nucleotide 1376G→T mutation in G6PD-deficient Chinese in Malaysia, *Malays J Pathol* 1995; **17**(2): 61–65.

17. Vulliamy T, Beutler E, Luzzatto L, Variants of glucose-6-phosphate dehydrogenase are due to missense mutations spread throughout the coding region of the gene, *Hum Mutat* 1993; **2**: 159–167.

18. Persico MG, Viglietto G, Martini G, Toniolo D, Paonessa G, Moscatelli C, Dono R, Vulliamy T, Luzzatto L, D'Urso M, Isolation of human glucose-6-phosphate dehydrogenase (G6PD) cDNA clones: Primary

structure of the protein and unusually 5' non-coding region, *Nucleic Acids Res* 1986; **14**: 2511,7822.

19. Martini G, Toniolo D, Vulliamy T, Luzzatto L, Dono R, Viglietto G, Paonessa G, D'Urso M, Persico MG, Structural analysis of the X-linked gene encoding human glucose-6-phosphate dehydrogenase, *EMBO J* 1986; **5**: 1849.

20. Takizawa T, Huang IY, Ikuta T, Yoshida A, Human glucose-6-phosphate dehydrogenase: Primary structure and cDNA cloning, *Proc Natl Acad Sci* USA,1986; **83**: 4157.

21. Roth Jr EF, Raventos-Suarez C, Rinaldi A, Nagel RL, Glucose-6-phosphate dehydrogenase deficiency inhibits *in vitro* growth of *Plasmodium falciparum*, *Proc Natl Acad* Sci USA 1983; **80**: 298–299.

22. Luzzatto L, Sodeinde O, Martini G, Genetic variation in the host and adaptive phenomena in *Plasmodium falciparum* infection, *Ciba Found Symp* 1983; **94**: 159–162.

23. Ninfali P, Baronciani L, Ruzzo A, Fortini C, Amadori E, Dall'ara G, Magnani M, Beutler E, Molecular analysis of G6PD variants in northern Italy: a study on the population from the Ferrara district, *Hum Genet* 1993; **92**: 139–142.

24. Saha N, Hong SH, Low PS, Tay JS, Biochemical characteristics of four common molecular variants in glucose-6-phosphate dehydrogenase-deficiency Chinese in Singapore, *Hum Hered* 1995; **45**(5): 253–257.

25. Quak SH, Saha N, Tay JS, Glucose-6-phosphate dehydrogenase deficiency in Singapore, *Am Acad Med Singapore* 1996; **25**(1): 45–48.

26. Hsia YE, Miyakawa F, Baltazar J, Ching NSP, Yuen J, Westwood B, Beutler E, Frequency of glucose-6-phosphate dehydrogenase (G6PD) mutations in Chinese, Filipinos and Laotians from Hawaii, *Hum Genet* 1993; **92**: 470–476.

27. Silao CL, Shirakawa T, Nishiyama K, Padilla C, Matsuo M, Molecular basis of glucose-6-phosphate dehydrogenase deficiency among Filipinos, *Pediatr Int* 1999; **42**(2): 138–41.

28. Padilla C, G6PD deficiency in the Philippines, in *Abstracts of Asian Symposium in Neonatology: G6PD Deficiency & Related Conditions*, 8–9 August 2000, Yogyakarta, Indonesia, p. 3.

29. Laosombat V, G6PD deficiency and neonatal jaundice in Thai patients, in *Abstracts of Asian Symposium in Neonatology: G6PD Deficiency & Related Conditions*, 8–9 August 2000, Yogyakarta, Indonesia, p. 15.

30. Panich V, Na-Nakorn S, Wasi P, G6PD variants in Chinese in Thailand, *Southeast Asia J Trop Med Public Health* 1980; **11**(2): 250–255.

31. Lai PS, Shirakawa T, Nishiyama K, Matsuo M, MPTP detection of G6PD molecular variants, G1376T and G1388A, in *Abstracts of the South East Asia Symposium on Neonatology*, 1–2 March 1999, Yogyakarta, Indonesia, p.31, Yogyakarta, March 1999.

32. Lai PS, G6PD deficiency in Singapore: screening by MPTP, in D Setyowireni, M Julia, S Wardita, and A Surjona (eds.), *South East Asia Symposium on Neonatology—Providing Optimum Neonatal Care to Reduce its morbidity and Mortality*, Yogyakarta: Center for Maternal-Perinatal Health, 1999: 91–99.

33. Joseph R, Mass screening, profile and clinical consequences for G6PD deficiency in Singapore, in *Abstracts of Asian Symposium in Neonatology: G6PD Deficiency & Related Conditions*, 8–9 August 2000, Yogyakarta, Indonesia, p. 9.

34. Shirakawa T, Nishiyama K, Lai PS, Matsuo M, A comprehensive method to scan for point mutation of the glucose-6-phosphate dehydrogenase gene, *Jpn J Human Genet* 1997; **42**: 417–423.

35. Shirakawa T, Nishiyama K, Lai PS, Yusoff NM, Dao NTN, Matsuo M, A simple and rapid mutation detection system for G6PD deficiency, in *Abstracts of Asian Symposium in Neonatology: G6PD Deficiency & Related Conditions*, 8–9 August 2000, Yogyakarta, Indonesia, p. 6.

36. Yusoff NM, Shirakawa T, Nishiyama K, Ghazali S, Choo KE, Isa MN, Matsuo M, Genetic variations in the glucose-6-phosphate dehydrogenase gene in neonatal jaundice in Kelantan Malays, Malaysia, in *Abstracts of Asian Symposium in Neonatology: G6PD Deficiency & Related Conditions*, 8–9 August 2000, Yogyakarta, Indonesia, p.7.

37. Suryantoro P, Formazan Ring test for G6PD deficiency identification in the newborn infants in Yogyakarta, *BIK* 1998; 21–27.

38. Soemantri AG, Saha S, Saha N, Tay JSH, Molecular variants of red cell glucose-6-phosphate dehydrogenase deficiency in central Java, Indonesia, *Hum Hered* 1995; **45**: 346–350.

39. Suryantoro P, Tasmini, Pramuji H, Patria SY, G6PD report from Yogyakarta, in *Abstracts of Asian Symposium in Neonatology: G6PD Deficiency & Related Conditions*, 8–9 August 2000, Yogyakarta, Indonesia, p. 12.

FRONTIERS IN HUMAN GENETICS
Diseases and Technologies

© 2001 by World Scientific Publishing Co. Pte. Ltd.

GLUCOSE-6-PHOSPHATE-DEHYDROGENASE (G6PD) DEFICIENCY: PRELIMINARY REPORT OF THE MULTIPLEX PCR TANDEM FORWARD PRIMERS (MPTP) FOR INDONESIAN YOGYAKARTA CASES

PURNOMO SURYANTORO*, TASMINI[†],
KAORU NISHIYAMA[‡], TAKU SHIRAKAWA[‡] and MASAFUMI MATSUO[‡]
**Pediatric Department, Faculty of Medicine, Gadjah Mada University*
[†]*Department of Biochemistry, Faculty of Medicine, Gadjah Mada University*
Yogyakarta, Indonesia
[‡]*Kobe School of Medicine, Kobe, Japan.*

G6PD is one of the enzymes needed by erythrocytes to regenerate ADP to ATP for recovering energy resource. Deficiency of this enzyme will lead to an increase of erythrocyte lysis. The incidence of this disease in Indonesia is considerably high possibly due to malarial selection. It is estimated that 2-6 % of the population are carriers. Routine sequencing of the 60 common mutations covering all 12 exons of G6PD gene is laborious, time consuming and expensive. Therefore the MPTP method was used to detect the most common mutations located in exons 6, 11 and 12.

Among 57 samples of normal newborn babies from Yogyakarta, Indonesia that were screened by the Formazan method, 5 were G6PD deficient whereas among 5 newborns with severe jaundice, 2 were G6PD deficient. Of these seven deficiency cases, one was shown to carry the Silent variant (1311 C→T mutation).

The MPTP method for detecting the mutation of G6PD gene was found to bed useful for screening G6PD cases from Yogyakarta, Indonesia.
Keywords: G6PD, deficiency, Indonesia, MPTP

1 INTRODUCTION

Glucose-6-phosphate dehydrogenase (G6PD) is an X-chromosome encoded enzyme that catalyze the first step in the pentose phosphate pathway. Deficient subjects are mostly asymptomatic, but clinical manifestations may include neonatal jaundice, acute hemolytic anemia which are stimulated by viral infection or drugs such as salycylate, sulfonamide, etc.

G6PD deficiency is common in Indonesia as it is a malaria endemic region. It is estimated to affect 400 million people worldwide. To date, more

* Corresponding author

than 300 different G6PD variants have been identified on the basis of biochemical parameters.

The G6PD gene is 65 kilobases in length and consists of 12 exons. Vast genetic heterogeneity has been suggested by Beutler and Kuhl [1]. However, Ganczakowski *et al* [2] reported that at the molecular level, only 60 different point mutations have been found. In certain geographical areas, only a few specific mutations affecting some exons have been found.

The most common mutation reported in Asia and the Middle East was the Mediterranean variant[1,3,4] whereas in Thailand, the most common variants are the Mahidol, Union, Canton and Anant-Dhon variants[5]. In Indonesia, the majority of mutations are found in exons 5, 6, 11 and 12.

The **M**ultiplex **P**CR with **T**andems forward **P**rimers (MPTP) suggested by Shirakawa *et al* [6] is a comprehensive screening system for mutations in X-linked recessive diseases. This polymerase chain reaction (PCR) amplification using a two-steps method is simple, rapid and can be carried out in any laboratory which has a PCR machine. In Singapore, the Canton (G1376T) and Kaiping (G1388A) G6PD variants were found by MPTP screening of exon 12[7].

The aim of this study is to identify the G6PD deficiency at the molecular level among the local population group around Yogyakarta City, using the MPTP technique described by Shirakawa *et al.* Specifically, the following mutations will be screened: 487G→A (Mahidol) and 563C→T (Mediterranean) mutations on exon 6; the 1311C→T (silent) and 1360C→T (Union) mutations on exon 11; and the 1376G→T (Canton) and 1388G→A (Kaiping) mutations on exon 12.

2 METHODS

Dried blood spots were collected on filter papers from patients admitted in the Pantirapih Hospital and Pediatric Department of Sardjito General Hospital in Yogyakarta, Indonesia. The blood spots were then screened for G6PD deficiency by the Formasan Ring test[8]. Genomic DNA was extracted from the spots from patients who were classified as deficient by the Formasan Ring test.

The sequences of primers, the target of genomic DNA and the length of PCR product for first and second PCR are listed in Tables 1 and 2.

Table 1. List of primers for first round PCR.

Exon	Primer code	Sequence	PCR product size
6	6,7-a 6,7-b	5' –caaaaccaatgaggaagcca 5'- tgcctcgtcacagatgggcc	1321 bp
11-12	10-a 13-b	5'- ggtgggatggtaggtgatgcc 5'- aggaatgtgcagctgaggtcaat	1054 bp

Table 2. List of primers for second round PCR.

Exon	Normal target primers		Mutation specific primers		G6PD variants
	Code	Sequence	Code	Sequence	
6	6-F	5'gtaacgcagctcc			
	6-487/493	5'agcagaggcctgga	6-487M	5'agcagaAgctgga	Mahidol 487 G→A
			6-493M	5'agcagaggctggG	Taiwan-Hakka 493A→G
	6-563	5'acatctcctccctg	6-563M	5'acatctTctccctg	Mediterranean 563 C→T
	6-R	5'gcagtgggccagg			
11	11-F	5'tccctcacagaac			
	11-1360	5'cacttcgtgcgc	11-360M	5'cacttcgtgTgc	Union 1360 C→T
	11-R	5'catagcccacaggt			
12	12-F`	5'ctatgtcccctcag			
	12-1376	5'acgagctccgtg	12-1376M	5'gacgagctccTtg	Canton 1376 G→T
	12-1388	5'aggcctggcgta	12-1388M	5'aggcctggcAta	Kaiping 1388 G→A
	12-R	5-ggcttctccagctc			

The PCR reactions were performed as follows: A 10μl of solution containing 8μl of a master mix (10mM Tris-HCl pH 8.3, 10 mM KCL 250 μM dNTPs, 2.5mM MgCL2 and 5% DMSO), 1 μl of primer mixture, 1 μl of the first PCR product and another 10 μl of solution containing of 8μl of the master mixture, 1.9 μl of distilled water and 0.1 μl of 5U AmpliTaq DNA

Polymerase Stoffel Fragment (Perkin-Elmer Co.,USA) were pre-heated to 85°C individually and then mixed together at 85 °C. PCR was then carried out with the following parameters: 25 cycles of denaturation at 95°C, annealing at 47°C and elongation at 70°C for 30 seconds per step. The products were electrophoresced on a 4% agarose gel (MetaPhor/Seakem HGT 3:1) containing 1.0μg/ml ethidium bromide and photographed under ultraviolet light. Figure 1 shows the locations of the primers in exons 11 and 12.

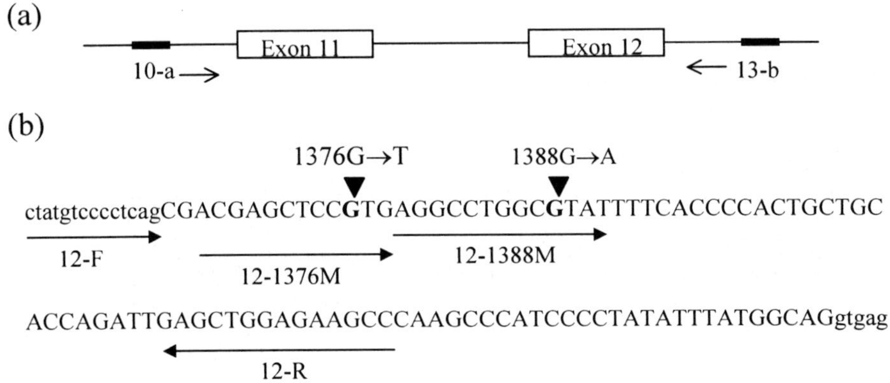

Fig. 1. Location of primers flanking exons 11 and 12
(1a) The boxes represent exons 11 and 12; and the lines represent introns 10, 11 and 12. The arrows indicate the location and direction of primers 10-a and 13-b.
(1b) MPTP using 4 primers in the exon 12 region as indicated by arrows. Three different PCR products are expected for the Canton and Kaiping G6PD deficiency variants

3 RESULTS

During the period between August to December 1998, we collected fifty-seven dried blood spots on the filter papers from healthy infants born at two hospitals in Yogyakarta, Indonesia. The Formazan tests were carried out in the laboratory of ICMR/Kobe School of Medicine, Japan. A total of twenty-one samples showed small rings but fifteen of them were not of suitable quality such that only five could be included in this study.

From another five newborn babies with severe bilirubinemia, two cases have G6PD enzyme levels of 113 and 81 μU/103 erythrocytes as compared to the normal level of 114-180 μU/ 1000 erythrocytes). Thus, there were a total of seven samples for further MPTP studies.

Figures 2 and 3 show results of the screening for exon 6 in which 3 bands are expected indicating the Mahidol, Taiwan Hakka and Mediterranean G6PD variants. In Figure 2, the third band of lane 4 was faint with the wild type primers (normal target primers) indicative of the Mediterranean variant. However, lane 16 showed that the first and second bands corresponding to the Mahidol and Taiwan Hakka variants were faint and clearly missing. Therefore this case could neither be classified as the Mahidol nor the Mediterranean variant.

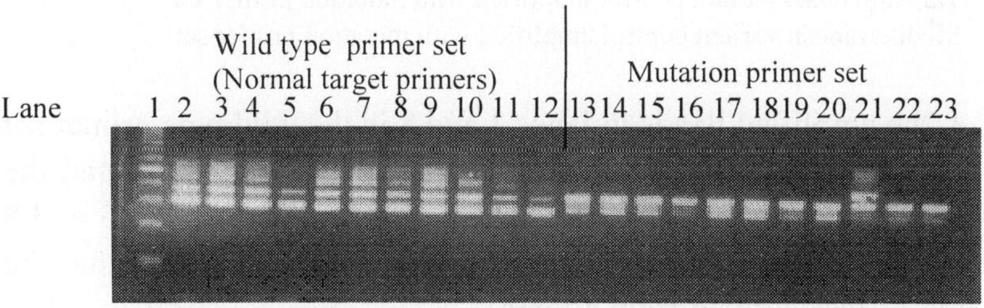

Fig. 2. Analysis of seven cases of G6PD deficiency for 487 G→A (Mahidol variant) and 563 C→T (Mediterranean variant) in exon 6

Lane 1: Markers

Lanes 2-8: Seven G6PD deficient samples amplified with the normal primer set

Lane 9: Normal control individual amplified with the normal primer set

Lane 10: Mahidol variant control amplified with the normal target primer set

Lane 11: Taiwan-Hakka variant amplified with the normal target primer set

Lane 12: Mediterranean variant control amplified with the normal target primer set

Lanes 13-19: Seven G6PD deficient samples amplified with the mutation primer set

Lane 20: Normal control individual amplified with the mutation primer set

Lane 21: Mahidol variant control amplified with the mutation primer set

Lane 22: Taiwan-Hakka variant amplified with the mutation primer set

Lane 23: Mediterranean variant control amplified with the mutation primer set

Since there is only a 6 bp difference between the first (Mahidol variant) and second (Taiwan-Hakka variant) amplification products, screening for the Taiwan-Hakka mutation was repeated. The results showed that the Taiwan-Hakka variant is absent for exon 6 (Fig. 3).

Mutation primer set

Lane 1 2 3 4 5 6 7 8 9 10 11 12

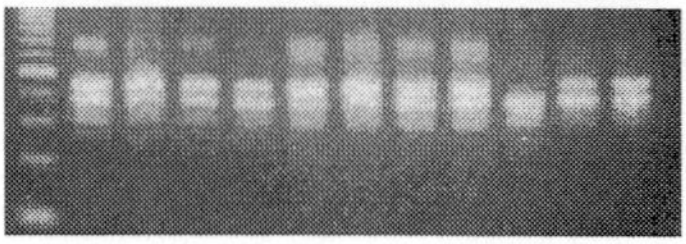

Fig. 3. Analysis of seven G6PD deficient cases for 493 A→G (Taiwan-Hakka) on exon 6.

Lane 1: Markers

Lanes 2-8: Seven G6PD deficient samples amplified with Taiwan-Hakka mutation primer set

Lane 9: Normal control individual amplified with mutation primer set

Lane 10: Mahidol variant control amplified with mutation primer set

Lane 11: Taiwan-Hakka variant control amplified with mutation primer set

Lane 12: Mediterranean variant control amplified with mutation primer set

In Fig. 4, the amplified bands in lanes 3 and 8 in the wild type primer set and lanes 13, 17, and 18 in the mutation primer set strongly indicated the presence of a silent variant (nucleotide 1311). The results for lanes 3 and 8 corresponded to that of lanes 13 and 18 respectively. It appeared that the sample in lane 17 was derived from a female.

← Wild type primer set → | ← Mutation primer set →

Lane 1 2 3 4 5 6 7 8 9 10 11 | 12 13 14 15 16 17 18 19 20 21

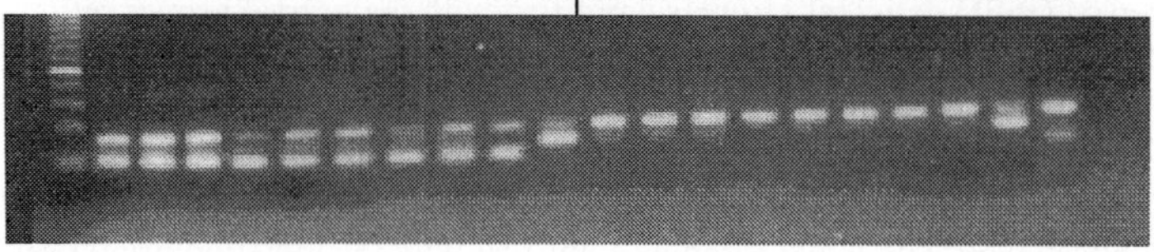

Fig. 4. Analysis of seven G6PD deficient cases for 1311 C→T (silent mutation variant) and 1360 C→T (Union variant) on exon 11.

Lane 1: Markers

Lanes 2-8: Seven G6PD deficient samples amplified with the wild type primer set

Lane 9: Normal control individual amplified with the wild type primer set

Lane 10: Silent variant control amplified with the wild type primer set

Lane 11: Union variant control amplified with the wild type primer set

Lanes 12-19: Seven G6PD deficient samples amplified with mutation primer set

Lane 20: Normal control individual amplified with mutation primer set

Lane 21: Silent variant control amplified with mutation primer set

Lane 22: Union variant control amplified with mutation primer set

By using a similar approach for mutation primer set for exon 12, the two common Canton and Kaiping variants, 1376C→T and 1388G→A, can be screened (Fig. 5).

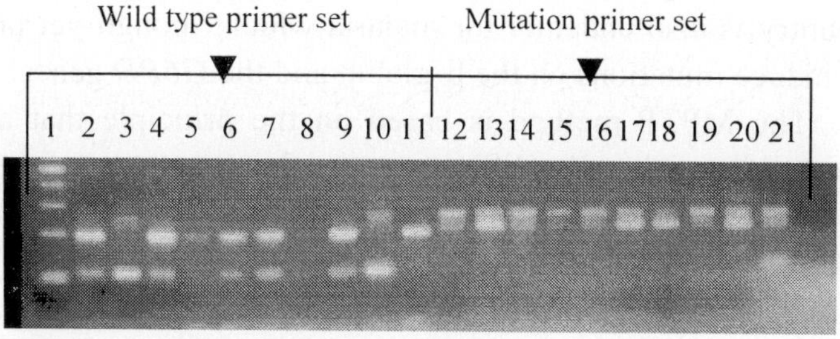

Fig. 5. Analysis of seven G6PD deficiency cases for 1376 C→T (Canton) and 1388 G→A (Kaiping variant) on exon 12.

Lane 1:	Markers
Lanes 2-8:	Seven samples amplified with the wild type type primer set
Lane 9:	Normal control individual amplified with the wildtype type primer set
Lane 10:	Canton variant control amplified with the wildtype type primer set
Lane 11:	Kaiping variant control amplified with the wildtype type primer set
Lanes 12-18:	Seven samples amplified with mutation primer set
Lane 19:	Normal control individual amplified with mutation primer set
Lane 20:	Canton variant control amplified with mutation primer set
Lane 21:	Kaiping variant control amplified with mutation primers

4 DISCUSSION

Indonesia has a heterogeneous population which traces its ancestry as originating from many population groups such as Cambodian, Thai, Indian, Mongol, Negroid etc. Population research thus offers a wide scope for understanding the molecular basis of diseases at the genetic level. This country is also endemic for malaria which, though yet unproven, is known to induce mutations on the β-globin and the *G6PD* genes.

The MPTP method is based on the principle that a short primer can recognize nucleotide differences in the target sequence by its failure to hybridize. Failure or inefficiency in annealing will result in the absence of the expected amplified product, and the reaction can be multiplexed to screen for several primer annealing sites simultaneously. This is useful if the sequence variations in the target region are localized to a narrow and short amplified region.

It has been reported previously that among 145 newborn babies in Yogyakarta, 6.2% of male babies have moderate G6PD enzyme activity and 1.4 % have low activity[7]. Female babies have a normal enzyme activity range. However, Tasmini *et al* (1995), using the Sigma kit to examine 111 neonates, found that none of them were identified as "deficient"[9]. When the erythrocyte hemolysis was correlated with the G6PD enzyme content the results were all negative.

Interesting results have also been reported by Sofro *et al.* (1994) who similarly used the Sigma kit[10]. Among 134 individuals of Batak descent, 10 males (43.5%) and 9 females (8.1%) were found to be G6PD deficient. Similar results were reported by Pramuji et al (1995) for the population around Palembang in Sumatra[11]. These results are interesting since the G6PD gene is X-linked and further research is needed to shed more light on these findings.

G6PD deficiency variants are very heterogeneous. In Thailand 14 different variants have been reported, with the Mahidol variant being the most common. This variant also exists among the Malay, Laotian, Cambodian, Vietnamese and Chinese populations[12]. In addition, 6 other variants have been reported – the Canton, Don (Taipei-Hakka), Union, A(-) Hong Kong, Pokfulam and Mediterranean. Among 1157 blood donors screened, there were 4 individuals who showed presence of G6PD variants.

Since erythrocytes in G6PD deficiency have higher hemolysis risk, it has been suggested that G6PD deficient individuals should be prohibited as blood donors.

Soemantri *et al.* (1995) reported that among the Javanese, the occurrence of the Mediteranean(563C→T), Canton (1376G→T), Mahidol (487G→A), Taipei-Hakka (493A→G), Kaiping (1388G→A) and silent (1311C→T) variants were 31%, 18%, 12%, 37% and 31%, respectively, with the remaining variants unknown[13].

Using the MPTP method, Lai PS (1999) reported successful identification of the Mediterranean, Kaiping and Canton variants from the Singapore population that consists predominantly of the Chinese ethnic group (77 %)[7]. In Yogyakarta we identified only one case of the Silent mutation variant and we conclude that this MPTP study should be repeated and/or extended to cover the other exons in the G6PD gene.

5 SUMMARY

In this study, five out of fifty-seven samples from healthy babies and 2 out of 5 newborns with jaundice were identified to have low G6PD activity by the formazan ring test. One out of seven patients was found to carry the 1311 C→T(silent variant) mutation by the MPTP method which could detect the 8 G6PD variants involving exons 5, 6 11 and 12. Further analysis of other exons using the MPTP method can be carried out for the Yogyakarta cases.

References

1. Beutler E, Kuhl W, The nt1311 polymorphism of G6PD: G6PD Mediterranean mutation may have originated independently in Europe and Asia. *Am J Hum Genet*. 1990 Dec; **47(6):**1008-1012.

2. Ganczakowski M, Town M, Bowden DK, Vulliamy TJ, Kaneko A, Clegg JB, Weatherall DJ, Luzzatto L, Multiple glucose 6-phosphate dehydrogenase-deficient variants correlate with malaria endemicity in

the Vanuatu archipelago (southwestern Pacific). *Am J Hum Genet*. 1995 Jan; **56(1)**:294-301.

3. Saha S, Saha N, Tay JS, Jeyaseelan K, Basair JB, Chew SE, Molecular characterisation of red cell glucose-6-phosphate dehydrogenase deficiency in Singapore Chinese. *Am J Hematol*. 1994 Dec; **47(4)**:273-277.

4. Kurdi-Haidar B, Mason PJ, Berrebi A, Ankra-Badu G, Al-Ali A, Oppenheim A, Luzzatto L, Origin and spread of the glucose-6-phosphate dehydrogenase variant (G6PD-Mediterranean) in the Middle East. *Am J Hum Genet* 1990; **47**:1013-1019.

5. Panich V, Na-Nakorn S, G-6-PD variants in Thailand, *J Med Assoc Thai*. 1980 Oct;**63(10)**:537-43.

6. Shirakawa T, Nishiyama K, Lai PS, Matsuo M, A comprehensive method to scan for point mutation of the glucose 6 phosphate dehydrogenase gene. *Jpn.J.HumanGenet*; 1997; **42:**417-423.

7. Lai PS, G6PD deficiency in Singapore: screening by MPTP, in Setyowireni D, Julia M, Wardita S and Surjona A (eds.), *South East Asia Symposium on Neonatology- Providing optimum Neonatal Care to reduce its morbidity and Mortality*, Yogyakarta: Center for Maternal-Perinatal Health, 1999:91-99.

8. Purnomo S, Formazan Ring test for G6PD deficiency identification in the newborn infants in Yogyakarta. **BIK** 1998:21-27.

9. Tasmini, Salam A, Ismadi M, Aktivitas ensim glukose 6 fosfat dehydrogenase (G6PD)dan fragilitas eritrosit bayi baru lahir. BPPS-UGM,**8(1B)**,February 1995:1-10.

10. Sofro AS, Lanni F, Tasmini, Madiyan M, Hastuti P, Sukiman I, Sucipto A, Defisiensi dehidrogenase glukosa 6 fosfat (G6PD) pada mahasiswa tenaga Kesehatan diMedan. BIK.1994:51-56.

11. Hastuti P, Sofro AS, Tasmini, Karakterisasi enzim glukosa 6 fosfat dehydrogenase pada penduduk di Palembang. Laporan Penelitian DPP-FK-UGM 1994-1995.

12. Panich V, G6PD variants in Southern Asia population, in Yoshida A and Beatler E (eds.), *Glucose 6 Phosphate Dehydrogenase,* Orlando:Academic Press, 1986:195-242.

13. Soemantri AG, Saha S, Saha N, Tay JSH, Molecular variants of red cell glucose 6 phosphate dehydrogenase deficiency in central Java, Indonesia. *Hum.Hered* 1995, **45**:346-350

FRONTIERS IN HUMAN GENETICS
Diseases and Technologies

© 2001 by World Scientific Publishing Co. Pte. Ltd.

CONGENITAL ADRENAL HYPERPLASIA: A REVIEW OF 13 CASES DETECTED BY NEWBORN SCREENING

JOY Y. LEE*,[†], CARMENCITA D. PADILLA*,[‡], CARMELITA FAGELA-DOMINGO*, SIOKSOAN C CUA[§] and LORNA R. ABAD*
Department of Pediatrics
University of the Philippines Manila-Philippine General Hospital
Manila, Philippines
[†]*E-mail: joyyaplito@hotmail.com*
[‡]*E-mail: carpadil@fastmail.i-next.net*
[§]*E-mail:sioksoan@mail.com*

Newborn screening for 21-hydroxylase deficiency, which is the most common type of congenital adrenal hyperplasia (CAH), was started in the Philippines in June 1996. CAH was screened using the Delfia 17-OHP time-resolved fluoroimmunoassay with a false positive rate of 0.02% and false negative rate of 0.002%. Radioimmunoassay was used as a confirmatory test. A total of 13 cases of 21-hydroxylase deficiency cases were detected out of 100,369 newborns screened. Females outnumbered the males (1.6:1). Mean age when screening was done was 13.8 days. The mean age when screening results were made available was 27.5 days. This was mainly attributed to the delay in sending samples to the screening laboratory which was on the average 11.9 days. The salt-losing type of 21-hydroxylase was the most common clinical form. All 13 cases are alive.
Keywords: congenital adrenal hyperplasia, 21-hydroxylase deficiency, newborn screening

* for the Newborn Screening Study group; † Corresponding author

1 INTRODUCTION

Congenital adrenal hyperplasia (CAH) is a group of autosomal recessive disorders involving an enzyme defect in the synthesis of cortisol. Clinical problems arise from the impaired production of the steroid hormones (glucocorticoids, mineralocorticoids or gonadal sex steroids) as well as from overproduction of certain steroids before the enzymatic block.

The majority (~90%) of CAH cases are due to a 21-hydroxylase deficiency which can be detected by newborn screening. Many of these screening programs were established in the mid-1980s. The justification for such programs lies in the increased mortality among missed cases of CAH who die of salt losing crisis and increased morbidity that results from adrenal crisis and incorrect sex assignment.

In the Philippines, the newborn screening for CAH, particularly 21-hydroxylase deficiency, was started in June 1996. Hospitals from across the country voluntarily participated in this multicenter study. Blood samples were collected on special filter paper. The screening method was the Delfia 17-OHP time resolved fluoroimmunoassay with a false positive rate of 0.02% and false negative rate of 0.002%.[1] Radioimmunoassay is used for confirmatory testing. A total of thirteen (13) CAH cases out of 100,368 newborns have been confirmed from June 1996 to December 1999.

This paper aims to present the clinical profile of the detected cases and to determine the role of newborn screening.

2 METHODOLOGY

All confirmed cases of congenital adrenal hyperplasia in the newborn screening laboratory located at the National Institutes of Health-University of the Philippines Manila were identified from records which were carefully filed and kept. A review of their sample cards was done. The following data were obtained: sex, birthweight, date of screening, date of availability of results, signs and symptoms, and other pertinent clinical information. The pediatric endocrinologists of some patients were also consulted regarding confirmatory test results, electrolyte results and outcome.

3 RESULTS

A total of 13 cases of congenital adrenal hyperplasia were detected in the newborn screening laboratory with a weighted incidence of 1:6,135. Almost all (12/13 or 92%) were term babies with birth weights appropriate for gestational age.

The sex ratio is 1.6:1 with 8 females and 5 males. The females had varying degrees of masculinization. Chromosomal studies done in 62% (5/8) of the cases showed a female (46XX) pattern. One case was wrongly assigned a male sex. Family history in 2 cases showed a similar condition. One case had a sibling who died due to symptoms similar to a salt-losing crisis and another patient had a cousin with ambiguous genitalia. There was no history of consanguinity in all cases.

CAH was not suspected in 7/13 or 54% of cases before newborn screening was done. Noteworthy is the absence of clinical suspicion in all male affected infants. Majority of cases (10/13 or 77%) are salt losers with signs and symptoms of electrolyte imbalance, vomiting, diarrhea and failure to thrive (see Table 1). One case is simple virilizing. The patient presented only with slightly enlarged clitoris.

The mean age when screening was done was 13.8 days. Screening results were available at a mean age of 27.5 days. Sample cards were received by the screening laboratory after 11.9 days from sample collection.

Almost all patients were being seen and followed up by their pediatric endocrinologists.

Table 1. Symptomatology

Sign / Symptom	Proportion
Hyperpigmentation	3/13 (23%)
Ambiguous genitalia	7/13 (54%)
Vomiting	3/13 (23%)
Failure to thrive	3/13 (23%)
Diarrhea	3/13 (23%)
Electrolyte imbalance	8/13 (62%)

4 DISCUSSION

Twenty-one hydroxylase deficiency results in adrenal insufficiency due to deficient glucocorticoid and mineralocorticoid production. Aside from these, an excess production of androgens which, in females, can cause varying degrees of masculinization will also be present.

Two clinical forms which correlate with the relative severity of the enzyme defect are seen in neonates: simple virilizing and salt losing. Simple virilizing form which has a partial enzymatic deficiency manifests as varying degrees of male differentiation of the external genitalia and urogenital sinus in females. The excess androgens can result in accelerated growth and skeletal maturation. These patients usually do not develop adrenal insufficiency unless exposed to major stress.

The salt losing form results from the severe deficiency of 21-hydroxylase. Affected females show more complete masculinization of the external genitalia. A significant impairment of cortisol and aldosterone production results in adrenal insufficiency and sodium wasting usually in the first weeks of life. Early detection of such cases is thus crucial to the ultimate outcome of affected babies.

The newborn screening program in the Philippines is only in its infancy. A total of 13 confirmed cases of CAH out of 100,368 newborns have been detected by the newborn screening laboratory from June 1996 to December 1999.

The sex ratio obtained was 1.6:1 with more females. A clinical suspicion of CAH was present in most of the female affected infants primarily because this group of patients presented with abnormal external genitalia at birth. Affected males, on the other hand, can be asymptomatic which can explain why CAH was not even suspected at the time screening was done.

The important role of newborn screening is highly illustrated by two cases in this series. In one case, a male infant, though born in a participating hospital, was not initially screened for CAH. Screening was later done at age 110 days when the infant had failure to thrive. Newborn screening expedited the diagnosis in this case.

In another case, an incorrect sex was assigned to an affected female. The result of the newborn screening aided in the rectification of the error.

The mean age when screening was done on affected cases was 13.8 days. This could have been earlier at 3.5 days had it not been for 2 cases in which screening was delayed at ages 110 and 31 days. Screening results were made available at a mean age of 27.5 days (range: 7-124 days). The screening laboratory was able to release results within three days. The two-week difference from sample collection to release of results, however, was mainly due to the time elapsed in sending the samples to the screening laboratory which was an average of 11.9 days. Screening practices are crucial in the early detection and treatment of affected cases since onset of adrenal crisis will usually be between 6 to 14 days for almost half of affected patients and 75% will manifest adrenal crisis by one month of age. Results of newborn screening would have been released earlier in this series if screening was done as recommended on the 2nd to 3rd day of life in all babies delivered and there was no delay in sending cards to the screening laboratory.

Newborn screening has played a significant role in the diagnosis of affected patients especially males who may be easily missed due to absence of ambiguous genitalia. It has also helped rectify incorrect sex assignment and reduce mortality and morbidity among the affected patients.

5 SUMMARY

Thirteen cases of 21-hydroxylase deficiency were confirmed. Females who presented with varying degrees of virilization outnumbered the males (1.6:1). The salt-losing type was the most common clinical form. All 13 cases are presently alive. Newborn screening has expedited the diagnosis of congenital adrenal hyperplasia especially among affected males and has helped in rectifying an incorrect sex assignment in one of the females. Mean age when screening was done was 13.8 days while the mean age when screening results were made available was 27.5 days.

6 RECOMMENDATIONS

A cost-benefit analysis of the program will be of great help in the decision of whether or not to implement it on a nationwide scale.

Blood samples should be collected on the 2nd to 3rd day of life and sent off to the laboratory as soon as possible.

Newborn screening for congenital adrenal hyperplasia must be made available to every newborn infant.

Acknowledgements

We would like to thank the consultant staff of the Department of Pediatrics and Department of Obstetrics and Gynecology, resident staff and nursing staff of all participating hospitals, namely, Capitol Medical Center, Cardinal Santos Medical Center, Children's Medical Center, Chinese General Hospital, De Los Santos Medical Center, Dr. Victor R. Potenciano Medical Center, Far Eastern University- Nicanor Reyes Memorial Foundation, Manila Center University-Felimon D. Tanchoco Medical Foundation, Manila Doctors Hospital, Mary Chiles Hospital, Medical Center Manila, Metropolitan Hospital, Ospital ng Maynila, Our Lady of Lourdes Hospital, Perpetual Help Medical Center, Philippine Children's Medical Center, Philippine General Hospital, Quezon City General Hospital, Quirino Memorial Medical Center, Rizal Medical Center, St. Luke's Medical Center, St. Martin de Porres Charity Hospital, United Doctors Medical Center, University of the East-Ramon Magsaysay Memorial Medical Center, Dr. Jesus Delgado Memorial Hospital, Hospital of the Infant Jesus, Makati Medical Center, Manila Sanitarium and Hospital, The Medical City, University of Santo Tomas Hospital, Mary Johnston Hospital.

We would like to express our sincerest thanks to Dr. Bridget Wilcken, Dr. Veronica Wiley, staff of the New South Wales Newborn Screening Program, University of the Philippines Manila, Philippine Pediatric Society and Philippine Obstetrics and Gynecologic Society for their support.

References

1. Lee JY, Padilla CD, Domingo CF, Screening for congenital Adrenal Hyperplasia in the Philippines. (unpublished)
2. Miller WL, Pathophysiology, genetics and treatment of hyperandrogenism. *Pediatr Clin North Am* 1997; **44**: 375–395.
3. Newborn Screening Fact Sheets. *Pediatrics* 1996; **98**: 473–501.
4. Donohue PA, Parker K, *et al.*, Congenital adrenal hyperplasia, in CR Scriver *et al.* (eds.), *The Metabolic and Molecular Basis of Inherited Diseases*, McGraw Hill Companies, New York, 1995: 2929–2961.
5. Laue L, Cutler G, 21-hydroxylase deficiency: overview of treatment. *The Endocrinologist* 1992; **75**: 291–299.

6. Pang C, Dean W, *et al.*, Congenital adrenal hyperplasia screening in a greater worldwide newborn population: Genetic frequency and the effectiveness of screening, in B Wilcken and D Webster (eds.), *Neonatal Screening in the Nineties*, The Kelvin Press, 1991: 208–209.
7. Pang S, Congenital adrenal hyperplasia. *Endocrinol Metab Clin North Am* 1997; **26**: 853–891.
8. Kohn B, Day D, Alemzadah R, Enerio D, Patel SV, Pelczar JV *et al.*, Splicing mutations in CYP 21 associated with delayed presentation of salt-wasting congenital adrenal hyperplasia. *Am J Med Genet* 1995; **57**: 450–454.
9. Virdi NK, Rayner PHW, Rudd BT, Green A, Should we screen for congenital adrenal hyperplasia? A review of 117 cases. *Arch Dis Child* 1987; **62**: 659–662.
10. al Saedi S, Dean H, Dent W, Stockl E, Cronin C, Screening for congenital adrenal hyperplasia: the Delfia Screening Test overestimates serum 17-hydroxyprogesterone in preterm infants. *Pediatrics* 1996; **97**: 100–102.
11. Cacciari E, Balsamo A, Cassio A, Piazzi S, Bernardi F, Salardi S *et al.*, Neonatal screening for congenital adrenal hyperplasia. *Arch Dis Child* 1983; **58**: 803–806.
12. Pang S, Wallace M, Hofman L, Thuline HC, Dorche C, Lyon IC *et al.*, Worldwide experience in newborn screening for classical congenital adrenal hyperplasia due to 21-hydroxylase deficiency. *Pediatrics* 1988; **81**: 866–874.

© 2001 by World Scientific Publishing Co. Pte. Ltd.

ANALYSIS OF DELETION BREAKPOINTS IN DYSTROPHIN TRANSCRIPTS

HH KHNG*, PS LOW*, WL LEE*,[†], HT ONG* and PS LAI*,[‡]
**Department of Paediatrics, National University of Singapore*
[†]*Department of Neurology, National Neuroscience Institute, Singapore*

Duchenne and Becker muscular dystrophies are allelic X-linked disorders resulting from defects in the gene coding for the dystrophin muscle protein. The dystrophin gene is more than 2300kb in size and consists of 79 exons. This large size and complexity presents a challenge to direct identification of point mutations and small deletions that cannot be identified by multiplex deletion testing or Southern blotting. One approach to this problem is to analyse the expression of ectopic dystrophin mRNA transcripts. Although the dystrophin gene transcript is distributed only over approximately 0.1% of the genome, analysis of such ectopic lymphocyte dystrophin transcripts can shed light on the pathogenic events at the transcriptional level.
Keywords: Duchenne muscular dystrophy, dystrophin, mRNA, RT-PCR, deletion, frameshift-mutation

1 INTRODUCTION

Duchenne's muscular dystrophy (DMD) is an X-linked recessive muscle degenerative disorder leading to severe disability and death.[1] It occurs with a frequency of approximately 1 in 3500 newborn males.[2] Becker's muscular dystrophy (BMD), the allelic form of DMD is milder, with patients surviving up to the fourth or fifth decade of life. Both DMD and BMD are caused by alterations or deficiency of a cytoskeletal protein called dystrophin. The gene coding for dystrophin protein located on chromosome Xp21 and spans more than 2300 kilobases in length.[2] The 14 kilobase

[‡] Corresponding author

transcript from this gene encompasses 79 exons.[3]

The majority of mutations in the dystrophin gene are intragenic deletions of one or more exons (65%)[1, 2] or duplications (5%).[4] The remaining mutations are either mutations in promoter elements, point mutations or micro-deletions in coding sequence or splice-site consensus sequences.[5] Gross deletions observed so far are mainly localised in two mutation "hotspots" — one in the proximal portion of the gene and one in the central portion.[2,6-9] DMD arises when the mutation results in premature termination of translation whereas BMD arises when the mutation produces a truncated but semi-functional dystrophin protein.[10]

The vast size of the gene and the large number of exons in the dystrophin gene present difficulties in molecular analysis. It is not practical to routinely screen for mutations in this gene by direct sequencing of all the coding exons. A number of approaches have been reported in the attempt to screen for mutations more efficiently. We report the results of a study using ectopic dystrophin mRNA from lymphocytes of patients to analyse gene mutations. Low levels of dystrophin mRNA are often detected in lymphocytes, lymphoblasts and fibroblasts as a result of 'illegitimate' or ectopic transcription.[11-12] These transcripts have been shown to reflect the true dystrophin mRNA in expressing tissues[13] and have been used for RNA studies in the dystrophin gene.[14-17] Being only fourteen kilobases in size, the mRNA transcript can be quickly analysed in a few overlapping amplification and sequencing reactions. In addition, analysis of ectopic transcripts allows the detection of pathological DMD/BMD transcripts in both hemizygous patients and heterozygous carriers.[15] Transcripts of abnormal sizes resulting from a genomic deletion or splice site mutation can be easily detected through RT-PCR analysis.

Direct sequencing of the aberrant products will also enable detection of deletion and duplication mutations by characterization of splice junctions in transcripts. Analysis of mRNA by this method will also be useful for determining the effects of genomic mutations on transcript structures and for the detection of alternative transcripts.

2 MATERIALS AND METHODS

2.1 RNA Extraction

Total RNA was extracted from EBV-transformed lymphocytes from five DMD patients using Trizol Reagent (GIBCO BRL). The lymphocytes were pelleted and 1ml of Trizol was added per culture of between 5-10 x 10^6 cells. RNA was extracted according to the manufacturer's instructions. The RNA pellet was stored as an ethanol precipitate at -70°C until use.

2.2 RT-PCR

RNA samples were reverse transcribed to cDNA using Superscript™ II Reverse Transcriptase (GIBCO BRL). 5μg total RNA was incubated in a volume of 8μl with 50ng random hexamers at 70°C for 10min. The mixture was snap-chilled on ice and made up to 20μl with a premix that contained 4μl 5X First Strand Buffer (GIBCO BRL), 1μl 1.0mM dNTPs, 2μl 0.1M Dithiotrietol and 200 units Superscript™ II Reverse Transcriptase (GIBCO BRL). The mixture was then incubated at 42°C for 1 hour. The cDNA synthesised was stored at -20°C prior to use.

Each cDNA sample was subjected to separate nested PCRs in ten overlapping reactions spanning the entire dystrophin gene using ten sets of primers (Table 1). The conditions of amplification and primer sequences have been previously reported[17]. First round of PCR amplification was done using GmdXa and GmdXb primers (x=1-10 for each of the set of 10 reactions). A second round PCR was performed using an inner set of primers GmdXc and GmdXd. For first round PCR, a PCR mix containing 1X PCR Buffer, 1.5mM $MgCl_2$, 0.2μM dNTPs, 0.2μM of each primer and 2.5U of Taq Polymerase (Pharmacia) was added to 2μl of cDNA. Thirty-five cycles of PCR (at 95°C for 1min, 50°C for 2min and 72°C for 3min) were performed. 2μl of the first round amplification product was then added to a 50μl mixture containing 1X PCR Buffer, 0.2mM dNTPs, 0.2μM of primers GmdXc and GmdXd and 2.5U of Taq Polymerase (Pharmacia). A further 35 cycles of PCR was performed. Negative and positive control reactions were performed by using water and normal cDNA sample

Table 1. Primers used for RT-PCR reactions.

PCR Reaction No.	Primers	PCR product size (bp)	Exons covered
1	Gmd1a + Gmd1b	1207	1-10
	Gmd1c + Gmd1d	1159	
2	Gmd2a + Gmd2b	1329	9-18
	Gmd2c + Gmd2d	1263	
3	Gmd3a + Gmd3b	1320	17-25
	Gmd3c + Gmd3d	1251	
4	Gmd4a + Gmd4b	1344	25-33
	Gmd4c + Gmd4d	1281	
5	Gmd5a + Gmd5b	1198	33-40
	Gmd5c + Gmd5d	1131	
6	Gmd6a + Gmd6b	801	40-44
	Gmd6c + Gmd6d	739	
7	Gmd7a + Gmd7b	1311	43-51
	Gmd7c + Gmd7d	1227	
8	Gmd8a + Gmd8b	1307	51-58
	Gmd8c + Gmd8d	1252	
9	Gmd9a + Gmd9b	1316	57-67
	Gmd9c + Gmd9d	1268	
10	Gmd10a + Gmd10b	1298	67-end
	Gmd10c + Gmd10d	1245	

respectively as the template for PCR. 8μl of the final PCR product was electrophoresced on a 1% agarose gel containing 1ug/ml ethidium bromide.

2.3 Sequencing Analysis

Sequencing of aberrant RT-PCR products was subsequently carried out to determine the deletion breakpoints or point mutations. RT-PCR products were purified using either Qiaquick Gel extraction kit (Qiagen) or Qiaquick PCR purification kit (Qiagen). The purified DNA was then sequenced either using the ABI Prism Bigdye Ready Reaction Kit on the Applied Biosystems 377 Sequencer (Applied Biosystems Inc.) or using the Cy™5 Thermo Sequenase Dye Terminator Kit on the ALFexpress™ DNA Sequencer (Amersham Pharmacia Biotech) according to the respective manufacturers' instructions.

3 RESULTS

It was found that amplification products of reverse-transcribed RNA from patients and normal controls could be clearly resolved on agarose gels. The analysis of the RT-PCR products showed aberrant-sized products in five cases (patients CL 334, CL402, CL636, CL681 and CL723). The PCR products were smaller than expected, suggesting the presence of either a deletion or a splice-site mutation (Fig. 1). Direct sequencing of these products was performed to determine the precise mutations.

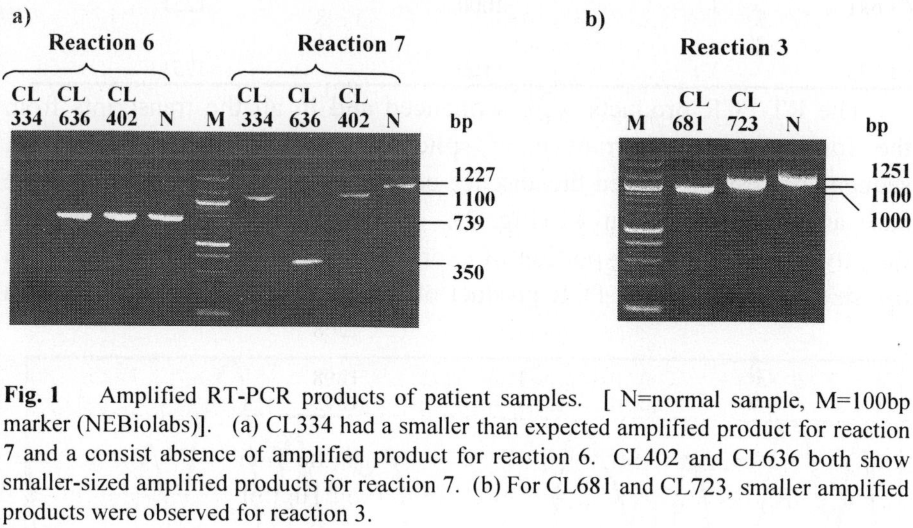

Fig. 1 Amplified RT-PCR products of patient samples. [N=normal sample, M=100bp marker (NEBiolabs)]. (a) CL334 had a smaller than expected amplified product for reaction 7 and a consist absence of amplified product for reaction 6. CL402 and CL636 both show smaller-sized amplified products for reaction 7. (b) For CL681 and CL723, smaller amplified products were observed for reaction 3.

3.1 Deletion Mutations

Patients CL334, CL402 and CL636 had smaller than expected amplified products for reaction 7 while patient CL681 had a smaller amplified product for reaction 3 (Table 2). In CL334, repeated PCR analysis of reaction 6 did not show any detectable amplified product.

Table 2 Summary of RT-PCR results from patients' samples which had RT-PCR products of abnormal size

Sample	PCR Reaction Number	Size of aberrant PCR products (bp)	Size of normal PCR product (bp)
CL334	6	absent	739
	7	1100	1227
CL402	7	1100	1227
CL636	7	350	1227
CL681	3	1000	1251
CL723	3	1100	1251

The RT-PCR products were sequenced and in all the transcripts from the four patients, aberrant exon splicing was observed. In CL334, sequencing results showed the absence of exon 44 and a deletion of a single base at the start of exon 45 (Fig. 2). The 3' end of exon 43 is spliced directly to the remaining portion of exon 45. This result is consistent with the size of the truncated PCR product of reaction 7 in CL334 which has a

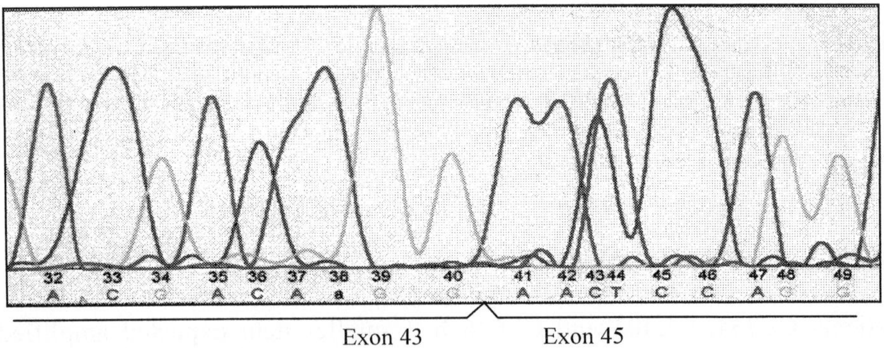

Fig. 2 Partial nucleotide sequence analysis of CL334. The sequences at the exon border of anomalous splicing events show that exon 44 was spliced out of the transcript together with the first base of exon 45. The 3' end of exon 43 is spliced to the remaining portion of exon 45. This led to a translational frameshift and the creation of a stop codon 46 nucleotides downstream, resulting in a truncated dystrophin protein.

loss of approximately 148bp corresponding to the size of exon 44. The deletion of exon 44 and the single base of exon 45 in the transcript caused a translational frameshift about 47 bases downstream from the splice junction.

Likewise, in CL402, CL636 and CL681, sequence analysis showed that exons were spliced out of the transcripts. Splicing out of exons and subsequent juxtaposition of the flanking exons resulted in translational frameshifts that were out-of-frame and which resulted in premature termination of translation (Fig. 3). These results were consistent with the intermediate to severe DMD phenotypes observed in the above patients.

To determine the exact nature of the mutation, genomic DNA from the patients were amplified using intronic primers described by Chamberlain *et al.*[18] which spans the region of the breakpoints. In all the four samples, the results showed the absence of the exons corresponding to those absent in their respective mRNA transcripts thus confirming that the mutations were due to intragenic deletions (Table 3).

Table 3 Summary of RT-PCR and Genomic DNA analysis results.

Patient	RT-PCR analysis Exons deleted	Genomic DNA analysis Exons deleted
CL334	44-45*	44
CL402	49-50	49-50
CL636	45-50	45-50
CL681	20	20
CL723	19	Nil

* Only the first base of exon 45 was deleted.

a. Patient CL334 (Ex 44 deletion)

```
          Exon 43              Exon 44
Normal:   ...VNKMYKDRQG|RFDRSVEKWRRFHYDIKIFNQW...
Patient:  ...VNKMYKDRQG|TPGWHA.........VRQKKE|ARRTKEYLVRISKRFK*
          Exon 43              Exon 45         Exon 46
```

b. Patient CL 402 (Ex 49-50 deletion)

```
          Exon 48              Exon 49
Normal:   ...YNQEGPFDVQ|ETEIAVQAKQPDVEEILSKGQHLY...
Patient:  ...YNQEGPFDVQ|LLLRLLLW*
          Exon 48              Exon 51
```

c. Patient CL636 (Ex 45-50 deletion)

```
          Exon 44              Exon 45
Normal:   ...EHAKYKWYLK|ELQDGIGQRQTVVRTLNATGEEIIQ...
Patient:  ...EHAKYKWYLK|LLLRLLLW*
          Exon 44              Exon 51
```

d. Patient CL681 (Ex 20 deletion)

```
          Exon 19              Exon 20
Normal:   ...RSAQALVEQMVN|EGVNADSIKQASEQLNS...
Patient:  ...RSAQALVEQMVN|G*
          Exon 19              Exon 21
```

e. Patient CL723 (Ex 19 deletion)

```
          Exon 18              Exon 19
Normal:   ...NFSDLKEKVN|AIEREKAEKFRKLQDASRSA...
Patient:  ...NFSDLKEKVN|RVLMQIASNKPQNN*
          Exon 18              Exon 20
```

Fig. 3 Predicted amino acid sequence analysis of patient samples at the splice junctions of anomalous splicing. Premature termination of translation was found in all the five patients.

3.2 Splicing Mutations

Patient CL723 had a smaller than expected PCR product of 1100bp for reaction 3 PCR. Sequencing results showed the absence of exon 19 from the transcript. This caused exon 20 to be spliced directly to exon 18 and created a stop codon 43 nucleotides downstream in exon 20. Amplification of exon 19 region from the genomic DNA of the patient using the exon 19 primer described by Chamberlain *et al.*[18] showed the presence of a normal PCR product of 459bp (Fig. 4). This indicated that the loss of exon 19 from the mRNA transcript was not due to a deletion but likely to be due to other causes, such as splice site errors, alternative transcripts, etc. Further characterisation of the exact nature of this splice site mutation is in progress.

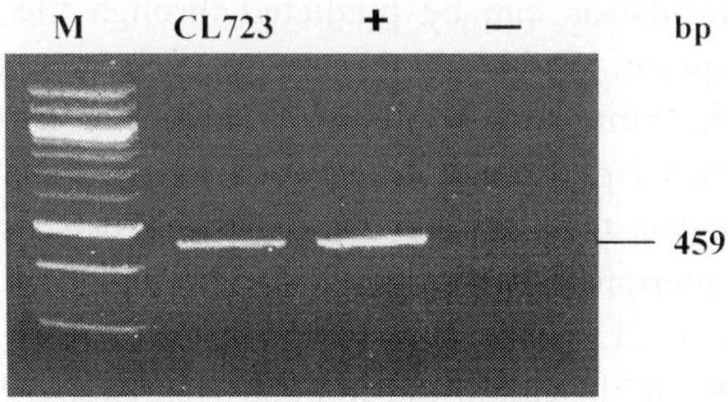

Fig. 4 Amplification of genomic DNA from patient CL723 in the exon 19 region. No deletion was observed. (M=100bp marker, + = normal, − = negative control).

4 DISCUSSION

Although reports showed that 65% of DMD/BMD patients have deletions in the dystrophin gene,[1, 3] deletions have previously been observed in only 56.5% of our DMD/BMD patients using multiplex PCR screening of the hotspot regions.[19] This lower frequency is likely to be due to the occurrence of single point mutations and splice-site mutations or deletions that are undetectable by the multiplex PCR amplification technique which screens for mutation in the deletion hotspot regions.

The approach of ectopic dystrophin mRNA analysis appears to be an efficient and effective method of mutation detection. In the five patients studied, analysis of the dystrophin mRNA not only identifies the type of mutations involved in aberrant and alternative transcripts but will also provide an understanding of the process of pre-mRNA processing, exon skipping etc. and allow further studies for therapy through targeted intervention to produce functional transcripts in patients.

Analysis of the dystrophin gene at the mRNA level allows the entire coding region of the gene to be screened. Further, as mRNA represents the actual template for translation, it is a better candidate for mutation screening than genomic DNA and for correlation of disease severity and type of mutation. It provides a direct description of the nature of the mutation and its likely effect on the protein, for example, a frameshift effect leading to translation termination can be predicted through the identification of the deletion breakpoint. This is in contrast to deletion studies at the genomic level where the frame-shift hypothesis is based on extrapolation from gene structure to transcript structure. In addition, splice-site mutations, which are not apparent at the DNA level of analysis, can also be detected by the presence of aberrant-sized transcripts resulting from abnormal splicing events. Patient CL723 is an example. Analysis of genomic DNA by multiplex PCR amplification did not show an exon 19 deletion. However, exon 19 was found to be deleted at the transcript level presumably due to aberrant splicing of the mRNA. This reflects some of the limitations of genomic DNA analysis for molecular diagnosis of DMD/BMD.

The present challenge is to efficiently screen for point mutations or micro-deletions in the remaining cases of DMD/BMD in which no gross mutations were detected by the multiplex PCR technique. These mutations can be detected by the screening of individual exons at the genomic level using methods such as SSCP,[20] heteroduplex,[21] DGGE[22] etc. The protein truncation test (PTT) in combination with RT-PCR has also been used for the detection of nonsense mutations.[23, 24] This method would in theory have detected all 5 mutants studied here. However, this approach does not detect pathogenic nucleotide substitutions or minor deletions, which might give rise to a normal sized but dysfunctional protein. In addition, in some samples alternative transcripts or non-specific amplification and PCR artefacts may be seen in the RT-PCR products. Usually, patient reactions are run alongside normal controls and only major transcript species are analysed as compared to controls to eliminate such problems. However, there is still a need to find a way to adapt the RT-PCR method and to develop more effective techniques to screen for such mutations more efficiently.

References

1 Koenig M, Hoffman EP, Bertelson CJ, MonacoAP, Feener C, Kunkel LM, Complete cloning of the Duchenne muscular dystrophy (DMD) cDNA and preliminary genomic organization of the DMD gene in normal and affected individuals. *Cell* 1987; **50**:509–517.

2 Den dunnen, JT, Grootscholten, PM, Bakker, E, Blonden, LAJ, Ginjaar, HB, Wapenaar, MC, van Passen HMB, van Broeckhoven C, Pearson P and van Ommen GJB, Topography of the Duchenne Muscular Dystrophy (DMD) gene: FIGE and cDNA analysis of 194 cases reveals 115 deletions and 13 duplications. *Am J Hum. Genet*. 1989; **45**:835–847.

3 Roberts RG, Coffey AJ, Bobrow M, Bentley DR, Exon structure of the human dystrophin gene. *Genomics* 1993; **16**:536–538.

4 Hu X,Ray PN, Murphy EG, Thompson MW, Worton R, Duplicational mutation at the Duchenne muscular dystrophy locus: its frequency, distribution, origin and phenotype/genotype correlation. *Am J Hum Genet* 1990; **46**: 682–695

5 Reninsland F, Reiss J, Microlesions and polymorphisms in the Duchenne/Becker muscular dystrophy gene. *Hum Genet* 1994; **94**:111–116.

6 Darras BT, Blattner P, Harper JF, Spiro AJ, Alters S, Francke U, Intragenic deletions in 21 Duchenne muscular dystrophy (DMD) families studied with Dystrophin cDNA: Location of breakpoints on Hind III and Bgl II exon-containing fragment maps, meiotic and mitotic origin of the mutation. *Am J Hum Genet* 1988; **43**:629

7 Baumbach, L.A., Chamberlain, J.S., Ward, P.A., Farwell, N.J., and Caskey, CT, Molecular and clinical correlations of deletions leading to Duchenne and Becker muscular dystrophies. *Neurol*. 1989; **39**:465–474.

8 Gillard EF, Chamberlain JS, Murphy EG, Duff CL, Smith B, Burghes AHM, Thompson MW, Sutherland J, Oss I, Budrug SE, Klamut HJ, Ray PN, Worton R, Molecular and phenotypic analysis of patients with deletions within the deletion-rich region of Duchenne muscular dystrophy (DMD) gene. *Am J Hum Genet* 1989; **45**:507–520.

9 Lindlof M, Kiuru A, Kaariainem H, Kalimo H, Lang H, Piltko H, Rapola J, Somer H, Somer M, Savontaus ML, de la Chapelle A, Gene deletions in X-linked muscular dystrophy. *Am J Hum Genet* 1989; **44**:682–695

10 Monaco, AP, Bertelson, CJ, Liechti-Gallati, S, Moser, H, and Kunkel, LM, An explanation for the phenotypic differences between patients bearing partial deletions of the DMD locus. *Genomics* 1988; **2**:90–95.

11 Chelly J, Concorder JP, Kaplan JC, Kahn A, Illegitimate transcription: transcription of any gene in any cell type. *Proc Natl Acad Sci USA* 1989; **6**:2617–2621

12 Sarker G, Sommer SS, Access to a messenger RNA sequence or its protein product is not limited by tissue or species specificity. *Science* 1989; **244**:331–334.

13 Chelly J, Gilgenkrantz H, Hugnot JP *et al.*, Illegitimate transcription. Application to the analysis of truncated transcripts of the Dystrophin gene in non-muscle cultured cells from Duchenne and Becker patients. *J Clin Inves* 1991; **88**:1161–1166.

14 Reninsland F, Hahn A, Niemann-Seyde S, Slomski R, Hanefeld F and Reiss J, Identification of a new DMD gene deletion by ectopic transcript analysis. *J Med Genet* 1992; **29**:647–651.

15 Schloesser M, Slomski R, Wagner M, *et al.*, Characterisation of pathological Dystrophin transcripts from the lymphocytes of a muscular dystrophy carrier. *Mol Biol Med* 1990; **7**:519–523.

16 Roberts RG, Bentley DR, Barby TFM, Manners E, Bobrow M, Direct diagnosis of carriers of Duchenne and Becker muscular dystrophy by amplification of lymphocyte RNA. *Lancet* 1990; **336**:1523–1526.

17 Roberts RG, Barby TFM, Manners E, Bobrow M, Bentley DR, Direct detection of dystrophin gene rearrangements by analysis of dystrophin mRNA in peripheral blood lymphocytes. *Am J Hum Genet* 1991; **49**:298–310.

18 Chamberlain JS, Gibbs RA, Rainier JE, Caskey CT, Deletion screening of the Duchenne muscular dystrophy locus via multiplex DNA amplification. *Nucleic Acids Res* 1988; **16**: 11141–11156.

19 Lai PS, Khng HH, Ho KL, Lee WL and Low PS, Spectrum of deletions in the dystrophin gene. *Proceedings of the Inauguration and International Conference of the Federation of Asia-Pacific Microbiology Societies* 1998, p.131.

20 Nigro V, Nigro G, Esposito MG *et al.*, Novel small mutations along the DMD/BMD gene associated with different phenotypes. *Hum Mol Genet* 1994; **3**: 1907–1908.

21 Prior TW, Papp AC, Snyder PJ, Sedra MS, Western LM, Bartolo C, Moxley RT, Mendell JR, Heteroduplex analysis of the dystrophin gene: application to point mutation and carrier detection. *Am J Med Genet* 1994; **50**:68–73.

22 Tubiello G, Carrera P, Soriani N, Morandi L, and Ferrari M, Mutational analysis of muscle and brain specific promoter regions of dystrophin gene in DMD/BMD Italian patients by denaturing gradient gel electrophoresis (DGGE). *Mol Cell Probes* 1995; **9**:441–446.

23 Roest PAM, Roberts RG, Sugino S, van Ommen GJB, den Dunnen JT. Protein truncation test (PTT) for rapid detection of translation-terminating mutations. *Hum Mol Genet* 1993; **2**:1719–1721.

24 Gardner RJ, Bobrow M, Roberts RG. The identification of point mutations in Duchenne Muscular Dystrophy patients by using reverse-transcription PCR and the protein truncation test. *Am J Hum Genet* 1995; **57**:311–320.

FRONTIERS IN HUMAN GENETICS
Diseases and Technologies

© 2001 by World Scientific Publishing Co. Pte. Ltd.

CONFIRMATION OF PREDICTED MUTATIONAL EFFECT BY *IN VITRO* PROTEIN TRANSLATION IN A PATIENT WITH DUCHENNE MUSCULAR DYSTROPHY

SKH TAY*,†, HH KHNG*, WL LEE*,‡, PS LOW* and PS LAI*
**Department of Paediatrics, National University of Singapore*
‡*Department of Neurology, National Neuroscience Institute, Singapore*

Duchenne muscular dystrophy (DMD) is an X-linked recessive lethal disorder affecting 3500 live born males. This disorder is caused by mutations in the gene that encodes dystrophin, a high molecular weight cytoskeletal protein. The large size and complexity of the gene pose limitations for detailed mutational analysis in patients, especially for non-deletion cases. In this report, we describe the characterization of a single nucleotide alteration in exon 37 from ectopic transcripts of immortalized lymphocytes from a DMD patient. This mutation is predicted to result in termination of translation of the dystrophin protein. *In vitro* translation of polypeptide using an amplified fragment of 2105 bases covering the nucleotide change was carried out from the extracted mRNA transcripts using the coupled reticulocyte lysate transcription/translation system. A truncated polypeptide of approximately 46 kDa was obtained, confirming premature chain termination had occurred. Demonstration of this mutational effect proves that the identified mutation is responsible for the pathogenic phenotype in the patient. This approach may be applied in future for direct identification of causative mutations resulting in truncated dystrophin in other DMD patients.
Keywords: Duchenne muscular dystrophy (DMD), reverse-transcription PCR (RT-PCR), protein truncation test (PTT), nonsense mutation, ectopic transcripts, translation

1 INTRODUCTION

Duchenne muscular dystrophy is an X-linked recessive disorder with an

---

† Corresponding author

estimated incidence of 1 in 3500 males.[1] It is a progressive neuromuscular disease characterized by proximal muscle weakness presenting at three to five years of age. The affected patients usually become wheel chair bound by twelve years of age, and die from respiratory or cardiac failure by the second decade.

This disorder is caused by mutations in a 2.4 Mb gene on Xp21 called dystrophin.[2,3] It is the largest known gene and encompasses seventy-nine coding exons. The mature mRNA is some 14 kb long and translates into a 427 kDa protein.[4] Dystrophin makes up part of a dystrophin-glycoprotein complex found in muscle that spans the sarcolemma, linking the subsarcolemmal cytoskeleton with the extracellular matrix.[5,6] It is postulated that the absence or deficiency of dystrophin may disrupt the stability of the sarcolemma and therefore affect calcium regulation of the cell. The increased calcium in the cell activates the intracellular calcium-dependent protease enzymes leading eventually to muscle cell necrosis and cell death.[7]

DNA analysis by multiplex polymerase chain reaction or Southern blot analysis may detect partial deletions in 65% of patients,[2,8] and partial duplications in another 5% of patients.[9] Local studies in Singapore have shown that a slightly lower deletion rate of 58.3–61% occurs in the population.[10,11,12] Hence, the molecular basis of the causative mutations in about 40% of patients in Singapore remains unknown. It is believed that for these patients, the mutations are likely to arise from very small deletions, insertions or substitutions involving single or a few nucleotides. For such cases, different approaches such as SSCP, heteroduplex analysis, and DGGE followed by DNA sequencing have been attempted for mutation detection. However, the number of point mutations identified by DNA-based studies is less than expected, with screening of 20% to 80% of coding region showing only 2-18% mutations.[13,14]

The limited detection rate of DMD gene mutations complicates genetic counseling in family studies and also diminishes the value of negative results in differential diagnosis. Indeed, the large size and complexity of the dystrophin gene poses a challenge for detailed mutational analysis of non-deletion cases. Recent work has focused screening the DMD lymphocyte mRNA for translation-terminating mutations by reverse transcription polymerase chain reaction (RT-PCR) and subsequent *in vitro* translation and transcription, as it is laborious and time-consuming using DNA-based analysis to sequence the entire dystrophin gene. In one such approach,[15-17] ectopic lymphocyte RNA is isolated, reverse transcribed and amplified by RT-PCR. Subsequently, nested PCR is performed for five to ten overlapping regions with modified primers containing a T7-promoter and a eukaryotic translation initiation sequence. *In vitro* transcription/translation of the PCR products is then performed. Abnormal fragments are readily detected and subjected to sequence analysis to characterize the mutation. However, mutations that do not result in aberrant fragments, larger or smaller than the expected size from the RT-PCR analysis, are not easily identified and sequencing of each nucleotide in the entire coding region has still to be carried out.

In this study, we describe the characterization of a single nucleotide substitution in exon 37 of the dystrophin gene in a seventeen-year-old DMD patient ELPL. Ectopic transcripts from immortalized lymphocytes from the patient were analyzed by reverse transcription of the mRNA. Subsequently, nested PCR was carried out in ten overlapping reactions.[18] The amplified products were then sequenced to detect any nucleotide alterations. A point mutation involving a C to T substitution in nucleotide number 5474 was detected from the sequencing of the nested PCR products covering exons 33 to 40. This is predicted to change a glutamine residue to a stop codon (Q95X), resulting in termination of translation of the dystrophin protein. To demonstrate that this sequence alteration is indeed responsible for the pathogenicity in this patient, protein analysis was carried out by *in vitro* translation studies using a separate set of RT-PCR primers.

2 MATERIALS AND METHODS

2.1 Patient Profile

The patient ELPL was a Chinese boy with a significant family history of two maternal uncles who had progressive muscle weakness from childhood and died in their twenties (Fig. 1). Patient ELPL presented with lower limb weakness at five years of age, and had difficulty in climbing stairs and running. Clinical examination at presentation showed pseudohypertrophy of

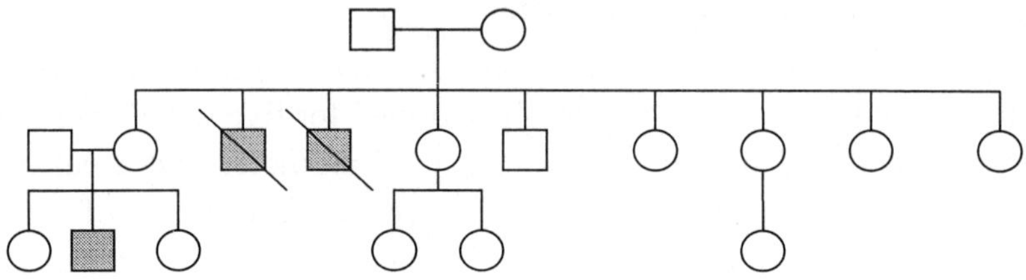

Fig. 1 Pedigree for patient ELPL

his calves, normal muscle tone and deep tendon reflexes, and proximal muscle weakness of the lower limbs. He had a waddling gait and positive Gower's sign. Subsequent follow-up revealed that he became chair bound at twelve years of age. Creatine kinase level was 47 793 U/l (Normal 35–290 U/l) at the age of seven years old. He was found not to have a gross deletion by multiplex PCR analysis and Southern blotting.

2.2 RNA Preparation

Total RNA was extracted from Epstein-Barr virus (EBV) transformed lymphocyte cell line from the patient using Trizol reagent (GIBCO BRL). The lymphocytes were pelleted and 1 ml of Trizol was added per culture of approximately 5–10 x 10^6 cells. RNA was extracted according to manufacturer's instruction and precipitated in 70% ethanol, and kept at –70°C until use. Pre-treatment with diethylpyrocarbonate was carried out for all reagents not provided by the manufacturer.

2.3 Reverse Transcription

Reverse transcription was performed on samples of total RNA in TE with the Superscript™ II Reverse Transcriptase (GIBCO BRL). An amount of 2–5 µg of total RNA was incubated in a volume of 8 µl with 50 ng random hexamers at 70°C for 10 minutes in an overlay of 50 µl of light paraffin. The reaction mixture was then snap-chilled on ice, and a pre-mix of 4 µl of 5X First Strand Buffer (GIBCO BRL), 1 µl of 1.0 mM dNTPs, 2 µl 0.1M DTT and 200 units of Superscript™ II Reverse Transcriptase (GIBCO BRL) was added. The mixture was then incubated at 42°C for one hour.

2.4 Nucleotide Sequence Analysis

Nested PCR of the ten overlapping reactions was performed as previously described.[18] PCR products from the RT-PCR reactions were directly sequenced on an ABI 377 automated fluorescent sequencer using the dye terminator method according to the manufacturer's instructions as previously described.[18]

2.5 Nested PCR for PTT

RT-PCR products for protein translation were obtained as follows: a PCR mixture containing 1X PCR buffer, 1.5 mM $MgCl_2$, 0.2 μM dNTPs, 0.2 μM of the NW3a and NW3b primer (see Table 1) and 2.5 U of Elongase enzyme (GIBCO BRL) was added to 5 μl of cDNA in a volume of 50 μl.

Table 1 Primers used in the exon 33 – 45 nested reaction

Primer	Sequence (5'→3')	Nucleotide position
NW3a	ATAAAAGTCTGAGTGAAGTGAAGTCTG	4734 – 4760
NW3b	TAGCAATGTTATCTGCTTCCTCCAACC	6885 – 6911
NW3c[1]	GTGGAAATGGTGATAAAGACTGGACGT	4763 – 4789
NW3d	CTGTCTGACAGCTGTTTGCAGACCTCC	6787 – 6813

[1]Prefaced by T7 bacteriophage promoter sequence and consensus eukaryotic translation initiation signal: GGATCCTAATACGACTCACTATAGGAACAGACCACCATG

30 cycles of PCR (at 94°C for 30 seconds, 55°C for 48 seconds and 68°C for 4 minutes) were performed on a HYBAID OMNIGENE machine. For the second round of PCR amplification, 2 μl of the first PCR product was added to a final volume of 50 μl containing 1X PCR buffer, 0.2 mM dNTPs, 0.2 μM nested NW3c[1] and NW3d primers and 2.5 U of Elongase enzyme (GIBCO BRL). Amplification was carried out under similar conditions as first round PCR. Negative and positive control reactions were included in the amplification reaction using water and normal cDNA sample respectively as templates for the nested PCR. A volume of 10 μl of the final PCR product was electrophoresed on an agarose gel containing 1 μg/ml of ethidium bromide.

2.6 *In Vitro* Transcription/Translation and Protein Analysis

The protein truncation test was performed with the TNT® Quick for PCR DNA (Promega) according to the manufacturer's instructions. 2 μl of nested PCR products was added to 20 μl of TNT® Quick master mix and 2 μl of ^{35}S-methionine to a volume of 25 μl. The reaction was performed at 30°C for 90 minutes. The translation products were separated by discontinuous SDS-PAGE through a 5% stacking gel and a 12% separating gel with a Tris-glycine buffer. A 10 cm x 10 cm x 0.75 mm vertical gel system (Mighty Small II, Hoeffer Scientific Instruments) was used and electrophoresis was performed at 15 mA constant current till the dye front moved to the resolving gel and then increased to 30 mA current for about five hours. High range Rainbow™ [^{14}C]methylated protein molecular weight marker was used. The gel was fixed in 10% glacial acetic acid for 30 minutes at room temperature, soaked overnight in 20% methanol and 3% glycerol, followed by soaking in Amplify™ (Amersham) for 30 minutes. The gel was dried under vacuum at 80°C for 90 minutes, and the signal detected by autoradiography on standard X-ray film.

3 RESULTS

We screened exons 1–79 by RT-PCR, using ten overlapping primer pairs on the immortalized cell line of the patient ELPL. No other family members were available for screening. Each fragment was between 1.0–1.4 kb. In this patient, analysis of the RT-PCR products on an agarose minigel did not show any aberrant products that differed from the expected sized-product from the normal control (Fig. 2).

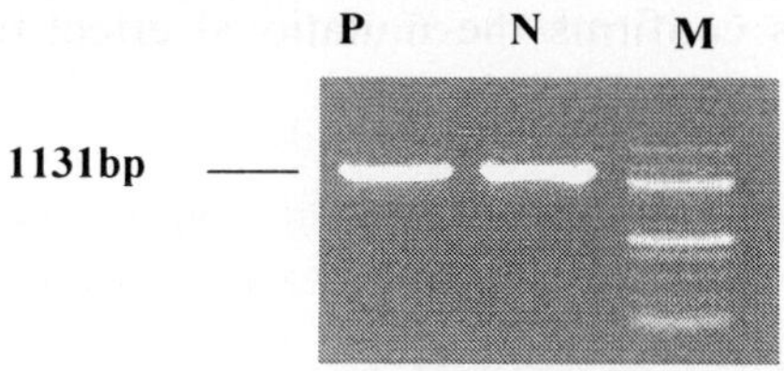

Fig. 2 Amplified RT-PCR products from reaction covering exons 33 to 40.[P= Patient ELPL, N= normal sample, M= 100 bp ladder (NEBiolabs)] Similar sized products were obtained from both normal control and patient ELPL sample.

All the products were subjected to nucleotide sequencing. Analysis of the nested reaction covering exons 33 to 40 showed a point mutation in nucleotide number 5474 involving a C to T substitution (Fig. 3). This is a nonsense mutation that is predicted to change a glutamine residue to a stop codon, resulting in termination of translation of the dystrophin protein.

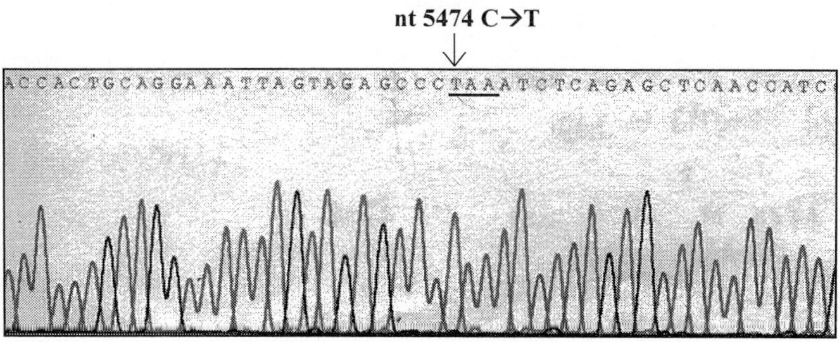

Fig. 3 Partial nucleotide sequence of exon 37 of the patient. The base change in nucleotide 5474 resulted in a nonsense mutation (CAA → TAA) which altered the glutamine residue to a stop signal. (Sequence numbering according to GenBank M18533)

In vitro protein synthesis using the reaction covering exons 33–45 showed a truncated band of about 46 kDa and not the normal full-length band (Fig. 4) of the expected size of 83 kDa. The lack of a normal full-length protein band and the presence of a shorter truncated band clearly indicate the presence of a chain-terminating mutation in this region. This protein truncation test (PTT) thus confirms the mutational effect of the abnormal sequence in exon 37.

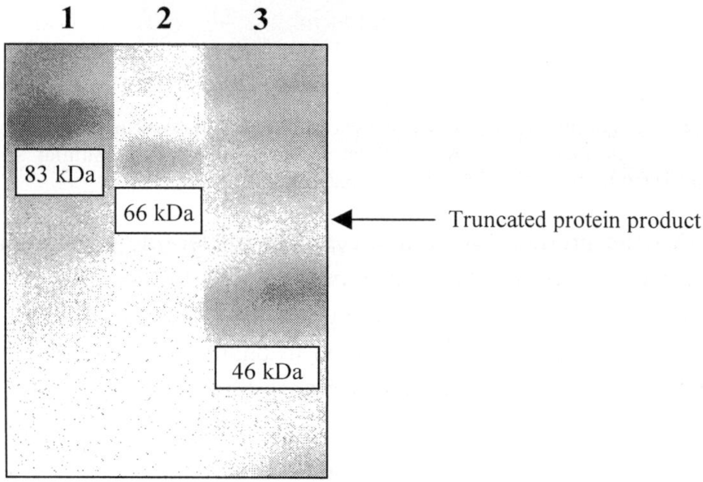

Fig. 4 Autoradiogram showing the protein products from the *in vitro* transcription/ translation analysis. Lane 1, normal control individual showing the production of normal sized protein (83kDa); Lane 2, Rainbow™ [^{14}C]-methylated Protein Molecular Weight Marker (Amersham Lifescience) [220kDa,97.4kDa, 66kDa, 46kDa, 30kDa, 21.5kDa, 14.3kDa] ; Lane 3, truncated protein product of 46kDa from patient.

4 DISCUSSION

The dystrophin gene is the largest gene known in man, and its size and complexity has complicated the molecular diagnosis of Duchenne muscular dystrophy. Various methods have been developed to identify the causative mutations. DNA-based methods such as multiplex PCR are able to identify the majority of deletions, but can only identify the mutations in 4% of non-deletion patients. RNA-based methods, however, by screening the total coding region, can identify the mutations in 82% of patients.[19] In Singapore, more than 50% of DMD patients do not have a deletion, and the use of RNA-based methods has opened a vista of possibilities in the accurate molecular diagnosis of this disease.

In this case, we were able to utilize RNA-based methods to characterize the mutation in a non-deletion patient, ELPL. RT-PCR methods and sequencing of the amplified product revealed a base substitution at nucleotide number 5474 in exon 37. This mutation is likely to be chain terminating as it changed a glutamine residue to a stop codon.

The results of the sequencing were further validated by the protein truncation test. In the segment containing exon 37, we were able to demonstrate a smaller truncated protein product. It has been shown previously that almost all the known DMD small mutations, such as gene rearrangements, result in premature termination of translation.[20,21] This case again serves to highlight the importance of this method in the molecular diagnosis of DMD. RT-PCR/PTT appears to be an efficient method and has certain advantages over techniques such as heteroduplex analysis or chemical cleavage mismatch techniques as products larger than four kilobases can be screened by RT-PCR/PTT in a single assay. In this case, we were also able to demonstrate that the nucleotide substitution in exon 37 was a pathogenic mutation resulting in abnormally truncated protein in the patient. This protein is truncated in the spectrin-like rod domain and would thus be non-functional in the patient.

Although the incidence of truncating mutations in the DMD gene is high, *in vitro* protein translation analysis such as the PTT may not elucidate some mutations that are either outside the coding region, or if the patient has a large duplication that lies outside of the amplification section, or if the patient is a mosaic such that the mutation not being represented in the lymphocyte transcripts.

It has been previously recommended by Whittock *et al*[17] that RNA from muscle biopsy samples be obtained to confirm the diagnosis of DMD by PTT because the number of dystrophin transcripts in muscle is about five hundred times higher than that of lymphocytes. He reported RT-PCR failure of some fragments when working with lymphocyte total RNA and this also appeared to be dependent on the age of the sample. In this patient, however, we were able to demonstrate the clinical utility of this method using immortalized lymphocytes of the patient. As far as we know there have been no other reports of the use of this method on immortalized lymphocytes. The use of fresh immortalized lymphocytes may thus overcome the problem of degradation of the transcripts that are produced in stored samples. The ability to perform the test on immortalized cell lines is a great advantage as repeated blood sampling and muscle biopsies may not be feasible. In addition, the use of immortalized cell lines is also of value if a molecular diagnosis is needed for the offspring of maternal relatives later. It should be noted, however, that careful culture of the lymphocytes prior to harvesting to give at least 3–5 μg of RNA per ml is necessary before performing reverse transcription.

5 CONCLUSION

A nonsense mutation (Q95X) was identified in RNA from immortalized lymphocytes of a DMD patient on sequencing. *In vitro* protein translation from the transcripts of immortalized lymphocytes of the patient showed a truncated product of 46kDa. This was smaller than the expected size of 83 kDa from normal controls. The results confirmed that this mutation resulted in premature polypeptide chain termination during dystrophin protein synthesis.

References

1. Emery AE, Population frequencies of inherited neuromuscular diseases – a world survey. *Neuromusc Disord* 1991; **1**:19-29.

2. den Dunnen JT, Grootscholten PM, Bakker E *et al*., Topography of the Duchenne muscular dystrophy (DMD) gene: FIGE and cDNA analysis of 194 cases reveals 115 deletions and 13 duplications. *Am J Hum Genet* 1989; **45**:835-847.

3. den Dunnen JT, Grootscholten PM, Dauwerse JG, *et al*., Reconstruction of the 2.4 Mb human DMD-gene by homologous YAC recombination. *Hum Molec Genet* 1992; **1**;19-28.

4. Hoffman EP, Brown RH Jr, Kunkel LM, Dystrophin: the protein product of the Duchenne muscular dystrophy locus. *Cell* 1987; **51**:919-928.

5. Ervasti JM, Campbell KP, Membrane organization of the dystrophin-glycoprotein complex. *Cell* 1991; **66**:1121-1131.

6. Ibraghimov-Beskrovnaya O, Ervasti JM, Leveille CJ, *et al.,* Primary structure of dystrophin-associated glycoproteins linking dystrophin to the extracellular matrix. *Nature* 1992; **355**:696-702.

7. Matsumara K, Ohlendieck K, Ionasescu VV, *et al,*. The role of the dystrophin-glycoprotein complex in the molecular pathogenesis of muscular dystrophies. *Neuromusc Disord* 1993; **3**;533-535.

8. Koenig M, Hoffman EP, Bertelson CJ, *et al*., Complete cloning of the Duchenne muscular dystrophy (DMD) cDNA and preliminary genomic organization of the DMD gene in normal and affected individuals. *Cell* 1987; **50**:509-517.

9. Hu XY, Ray PN, Murphy EG, Thompson MW, Worton RG, Duplication mutation at the Duchenne muscular dystrophy locus: its frequency, distribution, origin and phenotype-genotype correlation. *Am J Hum Genet* 1990; **40**:682-695.

10. Lai PS, Tay JSH, Low PS, *et al*., Deletion analysis of DMD/BMD children in Singapore using multiplex polymerase chain reaction (PCR) technique. *J Trop Pediatr* 1992; **38**(5):224-227.

11. Low PS, Lai PS, Lee WL, *et al.,* Molecular diagnosis of Duchenne muscular dystrophy in Singapore. *Ann Acad Med Singapore* 1996 ; **25**(1):84-89.

12. Lai PS, Khng HH, Ho KL, Lee WL and Low PS, Spectrum of deletions in the dystrophin gene, *Proceedings of the Inauguration and International conference of the Federation of Asia-Pacific Microbiology Societies* 1998, p. 131.

13. Roberts RG, Bobrow M, Bentley DR, Point mutations in the dystrophin gene. *Proc Natl Acad Sci USA* 1992; **89**:2331-2335.

14. Roberts RG, Gardner RJ, Bobrow M, Searching for the 1 in 2,400,000: A review of dystrophin gene point mutations. *Hum Mutat* 1994; **4**:1-11.

15. Roest PAM, Roberts RG, Sugino S, *et al.*, Protein truncation test (PTT) for rapid detection of translation-terminating mutations. *Hum Molec Genet* 1993; **2**:1719-1721.

16. Gardner RJ, Bobrow M, Roberts RG, The identification of point mutations in Duchenne muscular dystrophy patients by using reverse-transcription PCR and the protein truncation test. *Am J Hum Genet* 1995; **57**:311-320.

17. Whittock NV, Roberts RG, Mathew CG, Abbs SJ, Dystrophin point mutation screening using a multiplexed protein truncation test. *Genet Testing* 1997; **1**:115-123.

18. Khng HH, Low PS, Lee WL, Lai PS, Analysis of deletion breakpoints in dystrophin transcripts, in PS Lai, EPH Yap (eds.), Frontiers in Human Genetics: Diseases and Technologies; World Scientific Publishing, Singapore (in print)

19. Roest PAM, Bout M, van der Tujin AC, *et al.*, Splicing mutations in DMD/BMD detected by RT-PCR/PTT: Detection of a 19 AA insertion in the cystein rich domain of dystrophin compatible with BMD. *J Med Genet* 1996; **33**:935-939.

20. Bulman DR, Gangopadhyay SB, Bebchuck KG *et al.*, Point mutation in the human dystrophin gene: identification through western blot analysis. *Genomics* 1991; **10**:457-460.

21. Roberts RG, Bobrow M, Bentley DR, Point mutations in the dystrophin gene. *Proc Natl Acad Sci USA* 1992; **89**:2331-2335.

FRONTIERS IN HUMAN GENETICS
Diseases and Technologies

© 2001 by World Scientific Publishing Co. Pte. Ltd.

MAPLE SYRUP URINE DISEASE: A REPORT OF 26 CASES IN THE PHILIPPINES

CARMENCITA D. PADILLA*, CATHERINE LYNN T. SILAO and JOY Y. LEE
Department of Pediatrics
University of the Philippines Manila-Philippine General Hospital
Manila, Philippines
**E-mail: carpadil@fastmail.i-next.net*

Maple syrup urine disease (MSUD) is an autosomal recessive disorder in the oxidative decarboxylation of branched chain alpha-ketoacids (BCKAs) derived from leucine, isoleucine, and valine. Impaired activity of the branched chain ketoacid dehydrogenase complex causes MSUD's heterogenous clinical phenotype. Severity of the disease ranges from classical to mild variant types, thus the diagnosis is often overlooked and considered least likely to occur in clinical practice. In the Philippines, the local incidence has not been reported as yet. Using two-dimensional thin layer paper chromatography, we describe here 26 cases of MSUD diagnosed in different hospitals in the country from 1992 to 1998. A review of their demographic data, symptomatology, diagnosis, management and outcome is presented. Our results depict heretofore the importance of early detection of MSUD to ensure reduced morbidity, mortality, and length of hospitalization for these patients.
Keywords: maple syrup urine disease, branched chain ketoacid dehydrogenase complex, two-dimensional thin layer paper chromatography

1 INTRODUCTION

Maple syrup urine disease (MSUD) or branched chain alpha-ketoaciduria is an autosomal recessively inherited deficiency in the mitochondrial branched chain alpha-ketoacid dehydrogenase (BCKAD) complex. The inability to metabolize branched chain alpha-ketoacids derived from the essential branched chain amino acids leucine, isoleucine, and valine causes their eventual accumulation thus producing severe clinical consequences including ketoacidosis, mental retardation, and neurological impairment. The genetic effects that could lead to this phenotype are diverse and could interfere with the production, assembly or function of the BCKAD complex.

* Corresponding author

As a clinical entity, MSUD was first described by Menkes, Hurst, and Craig in 1954, when four siblings developed neurologic symptoms within the first week of life and died within 3 months of the onset of these symptoms. The urine of these patients had a very characteristic maple syrup or burnt sugar odor.[1,2] In 1957, Westall, Dancis and Miller identified it as a disorder of amino acid metabolism.[3]

Based on routine screening data from 26.8 million newborns, the worldwide frequency is pegged at approximately 1 in 185,000.[4] In the urban white population, frequency of MSUD ranges from 1 in 290,000 to 1 in 176 births among Old Order Mennonites of Eastern Pennsylvania. No local incidence of MSUD in the Philippines has been established as yet. However, 26 cases have been diagnosed during the past six years.

Early diagnosis and treatment of MSUD cannot be overemphasized since these are very crucial determinants to their outcome. Patients with MSUD have often developed metabolic decompensation in their neonatal period or later. These episodes have occasionally been fatal or have precipitated psychomotor developmental delay thereafter. In efforts to increase the awareness of clinicians regarding MSUD, we describe here the clinical profile of the cases diagnosed in the Philippines. This report clearly demonstrates the existence of maple syrup urine disease in clinical practice. Therefore when dealing with sick newborns, a high index of suspicion for inborn errors of metabolism must always be present.

2 METHODOLOGY

The present study is based on twenty-six children aged 0 to 2 years (mean 2.1 months) who were diagnosed with maple syrup urine disease in the 6-year interval from 1992 to 1998. The following data were collected retrospectively from their medical records: sex, age on admission, age at onset of symptoms, birth history, family history, symptomatology, initial impression, diagnostic examinations done, management and outcome. Supervising pediatricians were also interviewed when possible.

3 RESULTS

3.1 Demographic Features

Twenty-six cases were diagnosed as having MSUD during the interval between 1992 to 1998. There were 12 females and 14 males. Of these 26 diagnosed cases, 5 were previously reported by Padilla and Masangkay in 1992.[5] Onset of symptomatology was noted as early as 2 days in two patients. However, the mean age at onset of symptoms was 5.65 days (range: 2 - 10 days). The age when the patients were brought to medical attention ranged from birth to 2 years with a mean age of 2.1 months. There was normal birth history and non-consanguinity in all cases. Notably, three patients had siblings with the same condition.

Table 1. Patient profile

Total number of patients	26
Male	14
Female	12
Age on admission	
Mean	2.1 months
Range	0-2 years old
Age at onset of symptoms	
Mean	5.65 days
Range	2-10 days

3.2 Clinical Presentation

All 26 children with MSUD were phenotypically normal at birth. However, within 5.65 days (range: 2-10 days) non-specific symptoms such as poor suck with vomiting, seizures, weak cry and neurologic deterioration were common features. In all patients, ketosis and a sweet smelling urine were present.

3.3 Initial Diagnosis

The clinical features of the 26 patients overlapped with those of sepsis thus making diagnosis difficult as proven by 69% (18/26) of cases with an initial diagnosis of Sepsis Neonatorum and one case diagnosed as Tetanus Neonatorum. Only 7 patients (27%) had an initial diagnosis of MSUD.

3.4 Diagnostic Examination

The suspicion of MSUD arose on clinical grounds in most cases. An incidental diagnosis of MSUD was made by newborn screening for 1 patient whose blood sample was retested for another disorder using thin layer paper chromatography. Results showed increased concentrations of branched chain amino acids. Confirmation of the diagnosis by urine metabolic screening for the first 5 cases was done at the Biochemical Genetics Service of the Royal Alexandra Hospital for Children, New South Wales, Australia. The subsequent cases were confirmed at the Department of Biochemistry, University of the Philippines Manila. Both ferric chloride and dinitrophenylhydrazine (DNPH) tests were positive for all cases. Two dimensional thin layer paper chromatography of their urine samples revealed highly dense bands for leucine, isoleucine, and valine. Serum ammonia levels were significantly increased in 18 patients (69%) and 14 of the 26 cases (54%) had growths on septic work-up.

3.5 Acute Phase Treatment

Concomitant bacterial infections were present in most cases thus their consequential use of antibiotics. Twenty-one of the patients required ventilatory support.

In 58% (15/26) of patients, metabolic decompensation was managed by peritoneal dialysis which is one of the primary modes of therapy in the Philippine setting. No significant improvement was noted in a patient where exchange transfusion was done.

Patients were initially started on a zero protein diet and regular protein was gradually introduced in the diet at increments of 0.25 gm/kg/day until a maximum of 1 gm/kg/day was reached. The rest of the requirements were provided using a special formula milk called Maple Syrup Urine Disease (MSUD) milk which is a BCAA-free formula.

3.6 Outcome

Of the 26 patients included in the study, 7 (27%) died. Metabolic crisis or overwhelming infection were the most important events heralding death. The rest of the 19 patients responded well to the provided acute phase management and were discharged improved immediately thereafter. Unfortunately, of the 19 cases, five patients (26%) were lost to follow-up. Eight of the initial 26 cases were alive at the time of the survey. All surviving patients, however, have neurologic sequelae and psychomotor developmental delay.

4 DISCUSSION

The biochemical basis of MSUD is the inability to metabolize BCKAs derived from the essential BCAAs leucine, isoleucine, and valine. The elevated BCAAs and BCKAs may have severe clinical consequences including ketoacidosis, mental retardation, and neurological impairment. Variations in clinical presentations have led to the classification of MSUD into five clinical phenotypes : classic (0-2% enzyme activity), intermediate (3-30%), intermittent (5-20%) thiamine-responsive (2-40%) and E3 deficient (0-25%). The classic form, which accounts for 75% of MSUD patients, is manifested within the first 2 weeks of life by poor feeding, lethargy, failure to thrive, respiratory irregularities, coma, and death if left untreated. These symptoms are non-specific and may easily be mistaken for sepsis by the unsuspecting clinician.

In other clinical forms of MSUD, onset of symptoms appear 12 24 months later and is usually triggered by infections, immunizations, operations, or excesses in regular protein which places the affected individual in a catabolic state.

In this series, all patients had early onset of symptoms with a mean age of 5.65 days while the mean age of diagnosis was 2.1 months. The delay in diagnosis is attributed to the overlap of symptomatology between MSUD and sepsis. It is often the case when metabolic disorders come to mind only if the expected clinical improvement is not observed. A high index of suspicion when newborns fail to thrive and develop signs of ketoacidosis is thus needed to prompt the clinician to determine the levels of plasma and urinary metabolites. Marked elevations of BCAAs and BCKAs with positive alloisoleucine suggests the diagnosis of MSUD. Diagnosis of

MSUD is actually dependent on clinical findings with ketosis and a burnt sugar odor in body secretions acting as hallmarks of the disease. Noteworthy is the finding that all 26 patients presented with this characteristic smell especially in their urine.

Though two-dimensional thin layer chromatography was used in our series as a confirmatory diagnostic technique, diagnosis can be made using other techniques such as gas chromatography to measure BCKA levels in blood and urine and amino acid analyzers or mass spectrophotometers to measure BCAA plasma levels.

Management of an acute metabolic crisis has three goals: (1) rapid removal of toxic metabolites, (2) nutritional support, and (3) reducing the catabolic state and promoting anabolism.[4] During an acute metabolic crisis, it is imperative to eliminate the elevated BCAA levels and their metabolites particularly alpha-ketoisocaproic acid which is the alpha-ketoanalogue of leucine and the main neurotoxin. This may be done using continuous venovenous hemoperfusion (CVVH), hemodialysis, or peritoneal dialysis. The initial two modalities are considered the treatments of first choice and should be employed if possible since both have been shown to have a higher clearance rate for the toxic metabolites.[6] In the Philippines, however, peritoneal dialysis is the most practical procedure used due to its effectiveness and availability. Of the 26 cases, 15 (58%) underwent this procedure with noted improvement.

The mainstay of management is dietary modification. Management of MSUD requires the restriction of the three essential branched chain amino acids (BCAAs) to a sufficient intake for anabolism to optimize normal growth patterns and minimize the accumulation of toxic metabolites, excess BCAAs and BCKAs. Also, a small amount of normal protein is needed to promote growth. It is imperative that dietary treatment be continued throughout the patient's life.[4] The eight surviving patients are presently maintained on a special MSUD milk formula with frequent monitoring and dietary adjustments necessarily done to ensure adequate caloric and protein requirements. Thus, a clinical dietician is an important partner in the management.

Minimizing the catabolic state and/or promoting anabolism combined with the other two aspects of management produces the best results. Therefore, infections, excesses in regular proteins and other stressful events must be avoided.

Without prompt, appropriate and optimum therapy reducing and preventing plasma levels of the toxic metabolites, unwanted complications i.e. progressive neurologic deterioration, apnea, coma, and death ensue. Untreated patients usually die within the first few months of life of metabolic crisis and neurologic deterioration often followed by infection or other stresses. Surviving patients who begin therapy late suffer severe neurologic damage including mental retardation, spasticity or hypotonia as is the case with the surviving patients in this study. It is of utmost importance, therefore, to consider that inborn errors of metabolism may co-exist with other clinical conditions such as sepsis.

5 CONCLUSION

We have described the 26 cases of MSUD diagnosed from 1992 to 1998 in the Philippines. Clinical features presented were non-specific and easily mimicked those of sepsis neonatorum. Noteworthy was the presence of the characteristic maple syrup odor in all 26 cases. This clinical hallmark should give clinicians the first clue in considering MSUD. Because of the diversities in the genetically determined severity of certain diseases, it must be emphasized that inborn errors of metabolism such as MSUD be considered along with sepsis. To expedite the identification of future patients, every neonate should be given the benefit of a newborn screen and if needed, a urine metabolic screen. It cannot be overemphasized that early diagnosis and treatment may lead to improved outcomes for future MSUD patients. It is the authors hope, therefore, that the long term outlook for future affected patients be brighter with increased awareness among clinicians.

Acknowledgements

We would like to thank Ms. Raquel Zafra of the Department of Biochemistry, University of the Philippines Manila for her assistance; Dr. Bridget Wilcken and her staff at the Biochemical Genetics Service of the Royal Alexandra Hospital for Children, New South Wales, Australia.

References

1. Menkes JH, Hurst PL, Craig JM, A new syndrome: Progressive familial infantile cerebral dysfunction associated with an unusual urinary substance. *Pediatrics* 1954; **14**: 462.
2. Clarke, JTR, *A Clinical Guide to Inherited Metabolic Diseases*. Cambridge University Press, U.K., 1996: 181–182.
3. Westall RG, Dancis J *et al.*, Maple syrup urine disease — A new molecular disease. *Am J Dis Child* 1952; **94**: 571.
4. Chuang DT, Shih VE, Disorders of branched chain amino acid and ketoacid metabolism, in CR Scriver *et al.* (eds.), *The Metabolic and Molecular Basis of Inherited Disease*. McGraw Hill Companies, New York, 1995: 1239–1265.
5. Padilla CD, Masangkay SA, Maple syrup urine disease: Report of five cases. *The Philippine Children's Medical Center Journal* 1992; **1**: 111–114.
6. Thompson GN, Francis D, Halliday D, Acute illness in maple urine disease: dynamics of protein metabolism and implications for management. *J Pediatr* 1991; **119**: 35–41.
7. Tanaka K, Rosenberg LE, Disorders of branched chain amino acids and organic metabolism, in JB Stanbury *et al.*(eds.), *The Metabolic Basis of Inherited Diseases*. McGraw Hill Book Company, New York, 1983: 440.
8. Dancis J, Hutzler J, Snyderman SE, Cox R, Enzyme activity in classical and variant forms of maple syrup urine disease. *J Pediatr* 1972; **81**: 312–320.
9. Nyhan WL, Sakati ND, Disorders of amino acid metabolism, in *Diagnostic Recognition of Genetic Diseases*. Lea and Febiger, Philadelphia, 1987: 34.
10. Clow CL, Reade TM, Scriver CR, Outcome of early and long term management of classical maple syrup urine disease. *Pediatrics* 1981; **68**: 856–862.

11. Snyderman SE, The therapy of maple syrup urine disease. *Am J Dis Child* 1967; **113**: 68–73.
12. Naughten ER, Jenkins J, Francis DE, Leonard JV, Outcome of maple syrup urine disease. *Arch Dis Child* 1982; **57**: 918–921.
13. Kaplan P, Mazur A, Field M, Berlin GA, Berry GT, Heidenreich R *et al.* Intellectual outcome in children with maple syrup urine disease. *J Pediatr* 1991; **119**:46–50.
14. Synderman SE, Treatment outcome of maple syrup urine disease. *Acta Paediatr Jpn* 1988; **30**: 417–424.

FRONTIERS IN HUMAN GENETICS
Diseases and Technologies

© 2001 by World Scientific Publishing Co. Pte. Ltd.

THE HETEROGENEITY OF THALASSEMIA IN SOUTHEAST ASIA

SUTHAT FUCHAROEN* and PRANEE WINICHAGOON
Thalassemia Research Center, Institute of Science and Technology for Research and Development, Mahidol University, Salaya Campus,
25/25 M.3, Puttamonthon 4 Rd., Puttamonthon
Nakornpathom 73170, Thailand
**E-mail : grsfc@mahidol.ac.th*

Southeast (SE) Asia has high frequencies of thalassemias. The most common abnormal genes are α-thalassemia, β-thalassemia, hemoglobin (Hb) E and Hb Constant Spring (CS). The severe form of α-thalassemia, α-thalassemia 1, is most common in this region and HbE is the hallmark of SE Asia. β-thalassemia is very heterogeneous at the molecular level. They are grouped into β^{o}-thalassemia which is associated with absence of β-globin chain and β^{+}-thalassemia with partial production of the β-globin chain.

The two types of α-thalassemia genes (α-thalassemia 1 and α-thalassemia 2), β-thalassemia, HbE, HbCS and some other mutations, in different combinations, lead to more than 60 thalassemia syndromes. HbBart's hydrops fetalis (homozygous α-thalassemia 1) associated with total lethality is the most severe thalassemic disease. HbH disease results from interaction between α-thalassemia 1 and α-thalassemia 2 or between α-thalassemia 1 and HbCS. Compound heterozygosity between β-thalassemia and HbE leads to β-thalassemia/HbE disease.

α-Thalassemias are most often due to gene deletions. While HbCS occurs from the mutation at the termination codon of the α2-globin gene. β-thalassemia results from a variety of molecular mechanisms, most of which are single base substitutions, deletions or insertions. Hemoglobin E occurs from a Glu→Lys substitution at position 26 of the β-globin chain. The abnormal gene also results in reduced amounts of β^{E}-mRNA. Therefore, HbE has a mild β^{+}-thalassemia phenotype. Despite the same genotype, a remarkable variability in severity of β-

* Corresponding author

thalassemia is observed. Genetic factors which influence the severity of anemia include: nature of the mutation, coinheritance of α-thalassemia gene and association with increased HbF.

Keywords: thalassemia, Southeast Asia, heterogeneity

1 INTRODUCTION

Thalassemia (thal) and hemoglobinopathies are recessive inherited diseases that are widely distributed. Thalassemias are also common in Southeast Asia. α-thalassemias are very prevalent and is present throughout Southeast Asia. However, its distribution is heterogeneous.[1] The frequencies are higher in the northern than in the southern part of the region (30.6% in Chiengmai province of Thailand and 0.5% in Indonesia, respectively), and in Laos than in Khon Kaen Province in Northeastern Thailand (42.8% and 5.5%, respectively). The figures represent the combined frequencies of α-thal 1, α-thal 2, and Hb Constant Spring (CS). α-thal 2 is equally distributed, whereas α-thal 1 is most frequent in Chiengmai in Northern Thailand (12.2%).

High frequencies of β-thal are found in Northern Thailand (5–9%) and Laos (3–9%), in Indonesia (6–10%), and in Myanmar (4.3%).[1–3] It is interesting to note that minority ethnic groups in many countries in Southeast Asia also have a high incidence of β-thal.[1] HbE is the hallmark of Southeast Asia attaining frequencies of 50–60% and is almost limited to this region. It arises from a mutation at position 26 of the β-globin chain which replaces glutamic acid by lysine. The abnormal gene also results in reduced amounts of β^E-mRNA and reduced synthesis of β^E-globin chains, producing a mild β^+-thal phenotype.[4] Occurrence of HbE is highest on the Southeast Asian mainland, especially in the border areas joining Thailand, Laos, and Cambodia. It does not seem to occur at a significant frequency in the Chinese and the immigrant minority groups living in the northern part of Thailand.

Abnormal genes in different combinations lead to the occurrence of numerous complex thalassemic syndromes. The thalassemia disorders were originally confined to the tropics and subtropics. However, Southeast

Asians are now scattered around the world and have carried with them most of the complex thalassemic conditions. Familiarity of physicians in nontropical countries with the complexity of the problems will help in the management of these chronic diseases.

2 CLINICAL HETEROGENEITY

Clinically, thalassemia syndromes can be classified into three groups:

2.1 Asymptomatic Thalassemia (Thalassemia Minor)

This group includes all the heterozygotes or thal traits such as α-thal trait, β-thal trait, HbCS and HbE traits. However, many homozygous, compound and doubly heterozygous subjects are also symptom-free, e.g. homozygotes for HbE or α-thal 2, α-thal 2/HbCS, α-thal/β-thal, α-thal/HbE.

Hematologic examination in the adult is not reliable to establish definite diagnosis of α-thal traits since α-thal 2 trait is often indistinguishable from normal, and iron deficiency anemia leads to microcytosis which also occurs with α-thal 1 trait and homozygous α-thal 2. The amount of HbBart's (γ_4) in the umbilical cord blood may predict the genotypes of α-thal.

HbBart's	Genotypes
1–2%	α-thal 2 trait
5–8%	α-thal 1 trait or α-thal 2/α-thal 2
25%	HbH disease
80–85%	Homozygous α-thal 1 (HbBart's hydrops fetalis)

Accurate diagnosis of various α-thal genotypes can be performed by either Southern blot or polymerase chain reaction (PCR) techniques.[5]

In contrast, adults with β-thal trait usually have small MCV with elevated HbA_2 to about 5%. α-thal/β-thal cannot be differentiated from β-thal trait by screening. The Hb type A+E is the characteristic of HbE trait and the percentage of HbE reflects concomitant inheritance of α-thal genes.[6] HbE constitutes 25–30% in simple HbE trait while 19–21% HbE is characteristic of α-thal 1/HbE and 13–15% for α-thal 1/α-thal 2--β^A/β^E (AE Bart's disease). α-thal 2/HbE cannot be differentiated from simple HbE trait by screening methods.

2.2 Thalassemia Intermedia

This group consists of thalassemic diseases with mild-to-moderate anemia; the average Hb level in the steady state is 7–8 g/dl. It is also associated with mild-to-moderate jaundice and hepatosplenomegaly. Generally, the patients have mild symptoms or are symptom-free, do not require blood transfusions and do not have thalassemic facies. The major disorder in α-thal intermedia is HbH disease, either α-thal 1/α-thal 2 or α-thal 1/HbCS genotypes. Other α-thal syndromes included in this category are homozygous HbCS, HbAE Bart's disease (α-thal 1/α-thal 2 or α-thal 1/HbCS with HbE heterozygote) and EF Bart's disease (HbH with homozygous HbE or HbH with β-thal/HbE). For β-thal intermedia there are two major genotypes. First is β^{+}-thal, either in the homozygous state or in compound heterozygosity with β^{o}-thal. However, the latter may have severe thalassemia depending on the type of β^{+}-thal genes. Second is β^{o}-thal/HbE in which a remarkable variability in severity was observed. Approximately half of β^{o}-thal/HbE manifest as thal intermedia and a number of genetic determinants was found to ameliorate the severity of the patients.

2.3 Severe Thalassemia (Thalassemia Major)

This group consists mainly of two categories: homozygous α-thal 1 (HbBart's hydrops fetalis) and numerous β-thalassemic diseases such as homozygous β^{o}-thal and β^{o}-thal/HbE. Homozygous α-thal 1 is the most severe form of thalassemia. Because of the absence of α-globin synthesis, the fetus does not have either HbF or HbA, and is thus incompatible with life. The baby either die *in utero* or soon after birth. Patients with severe β-thal usually die early, in the first two decades of life.

3 HETEROGENEITY IN MOLECULAR MECHANISMS

The thalassemia syndromes are due to a decrease or absence in production of α- or β-globin chains resulting in α-thal and β-thal, respectively. The abnormalities of α- and β-globin gene expression in α-thal and β-thal detected so far have been due either to deletions or to changes in a single or small number of nucleotides within or close to the gene. Deletions of a large segment of the α-globin gene are the major cause of α-thal whereas a small insertion or deletion of nucleotides and base substitutions mostly cause β-thal.

3.1 Molecular Biology of α-Thalassemia

There are two common abnormal α-thal alleles.[7] One is the severe or α-thal 1 gene resulting from deletion of both of the duplicated α-globin genes (--/αα; normal = αα/αα). The deletion is about 17.5–20 kb long. The other allele is the mild or α-thal 2 gene resulting from deletion of one of the two adjacent α-globin genes (-α/αα). Two types of α-thal 2 have been described: one involving a deletion of 4.2 kb of DNA that includes the α2 globin gene (the leftward or $-\alpha^{4.2}$ kb type); the other involves a 3.7 kb deletion of DNA between the duplicated α-globin genes (the rightward or $-\alpha^{3.7}$ kb type). The latter is more common in Southeast Asia. HbCS results from an elongated α-globin chain. A mutation at the termination codon (position 142) of the α2-globin gene results in a codon for glutamine; as a consequence mRNA is translated to the next inphase termination codon to give a globin with 172 rather than the normal 141 amino acids.[8] The CS-mRNA has been shown to be unstable and only a small amount of HbCS is produced from it.[9] This results in an α-thal-2-like effect. Other forms of α-thal including nondeletional types and α-chain structural variants are rare in Southeast Asia.

3.2 Molecular Biology of β-Thalassemia

β-thal is a very heterogeneous disorder caused by many different defects in the β-globin gene. Gene deletions are not the major cause of β-thal and have been found in a small number of patients in certain population such as the 105-bp and 3.4 kb deletion in Thais, and the Filipino deletion in Filipino descent and South Celebes in Indonesia.[10–13] Point mutations and small deletions or insertions in the nucleotide sequence are the main molecular defects responsible for most β-thalassemias, and different parts of the country usually have a different incidence of the common β-thal mutations (Table 1).[14–18] This probably reflects differences in the ethnic population residing within these regions. For example, a G→C mutation at IVS-1-5 occurs in about 4.3% of patients from Central Thailand but appears to be much more frequent in the south, where a spillover from Malaysia is suspected. A higher proportion of the mutation A→G at codon 19 is also observed in patients originating from Southern Thailand and in Sumatranese (50%) in Indonesia.[13,18] Patients with Chinese origin in Malaysia, Singapore and Thailand are almost exclusively from the southeastern provinces of China, especially Guangdong, Fukien, and Guangxi. The common β-thal mutations in these people are: 4 bp deletion in codons 41/42 (-TTCT), A→T at codon 17, C→T at IVS-II-654, and A→G at -28 of the ATA box.[18] In Myanmar, the common β-thal mutations are: G→T in IVS-I-1, G→C in IVS-I-5, and the 4 bp deletion in codons 41/42. However, Karens and those of mixed Burman/Karen ethnicity have a predominance of G→T in IVS-I-1.

4 GENOTYPIC–PHENOTYPIC INTERACTION

The clinical severity of thalassemia is usually influenced by the particular mutation(s) of the globin gene which are present in each patient. The severity of α-thalassemia phenotype relates to the number of genes affected. The four α-globin genes deletion results in the most severe thalassemia, HbBart's hydrops fetalis, which is incompatible with life. Whereas the three α-globin genes deletion (α-thal 1/α-thal 2) or interaction of α-thal 1 and the nondeletion defects, such as HbCS, result in HbH disease. Since the expression of the two α-globin genes, α1 and α2 are unequal and the α2-globin mRNA was present at a higher level than the α1-globin mRNA; the abnormality on the α2-globin gene will result in a more defective α-thal.[19] In Southeast Asia, HbH-CS (α-thal 1/HbCS) is also the common HbH disease and the patients have more severe clinical symptoms than the classical HbH.[20]

In β-thal, β^{o}-thal is usually more severe than β^{+}-thal in which a wide range of β-globin production is observed.[18] But it is also known to be influenced by other factors which affect either α- or γ-globin gene expression.[21] In a study of 116 Thai patients with β-thal disease, 38 were found to have a mild clinical symptom with the average Hb of 7.5 g/dl and 30 were found to be severe thalassemic. Six out of 38 patients with mild clinical symptoms were β^{+}-thal/β^{+}-thal or β^{+}-thal/HbE and 16 out of the remaining 32 mild patients who were β^{o}-thal /β^{+}-thal or β^{o}-thal/HbE also had coinheritance of α-thal gene (Table 2). Thus the presence of two β^{+}-thal alleles and coinheritance of α-thal indicate the mildness of the disease in 58% of the mild cases. In the 30 patients with severe clinical symptoms, 15 were β^{o}-thal/β^{o}-thal, 4 were β^{o}-thal/β^{+}-thal and 11 were β^{o}-thal/HbE. Concomitant inheritance of α-thal was found in three cases of β^{o}-thal/β^{o}-thal but it cannot ameliorate the severity of the disease.

5 CONCLUSION

Southeast Asia is witness of a very big problem of thalassemia and hemoglobinopathies. Problems occur not only from a high frequency of these abnormal genes but also from its heterogeneity and complexity of multiple gene-gene interactions. It is not simple to cope up with the magnitude of these single gene disorders. It will need a good strategy of public education, screening, training of personnel, technology transfer etc. to prevent and control of the diseases.

Acknowledgements

This work was supported in part by the Prajadhipok Rambhai Barni Foundation, the National Research Council of Thailand and National Science and Technology Development Agency.

References

1. Fucharoen S, Winichagoon P. Hemoglobinopathies in Southeast Asia. *Hemoglobin* 1987; **11:** 65–88.
2. Sofro ASM. Molecular pathology of β-thalassemia in Indonesia. *Southeast Asian Journal of Tropical medicine and Public Health* 1995; **26:** 221–223.
3. Aung-Than-Batu, U-Hla-Pe. Haemoglobinopathies in Burma. *Tropical Geographical Medicine* 1971; **23:** 15–29.
4. Orkin SH, Kazazian HH, Jr, Antonarakis SE, Ostrer H, Goff SC, Sexton JP. Abnormal RNA processing due to the exon mutation of the β^E-globin gene. *Nature* 1982; **300:** 768–769.
5. Winichagoon P, Thonglairuam V, Fucharoen S, Tanphaichitr VS, Wasi P. α-thalassemia in Thailand. *Hemoglobin* 1988; **12:** 485–498.
6. Fucharoen S, Winichagoon P, Thonglairuam V. β-Thalassemia associated with α-thalassemia in Thailand. *Hemoglobin* 1988; **12:** 581–592.
7. Winichagoon P, Higgs DR, Goodbourn SEY, Clegg JB, Weatherall DJ, Wasi P. The molecular basis of α-thalassemia in Thailand. *The EMBO Journal* 1984; **3:** 1813–1818.
8. Clegg JB, Weatherall DJ, Milner PG. Haemoglobin Constant Spring: a chain termination mutant. *Nature* 1971; **234:** 337–340.
9. Hunt DM, Higgs DR, Winichagoon P, Clegg JB, Weatherall DJ. Haemoglobin Constant Spring has an unstable α chain messenger RNA. *British Journal of Haematology* 1982; **51:** 405–413.
10. Sanguansermsri T, Pape M, Laig M, Hundrieser J, Flatz G. β^o-Thalassemia in a Thai family is caused by a 3.4 kb deletion including the entire β-globin gene. *Hemoglobin* 1990; **14:** 157–168.
11. Nopparatana C, Panich V, Saechan V *et al.* The spectrum of β-thalassemia mutations in Southern Thailand. *Southeast Asian Journal of Tropical medicine and Public Health* 1995; **26:** 229–234.
12. Waye JS, Eng B, Hunt JA, Chui DHK. Filipino β-thalassemia due to a large deletion: identification of the deletion endpoints and polymerase chain reaction (PCR)-based diagnosis. *Human Genetics* 1994; **94:** 530–532.

13. Setianingsih I, Williamson R, Marzuk S *et al.* Molecular basis of β-thalassemia in Indonesia: application to prenatal diagnosis. *Molecular Diagnosis* 1998; **3:** 11–20.
14. Ng ISL, Ong JBK, Tan CL, Law HY. β-Thalassemia mutations in Singapore — a strategy for prenatal diagnosis. *Human Genetics* 1994; **94:** 385–388.
15. George E, Li H-J, Fei Y-J *et al.* Types of thalassemia among patients attending a large university clinic in Kuala Lumper, Malaysia. *Hemoglobin* 1992; **16:** 51–66.
16. Lie-Injo LE, Cai SP, Wahidiyat I *et al.* β-Thalassemia mutations in Indonesia and their linkage to β-haplotypes. *American Journal of Human Genetics* 1989; **45:** 971–975.
17. Brown JM, Thein SL, Weatherall DJ, Khin Mar Mar. The spectrum of β-thalassemia in Burma. *British Journal of Haematology* 1992; **81:** 574–578.
18. Fucharoen S, Winichagoon P. Hemoglobinopathies in Southeast Asia: Molecular biology and clinical medicine. *Hemoglobin* 1997; **21:** 299–319.
19. Liebhaber SA, Cash FE, Ballas SK. Human α-globin gene expression. The dominant role of the α2-locus in mRNA and protein synthesis. *Journal of Biological Chemistry* 1986; **261:** 15327–15333.
20. Fucharoen S, Winichagoon P, Pootrakul P *et al.* Differences between two types of Hb H disease, α-thalassemia 1/α-thalassemia 2 and α-thalassemia 1/Hb Constant Spring. *Birth Defects* 1988; **23** (5A): 317–326.
21. Winichagoon P, Thonglairuam V, Fucharoen S, Wilairat P, Fukumaki Y, Wasi P. Severity differences in β-thalassaemia/haemoglobin E syndromes: Implication of genetic factors. *British Journal of Haematology* 1993; **83:** 633–639.

FRONTIERS IN HUMAN GENETICS
Diseases and Technologies
© 2001 by World Scientific Publishing Co. Pte. Ltd.

LINKAGE ANALYSIS OF MESOMELIC DYSPLASIA, KANTAPUTRA TYPE

P.N. KANTAPUTRA*, M. FUJIMOTO†, S. KONDO†,§§, H. TOMITA†, N. NIIKAWA†, S. IKEGAWA‡, T. TAKAGI‡, Y. NAKAMURA‡, Y. FUKUSHIMA§, S. SONTA¶, M. MATSUO**, T. ISHIDA††, T. MATSUMOTO‡‡, H.-X. DENG¶¶, M. D'URSO***, V. VENTRUTO***, M.M. RINALDI†††, L.O. LANGER JR. ‡‡‡ and R.J. GORLIN§§§

**Department of Pediatric Dentistry, Faculty of Dentistry, Chiang Mai University, Chiang Mai 50200, Thailand*
**E-mail: dnpdi001@chiangmai.ac.th*
†Department of Human Genetics, Nagasaki University School of Medicine, 1-12-4 Sakamoto, Nagasaki 852, Japan.
‡Human Genome Center, Institute of Medical Science, University of Tokyo, Tokyo, Japan.
§Department of Hygiene and Medical Genetics, Shinshu University School of Medicine, Matsumoto, Japan.
¶Department of Genetics, Institute of Developmental Research, Aichi Human Service Center, Kasugai, Japan.
***Division of Genetics, ICMR, Kobe University, Kobe, Japan.*
††Unit of Human Genetics, Department of Biological Science, Faculty of Science, University of Tokyo, Tokyo, Japan.
‡‡Department of Pediatrics, Nagasaki University School of Medicine, Nagasaki, Japan.
§§Department of Plastic and Reconstructive Surgery, Nagasaki University School of Medicine, Nagasaki, Japan.
¶¶Neurogenetic Laboratory, Department of Neurology, Northwestern University Medical School, Chicago, IL, USA.
****Istituto Internazionale di Geneica e Biofisica-Area di Ricerca C.N.R., Napoli, Italy.*
†††Servizio di Genetica Medica Ospedale A. Cardarelli, Napoli, Italy.
‡‡‡Department of Radiology, School of Medicine, University of Minnesota, Minneapolis, USA.
§§§Department of Oral Pathology and Genetics, School of Dentistry, University of Minnesota, Minneapolis, USA.

---

* Corresponding author

Mesomelic dysplasia Kantaputra type (MDK) (MIM *156232) is a new autosomal dominant skeletal dysplasia characterized by bilateral symmetrical marked shortening of the ulnae and shortening and bowing of the radii. The proximal fibula is usually short and synostoses are present between the tibia and fibula and the small malformed calcaneus and talus. The prominent calcanei on the ventral surfaces of the distal fibulae are a characteristic feature of this new syndrome. Carpal and tarsal synostoses are present in some affected people. We studied this Thai family in which 15 members in 3 generations were affected. With reference to the breakpoints of a balanced translocation [t(2;8)(q31;p21)] in patients from a previously reported Italian family with a skeletal dysplasia that appears similar to MDK, a linkage analysis was performed using 50 CA-repeat markers mapped to nearby regions (2q22-q34 and 8p24-p21) of the translocation breakpoints. The results clearly ruled out a linkage of MDK to marker loci at the 8p24-p21 region, whereas all nine affected members available for the study shared a haplotype at four loci (*D2S2284, D2S326, D2S2188,* and *D2S2314*) spanning about 22.7 cM in the 2q24-q32 region. The computer-assisted two-point linkage analysis revealed maximum logarithm of odds (lod) scores of 4.82, 4.21, 4.82, and 4.21 ($\theta = 0$) at these loci, respectively. These data indicated that the MDK locus is in the vicinity of *D2S2284* and *D2S2188* loci that are the most likely mapped to 2q24-q32.

Keywords: mesomelic dysplasia, dyschondrosteosis, hypoplastic radius, hypoplastic ulna, carpal synostosis, tarsal synostosis, hypoplastic fibula, linkage analysis, human *HOXD* genes

1 INTRODUCTION

This new disorder was transmitted from males supposed to be identical twins their sons, daughters, and three grandchildren. This disorder has never bee reported before in the literature. Their clinical and radiographic appearanc were distinctly different from previously described types of mesomel dysplasia. The affected were of Thai origin, born to unrelated parents. Most the members of this family live in Chiang Mai and Sawunkalok, Sukotha Fifteen individuals were affected but only nine were available for the stu (Fig. 1). [1]

Fig. 1. Nine affected members and the first author (PN Kantaputra) sitting at the upper right corner.

2 CLINICAL FINDINGS

Affected individuals shared common clinical appearances. They were c normal intelligence. Craniofacial structures were unremarkable. Panorami films showed normal dental development. The forearms were severely shor radially bowed and the hands deviate ulnarly. They had moderately sho stature as the legs were somewhat short. The ankle joints were absent. Th long axes of the shanks, hind and mid foot were essentially parallel. Th affected stood on the tips of their toes with the dorsal feet deviated laterally The soles faced upwards. The feet appeared of normal length (Figs. 2a and b) The dorsum of the feet in the weight bearing areas appeared hyperkeratotic The average height and weight of affected male adults were 152.5 cm and 56. kg, respectively. An affected adult female was 135 cm tall and weighed 40 kg.

3 RADIOGRAPHIC FINDINGS

All affected individuals shared common radiographic appearances with son variation. The identical twin brothers (II-5 and II-8) had almost identic radiographic appearances. The radiographic findings of the youngest affect individual (IV-2) were somewhat different from those in the rest of the family.

3.1 Upper Limbs

The radii and ulnae were very short. The ulna was very broad an disproportionately shorter than the radius. The radius was markedly bow radially and broad, but less so than the ulna. The changes in the radius and ul were bilateral and symmetrical. The distal radius articulated with the scaphoi often articulated with the lunate and sometimes with the triquetrum. The dis ulna articulated with the triquetrum. The distal radius and ulna did not articula as the triangular-shaped proximal carpal row with the lunate at the apex l between them. The pisiform was absent. Carpal synostoses were present some individuals.

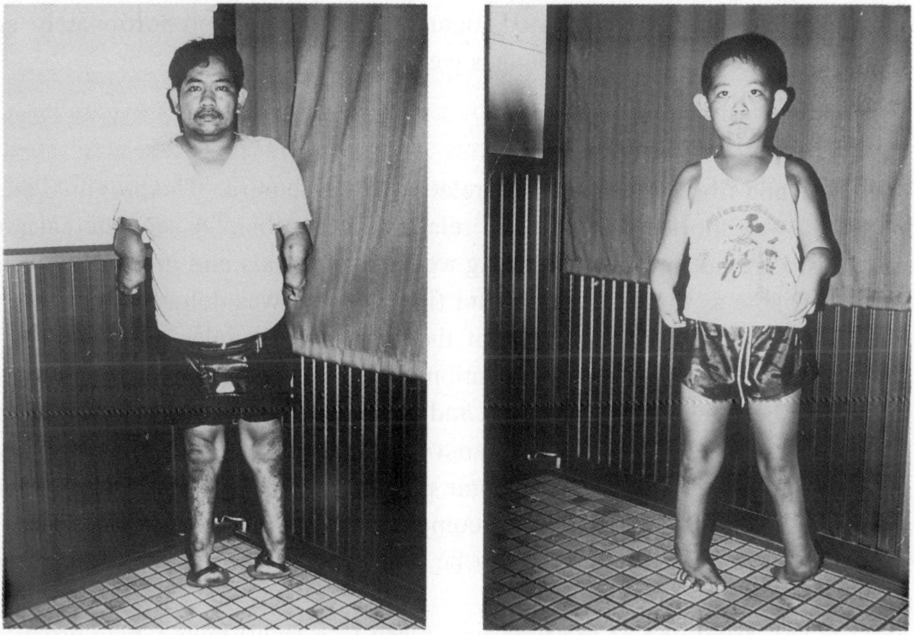

(a) (b)

Fig. 2. (a) Individual III-2 and (b) his son (IV-2). Forearms are severely short and radially bowed. Moderate short stature. Legs are short. They stand on the tips of their toes.

Scaphoid-trapezium fusion is the commonest, but capitate-trapezoid capitate-lunate, and in one individual, trapezium-scaphoid-trapezoid-capitate lunate fusion were present. This same individual (III-2) also had a fusio between the fibulo-calcaneal complex and the tibio-talar complex and fusions i the feet. The hand was deviated ulnarly due to the disproportionately shor ulna. The metacarpals and phalanges were normal (Fig. 3).

3.2 Lower Limbs

The tibiae and fibulae were short in relation to the femora. The proximal end c the fibula was shorter than usual in relation to the tibia. A separate calcaneu and talus were not present. The long axis of the tarsals and metatarsals was i the same plane as that of the tibia and fibula. There was deformity of the dist part of the fibula which, because of the prominent angular protrusion on th ventral surface and the distal articulation with the cuboid, appeared to represe synostosis between the distal fibula and calcaneus. This will be referred to a the fibulo-calcaneal complex. No transverse lines residual from this fusion we seen in any of the individuals. In some patients a small exostosis off the medi aspect of the distal fibulo-calcaneal complex was seen. This was at the level c the above described fusion and may be related to it. No lateral malleolus w present (Fig. 4).

The distal end of the tibia was fused to a small talus which articulate with the laterally placed navicular. This is referred to as the tibio-talar comple There was a more or less conspicuous narrow lucency between parallel den lines in the expected region of a tibio-talar synostosis. This represent incipient, incomplete, or residual changes related to the fusion. It was present all adult individuals. It was least apparent in the oldest individuals, the tw brothers. The medial malleolus was present.

Ventruto *et al.*[2] reported a family of which a father and his thr children were affected with skeletal abnormalities consisting of short forear cubitus valgus, curved radius, Madelung deformity, fusion of the first a second cervial vertebrae, and cleft of L5 and S1. All of the affected individua

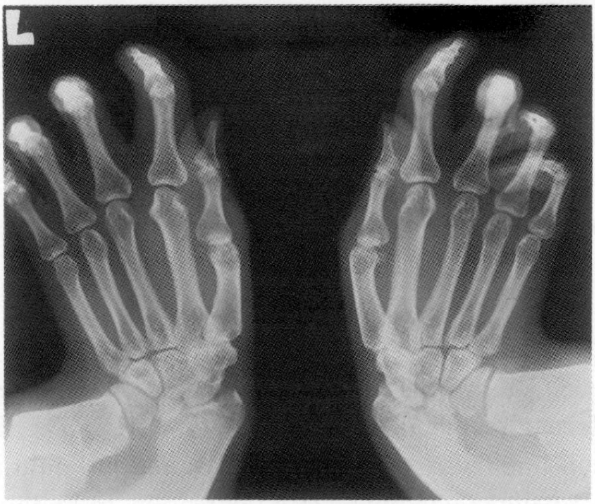

Fig. 3. Hand wrist radiographs of III-2. Radii are markedly short and bowed radially. The ulnae are shorter than the radii. Note carpal synostosis.

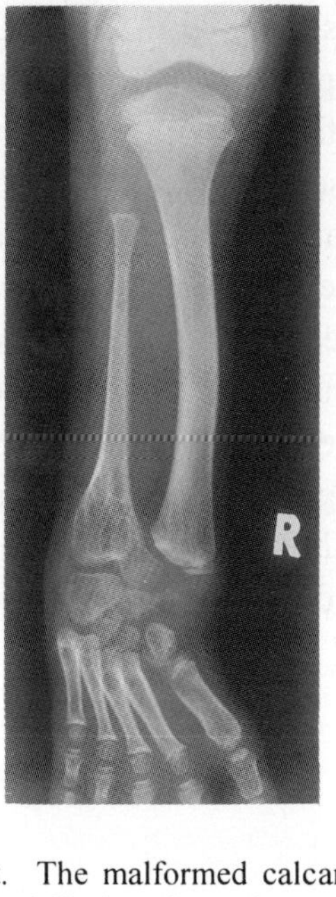

Fig. 4. Proximal fibula is short. The malformed calcaneus protruded form the medial aspect of ventral surface of the distal fibulo-calcaneal complex. The long axes of the tibia, fibula, and those of tarsals, and metatarsals are almost parallel. (Reprinted with permission from P.N. Kantaputra *et al. Am J Med Genet* (1992) **44**: 730–737; Copyright © 1992, Wiley-Liss, Inc., a subsidiary of John Wiley & Sons, Inc.)

had a reciprocal balanced translocation 2;8(q32;p23). Normal sibs had norma chromosomes. One affected female had two spontaneous first trimeste abortions.

Regarding mesomelic dysplasia, Langer type, it has been stated t represent the homozygous state of the autosomal dominant dyschondrosteosi gene which in it heterozygous state results in dyschondrosteosis.[3-7]

4 MATERIALS AND METHODS

Peripheral blood (10 ml) was drawn from the consenting nine affected membe and 13 unaffected members or their spouses (Fig. 1). Ebstein-Barr viru transformed cell lines were established and stored in liquid nitrogen until use High-molecular-weight genomic DNA extraction from those lymphoblasto cell lines were performed by standard method. Fifty pairs polymorph microsatellite (CA repeat) markers that are mapped at around the breakpoints the Italian patients were selected from Genethon collection.[8] The marke consisted of 24 CA repeats mapped to 2q22-q34 and 26 CA repeats assigned 8p24-p21, which correspond to nearby regions of the t(2;8)(q31:p2 breakpoints in the Italian family with dyschondrosteosis.[2] One of each prim was labeled with a fluorescence dye, Cy5, whereas the other pair was witho labeling. Genomic DNA of each individual was amplified by polymerase cha reaction (PCR) using primer sets for the above mentioned CA-repeat markers. The PCR conditions were previously described.[10] PCR products we sequenced using an automated DNA-sequencer (ALFexpress DNA sequence Pharmacia). Genotype of electrophoretic patterns was analyzed using comput software (Fragment ManagerTM version 1.2 software; Pharmacia). Lod scc was calculated using MLINK program of the LINKAGE package. Haploty analysis was performed and the disease locus was then mapped.

5 RESULTS

The genotypes were successfully determined in all loci and all the 21 family members examined. The resultant alleles at each locus were symbolized as numbers from 7 to 1, according to their size. Since none of the alleles at the region 8p24-p21 appeared consistently among the affected individuals, the region thereby being ruled out from the disease locus. All nine affected members shared a 1-3-5-4 haplotype at the *D2S2284, D2S326, D2S2188,* and *D2S2314* loci in the 2q22-q34 region (Fig. 5). The computer-assisted two-point linkage analysis revealed maximum logarithm of odds (lod) scores of 4.82, 4.21, 4.82, and 4.21 ($\theta = 0$) at these loci, respectively (Table 1). Haplotype analysis suggested that the MDK locus is localized between the *D2S2345* and *D2S118* loci with genetic distance of 22.7cM, since recombinations were not observed in a *D2S2284, D2S326, D2S2188,* and *D2S2314* interval, while a recombination was observed between *D2S2345* and *D2S2284* in IV-6 and between *D2S2314* and *D2S118* in IV-7 respectively (Fig. 5). Haplotypes of the proband and his twin-brother were deduced from those in their children and grandchildren. As the four marker loci correspond to 2q24-q34[8], our study indicates that the MDK disease locus is mapped to the region.

6 DISCUSSION

This new autosomal dominantly inherited mesomelic dysplasia is distinctly different from those previously described in the literature. The disorder started from the twins, both interestingly died of diabetic mellitus at age 61. The disorder had some similarities to mesomelic dysplasia, Langer type in the upper limbs. In both disorders, there are short and bowed radii, short ulna deviation of hands, and hypoplasia of fibulae. The shortening of the humerus, tibia, and fibula is greater in the Langer type. Madelung deformity of the wrists is a pathognomonic feature of the Langer type and is the result of growth retardation of the medial portion of the distal epiphyseal line of the radius. This results in

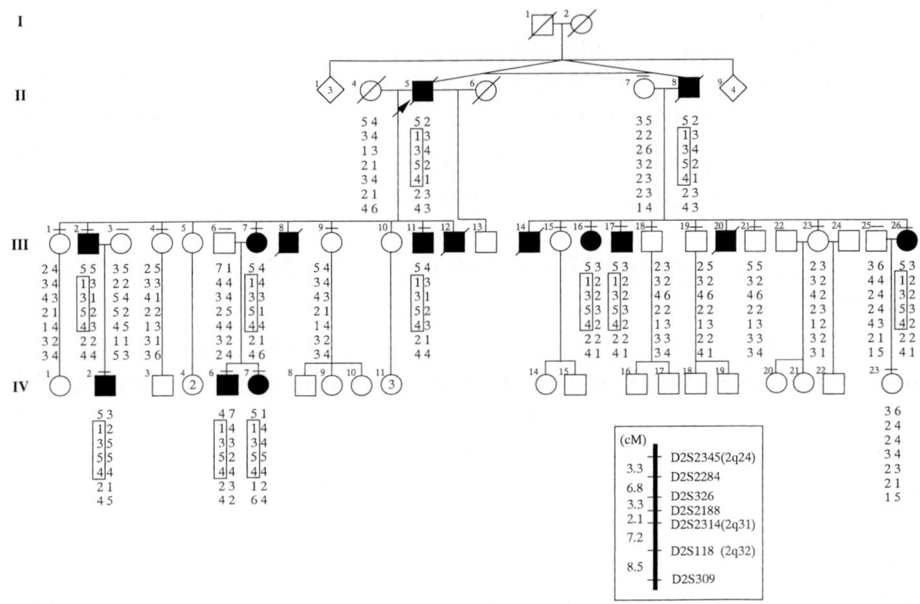

Fig. 5. Pedigree of the Thai family affected with mesomelic dysplasia, Kantaputra type. Individuals examined are marked with short bars above their symbols. Numbers in boxes depict haplotypes common among affected members (filled symbols). The inset shows the arrangement of the loci in the region examined. (Reprinted with permission from M. Fujimoto *et al. J Hum Genet* (1998) **43**: 32–36; Copyright © 1998, The Japan Society of Human Genetics and Springer-Verlag Tokyo.)

Table 1. Two-point logarithm of odds (lod) scores between the MDK and marker loci. (Reprinted with permission from M. Fujimoto *et al. J Hum Genet* (1998) **43**: 32–36; Copyright © 1998, The Japan Society of Human Genetics and Springer-Verlag Tokyo.)

Locus symbol	Distance (cM)	Lod score (Θ)										Zmax	Θmax
		0.00	0.05	0.10	0.15	0.20	0.25	0.30	0.35	0.40	0.45		
D2S2345		−∞	0.32	0.92	1.13	1.16	1.09	0.93	0.72	0.45	0.17	1.16	0.20
D2S2284	3.3	4.82	4.44	4.04	3.62	3.17	2.69	2.18	1.64	1.05	0.43	4.82	0.00
D2S326	6.8	4.21	3.88	3.53	3.16	2.76	2.34	1.89	1.41	0.89	0.35	4.21	0.00
D2S2188	3.3	4.82	4.44	4.04	3.62	3.17	2.69	2.18	1.64	1.05	0.43	4.82	0.00
D2S2314	2.1	4.21	3.88	3.53	3.16	2.76	2.34	1.89	1.41	0.89	0.35	4.21	0.00
D2S118	7.2	−∞	1.04	1.36	1.42	1.35	1.21	1.01	0.76	0.46	0.16	1.42	0.15
D2S309	8.5	−∞	1.61	1.87	1.88	1.76	1.56	1.30	0.98	0.62	0.23	1.88	0.15

ulnar and palmar inclination of the articular surface of the radius at the wrist, accompanied by lateral and dorsal bowing of the shaft.[11]

The strategy of our linkage analysis was based on the assumption that MDK in this Thai family was allelic to the disease in the Italian family.[2] The candidate positional approach was sucessful and provided evidence that the MDK locus is confined to a segment between the *D2S2345* and *D2S118* loci. However, the results of our linkage analysis do not rule out the possibility that the disease locus accidentally corresponds to the 2q31 breakpoint region in the Italian patients.

There have been several genes already mapped around 2q31 region. A candidate is the human homeotic gene, *HoxD* gene family.[12] Among the *Hoxd* genes, *Hoxd-11* is most plausible becuase it plays a important role in the development of the forearm in the chickens and mice.[13] Mice lacking *Hoxa-11* and *Hoxd-11* have been reported to be absence of radius and ulna.[14] However, its role in human development remains obscure because *HOXD11* has not been isolated. Therefore we are now isolating the gene. Once the mutation is found in the patients who have mesomelic dysplasia, Kantaputra type, we can identify it as the causative gene for this disorder.

Acknowledgements

We would like to thank all the members of the Thai and Italian families for their cooperation. We also would like to express our gratitude to the following; Dr. and Mrs. Montri Kantaputra for their generous help with contacting the Thai patients and arrangements; Prof. Robert Vickers for his support throughout the project; Dr. Lai Poh San for her great efforts in searching for a patient residing in Singapore; Dr. Osamu Miyoshi, Dr. Masayaki Egashira, Dr. Nobutomo Miwa, Dr. Kohki Yamada, Ms. Hiroko Iwaki, Ms Keiko Wakui, and Mr. Naoki Harada for their technical assistance. Assistance of the staff at Lanna Medical Lab in obtaining the blood from the patients is deeply appreciated. This work was supported in part by a grant-in-aid for scientific research A (No. 08307019) from the Ministry of Education, Science, Sport, and Culture of Japan.

References

1. Kantaputra PK, Gorlin RJ, Langer LO Jr, Dominant mesomelic dysplasia, ankle, carpal, and tarsal synostosis type: A new autosomal dominant bone disorder. *Am J Med Genet* 1992; **44**:730 - 737.

2. Ventruto V, Pisciotta R, Renda S, Festa B, Rinaldi MM, Stabile M, Cavaliere ML, Esposito M, Multiple skeletal familial associated with balanced reciprocal translocation 2;8(q32;13). *Am J Med Genet* 1983; **16**:589 - 594.

3. Silverman FN, Mesomelic dwarfism. *Progr Pediatr Radiol* 1973; **4**:546 - 562.

4. Espiritu C, Chen H, Wooley PV, Mesomelic dwarfism as the homozygous expression of dyschondrosteosis. *Am J Dis Child* 1975; **129**:375 - 377.

5. Kaitila II, Leisti JT, Rimoin DL, Mesomelic skeletal dysplasias. *Clin Orthop* 1976; **114**: 94 - 106.

6. Fryns JP, Van den Berghe H, Langer type of mesomelic dwarfism as the possible homozygous expression of dyschondrosteosis. *Hum Genet* 1979; **46**:21 - 27.

7. Kunze J, Klemm T, Mesomelic dysplasia, type Langer-A homozygous state for dyschondrosteosis. *Eur J Pediatr* 1978; **134**:269 - 272.

8. Dib C, Faure S, Fizames C, Samson D, Drouot N, Vignal A, Millasseau P, Marc S, Hazan J, Seboun E,, Lathrop M, Gyapay G, Morissette J, Weissenbach J, A comprehensive genetic map of the human genome based on 5264 microsattellites. *Nature* 1996; **380**:152 - 154.

Mansfield DC, Brown AF, Green DK, Carothers AD, Morris SW, Evans JH, 'right AF, Automation of genetic linkage analysis using fluorescent icrosatellite markers. *Genomics* 1994; **24**:225 - 233.

). Fujimoto M, Kantaputra PN, Ikegawa S, Fukushima Y, Sonta S, Matsuo M, hida T, Matsumoto T, Kondo S, Tomita H, Deng H, D'urso M, Rinaldi MM, entruto V, Takagi T, Nakamura Y, Niikawa N, The gene for mesomelic splasia, Kantaputra type is mapped to chromosome 2q24-q32. *J Hum Genet* 98; **43**:32 - 36.

. Langer LO Jr, Mesomelic dwarfism of the hypoplastic ulna, fibula, andible type. *Radiology* 1967; **89**:654 - 660.

. Apiuo F, Flagiello D, Cillo C, Malfoy B, Poupon MF, Dutrillaux B, Fine apping of human HOX gene clusters. *Cytogenet Cell Genet* 1996; **73**:114 - 5.

. Favier B, Dolle P, Developmental functions of mammalian Hox genes. *Mol uman Reprod* 1997; **3**:115 - 131.

. Davis AP, Witte DP, Hsieh-Li HM, Potter SS, Capecchi MR, Absence of lius and ulna in mice lacking hoxa-11 and hoxd-11. *Nature* 1995; **375**:791 - 5.

FRONTIERS IN HUMAN GENETICS
Diseases and Technologies
© 2001 by World Scientific Publishing Co. Pte. Ltd.

GENETICS AND SUSCEPTIBILITY TO TUBERCULOSIS : A REVIEW

EVA MARIA C. CUTIONGCO
Department of Pediatrics, College of Medicine
University of the Philippines - Manila

The role of genetic influences on the course of mycobacterial infections, particularly in tuberculosis is presented in this review. Studies have shown racial differences in susceptibility to tuberculosis, familial clustering of cases and a higher concordance rate of susceptibility among monozygotic twins compared to dizygotic twins. Animal models of infection have been used as tools in the genetic analysis of host susceptibility or resistance to tuberculosis. More recently, an autosomal dominant gene called *Bcg* has been identified which is believed to be responsible for a non-specific macrophage activation for bactericidal function. *Nramp1* or natural resistance associated macrophage protein 1 is the candidate gene for *Bcg*. It is suggested that *Nramp1* has a macrophage-specific membrane transport function. This review is presented to emphasize the important role of identifying genes regulating host innate and antigen-specific responses for better prevention and control of infections such as tuberculosis.
Keywords: tuberculosis, host susceptibility

Tuberculosis continues to be a major health concern worldwide. It has been estimated that 1.7 billion people are infected with *Mycobacterium tuberculosis* and tuberculosis is still responsible for as many as 3.3 million deaths a year.[1,2]

The role of genetic influences on the course of mycobacterial infections has been suspected but never proven, until recently.[3,4] According to Skamene, several lines of evidence suggesting a genetic influence and susceptibility to tuberculosis would include clustering of tuberculosis in certain ethnic groups, families and previously unexposed indigenous populations.[5]

Although an earlier report by Torchia found no evidence of racial difference in susceptibility to tuberculosis, more recent studies by Stead et al have proven otherwise.[6,7] In a study done among 53,000 residents of Arkansas nursing homes, data showed that black residents were more readily infected with tuberculosis compared to the white residents.[7] Among 25,398 residents who were tuberculin negative at the time of their entry into the nursing home, the infection rates were 13.8% and 7.2% in black and white residents, respectively.[7]

Crowle and Elkins investigated the cellular basis for these differences in susceptibility to tuberculosis. Their results showed that macrophages from black subjects had a reduced capacity to restrict the growth of *M. tuberculosis* compared to macrophages derived from white subjects.[8] The tuberculosis bacilli were found to grow consistently and significantly faster in infected macrophages from black donor phagocytes leading them to conclude their greater susceptibility compared to white donors.[8]

Strong evidence supporting the role of genetics and susceptibility to tuberculosis were provided by twin studies. A higher concordance rate of 70-80% was seen among monozygotic twins compared to 30% incidence among dizygotic twins.[9]

Association studies have been done between genetic markers and disease phenotype at the population level in different diseases including tuberculosis. Genetic markers, particularly of the major histocompatibility complex such as HLA-DR and HLA-DQ and its association with racial groups have been extensively studied; however, data showed divergent results.[10–12]

Several animal models of infection have been used as tools in the genetic analysis of host susceptibility or resistance.[13] Lurie and Danenberg in 1965 were the first to implicate the existence of simple genetic control of susceptibility to tuberculous pneumonitis in rabbits. They observed that after inhalation of a measured amount of virulent mycobacterium, there exists a genetic determinant of susceptibility segregating in a selective breeding program in rabbit families which seemed to control a process in the biochemical responses of the alveolar and other tissue macrophages to infection.[14,15]

Skamene and co-workers from the McGill Center for the Study of Host Resistance, identified a dominant autosomal gene called *Bcg* which plays a role in the control of susceptibility to tuberculosis. [5] They observed that inbred mouse strains could be separated into two distinct, non-overlapping groups with respect to their genetic resistance to the tuberculosis bacilli suggestive of the existence of a single locus with two alternative alleles, a resistant allele, which is the dominant one and susceptible allele as recessive.[15–17] These investigators believe that this *Bcg* gene is responsible for a non-specific macrophage activation for bactericidal function.[15] The

Bcg gene was found to map on the proximal portion of mouse chromosome 1.[18]

A candidate gene for *Bcg* known as *Nramp1* was isolated using the positional cloning approach.[19] *Nramp1*, which encodes an integral membrane protein, stands for natural resistance associated macrophage protein. It is expressed in phagocytes and although its function and mechanism of action remain largely unknown, it is suggested that it has a macrophage-specific membrane transport function.[19]

The suggestion that *Nramp1* is most likely the *Bcg* gene came from Vidal and co-workers. They observed that the *Nramp1* gene mapped within the *Bcg* candidate gene region and that Nramp expression is restricted to the reticulo-endothelial organs and macrophages.[19] They also noted that the predicted amino acid sequence of the Nramp protein is similar structurally to a eukaryotic nitrate transport system. [19] Data generated from the analysis of the *Nramp1* mRNA transcript identified a non-conservative gly-105 to asp 105 substitution in all susceptible strains.[19,20]

Experiments on mouse mutants bearing a null allele at *Nramp1* demonstrated that *Nramp1* and *Bcg* were allelic.[21] These mutant mice were susceptible to *M. tuberculosis* and in addition to other intracellular parasites such as *S. typhimurium* and *L. donovani*. [21]

More recent studies are trying to address the unresolved issues concerning *Nramp1* and its role in host resistance. Among these issues would be its unknown biochemical function, how its action affects the intracellular survival of microbes ingested by phagocytes, as well as how it affects the replication of antigenically unrelated microbes.

SMF1, a gene that codes for a hydrophobic protein shares homology with the *Nramp1* gene.[22] A mutation in this gene renders yeast cells sensitive to low manganese concentrations.[22] It was thus proposed that Nramp may also be involved in the transport of manganese from the extracellular milieu into the cytoplasm of a macrophage and after the generation of the phagosome, will remove manganese from the organelle. The depletion of manganese ions limits the ability of the mycobacteria to produce active enzymes and prevent propagation of ingested microorganisms.[22]

What do we gain from the study of genetics and susceptibility to tuberculosis? The molecular identification of these genes regulating host resistance to disease and their respective functions will have far reaching applications. In more developed countries, DNA based tests to assess the risk of infectious disease will be feasible, as well as therapy based on the functions of the genes.[13] In developing countries, such as in the Philippines, we rely heavily on vaccination programs to control infections; the identification of genes regulating host innate and antigen-specific responses are instrumental for vaccine development.

Genetics as it relates to the presence or absence of host resistance or susceptibility alleles will inevitably change public health approach towards prevention and control of infections.

References

1 Flynn JL, Chan J, Triebold KJ *et al.*, An essential role for interferon gamma in resistance to *Mycobacterium tuberculosis* infection. *J Exp Med* 1993; **178**:2249–2254.

2 Kochi A, The global tuberculosis situation and the new control strategy of the World Health Organization. *Tubercle* 1991; **72**:1–6.

3 Stead WW, Genetics and resistance to tuberculosis. *Ann Int Med* 1992; **116**:937–941.

4 Sorensen T, Nielsen GG, Andersen PK, Teasdale TW, Genetic and environmental influences on premature death in adult adoptees. *NEJM* 1988; **318**:727–732.

5 Skamene E, The Bcg gene story. *Immunobiology* 1994; **191**:451–460.

6 Torchia M, Tuberculosis among American Negroes: Medical research on a racial disease, 1830–1950. *J Hist Med Allied Sci* 1977; **32**:252–279.

7 Stead WW, Senner JW, Reddick WT, Lofgren JP, Racial differences in susceptibility to infection by *Mycobacterium tuberculosis*. *NEJM* 1990; **322**:422–427.

8 Crowle A, Elkins N, Relative permissiveness of macrophages from black and white people for virulent tubercle bacilli. *Infect Immun* 1990; **58**: 632–638.

9 Comstock GW, Tuberculosis in twins. *Am Rev of Respir Dis* 1978; **117**:621–624.

10 Selvaraj P, Reetha AM, Uma H *et al.*, Influence of HLA-DR and DQ phenotypes on tuberculin reactive status in pulmonary tuberculosis patients. *Tuber Lung Dis* 1996; **77**:369–373.

11 Pospelov L, Matrakshin AG, Chernousova LN *et al.*, Association of various genetic markers with tuberculosis and other lung diseases in Tuvinian children. *Tuber Lung Dis* 1996; **77**: 77–80.

12 Sanjeevi CB, Narayanan PR, Prabakar R *et al.*, No association or linkage with HLA-DR or DQ genes in South Indians with pulmonary tuberculosis. *Tuber Lung Dis* 1992; **73**:280–284.
13 Buschman Ė, Schurr E, Skamene E. Constitutional resistance to Mycobaterial infection: role of genetic factors in resistance to infection, in C Verduin *et al.*(eds), *Constitutional Resistance to Infection*, R.G. Landes Company, 1995:55–81.
14 Lurie M, Dannenberg A, Macrophage function in infectious disease with inbred rabbits. *Bacteriol Rev* 1965; **29**:466–476.
15 Skamene E, Population and Molecular genetics of susceptibility to tuberculosis. *Clin Invest Med* 1991; **14**:160–166.
16 Forget A, Skamene E, Gros P, Miailhe AC, Turcotte R, Differences in response among inbred mouse strains to infection with small doses of Mycobacterium bovis BCG. *Infect Immun* 1981; **32**:42–47.
17 Gros P, Skamene E, Forget A, Genetic control of natural resistance to Mycobacterium bovis BCG. *J Immunol* 1981; **127**: 2417–2421.
18 Schurr E, Skamene E, Forget A, Gros P, Linkage analysis of the Bcg gene on mouse chromosome 1: identification of a tightly linked marker. *J Immunol* 1989; **142**: 4507–4513.
19 Vidal SM, Malo D, Vogan K, et al. Natural resistance to infection with intracellular parasites: isolation of a candidate gene for Bcg. *Cell* 1993; **73**:469-485.
20 Malo D, Vogan K, Vidal S *et al.*, Haplotype mapping and sequence analysis of the mouse Nramp gene predict susceptibility to infection with intracellular parasites. *Genomics* 1994; **23**:51–61.
21 Vidal S, Tremblay ML, Govoni G *et al.*, The Ity/Lsh/Bcg locus: natural resistance to infection with intracellular parasites is abrogated by disruption of the Nramp 1 gene. *J Exp Med* 1995; **182**:655–666.
22 Supek F, Supekova L, Nelson H, Nelson N, A yeast manganese transporter related to the macrophage protein involved in conferring resistance to mycobacteria. *Proc Natl Acad Sci* 1996; **93**:5105–5110.

FRONTIERS IN HUMAN GENETICS
Diseases and Technologies
© 2001 by World Scientific Publishing Co. Pte. Ltd.

LOSS OF HETEROZYGOSITY OF THE CHROMOSOMAL REGION 11Q22-Q23 IN PRIMARY TUMOURS OF THE CENTRAL NERVOUS SYSTEM

ANN SG LEE*,†,**, TIMOTHY KY LEE‡, SUMANTY TOHARI*, ANGELA PC CHANG*, JIN WANG‡, YW SHEN‡,
TT YEO§ and JAMES KHOO¶

**Department of Clinical Research, Singapore General Hospital; †Division of Medical Sciences, National Cancer Centre, Singapore; ‡Department of Surgery, National University of Singapore;*
§Department of Neurosurgery, Tan Tock Seng Hospital;
¶Neurological Surgery, Mount Elizabeth Medical Centre, Singapore
***E-mail: dmslsg@nccs.com.sg*

Loss of heterozygosity (LOH) of the chromosomal region 11q22-q24 is common in several types of human solid neoplasms, including carcinomas of the breast, lung, colon, nasopharynx, cervix, ovary, uterus and malignant melanoma, suggesting the presence of one or several tumour suppressor genes in this region. To determine if LOH occurs in this chromosomal region in tumours of the central nervous system, we screened for LOH in matched normal and tumour DNA samples from 32 patients with primary tumours of the CNS, of which 20 were benign and 12 were malignant. Nine polymorphic microsatellite loci from the 11q22-q23 region, between D11S940 and D11S934 were analyzed by the polymerase chain reaction (PCR). Six of the 32 cases (18.8%) demonstrated LOH at one or more of the nine loci studied and no significant increase in the frequency of LOH was observed at any particular locus. Of these six cases with LOH, two were benign and four were malignant tumours. Since LOH was detected in both benign and malignant cases, we conclude that LOH at the chromosomal region 11q22-q23 in tumors of the CNS is a random event and may not play an important role in the pathogenesis of these tumours.
Keywords: loss of heterozygosity, microsatellite analysis, chromosome 11q22-q24, central nervous system, carcinogenesis, tumour suppressor genes

** Corresponding author

1. INTRODUCTION

Tumour suppressor genes have been shown to be important in the development of solid tumours. The detection of loss of heterozygosity (LOH) has been used to identify regions in the chromosome which might harbour tumor suppressor genes.

Studies on LOH of different solid tumours such as breast cancer[1], cervical cancer,[2] ovarian cancer,[3] malignant melanoma,[4] stomach adenocarcinoma,[5] lung adenocarcinoma[6] and nasopharyngeal adenocarcinoma[7] have suggested that a putative gene or genes on chromosome 11q22-q24 may be involved in the tumorigenesis of these diverse tumour types. There have been no previous reports on allelic loss of this region in brain tumours.

The aim of this study was to determine if LOH at chromosome 11q22-24 occurs also in tumours of the central nervous system. Microsatellite analysis of this region was done using nine highly polymorphic microsatellite markers.

2. MATERIALS AND METHODS

2.1 Samples

Tumour tissue of the brain or spine were snap-frozen in liquid nitrogen and stored at -70°C. Whole blood collected in EDTA tubes were also collected from each of the patients, and frozen at –70°C.

2.2 DNA Extraction

Cryostat sections were cut from each of the tumours and stained with H&E. DNA was extracted from tumour tissues using DNAzol (GIBCO/BRL) according to the manufacturer's instructions. DNA was extracted from frozen whole blood using sucrose lysis buffer and proteinase K digestion[8].

2.3 Microsatellite Analysis

Nine microsatellite markers from the chromosomal region 11q22-23 were selected: D11S940, D11S1340, D11S29, D11S924, D11S925, D11S1345,

D11S1328, D11S933 and D11S934. Polymerase chain reactions (PCR) were performed with 100-400ng of DNA in a 10μl volume. The sense primer was end-labelled with ^{33}P. The PCR reaction was carried out for 1 min at 94°C, 1 min at 60°C or 67°C, and 1 min at 72°C for 24 cycles. The PCR products were separated on a 8% polyacrylamide sequencing gel and exposed to X-ray film overnight and also exposed to the CS phosphor screens (BioRad) for 4 to 6 hours.

2.4 Assessment of LOH

DNA quantitation was performed using a BioRad Molecular Imager. LOH was determined by quantitating the signal intensity of each allele, and comparing the ratios of the intensity of the alleles from the tumour DNA with that of the constitutional (blood) DNA. A difference of over 30% between these ratios was scored as LOH. Samples with LOH were reanalyzed to confirm the result.

3. RESULTS

A total of 32 tumours from the central nervous system were analysed by microsatellite analysis at nine loci on chromosome 11q (Figure 1). Twenty were benign and comprised of 14 meningiomas, three neurofibromas and three pituitary tumours. Twelve tumours were malignant of which there were four astrocytomas, three gliomas, two oligodendrogiomas, one primitive neuroectodermal tumour (PNET), one ependymoma and one glioblastoma.

Representative examples of LOH are shown in Figure 2.

Table 1 shows the frequency of LOH at each of the nine microsatellite loci. Frequency of LOH ranged from 3.6% at D11S29 to 11.8% at D11S924.

Six of the 32 (18.8%) cases demonstrated LOH on at least one of the nine loci studied. Of these six cases with LOH, four tumours were malignant and two were benign (Table 2). Table 2 shows the results of microsatellite analysis for these six cases.

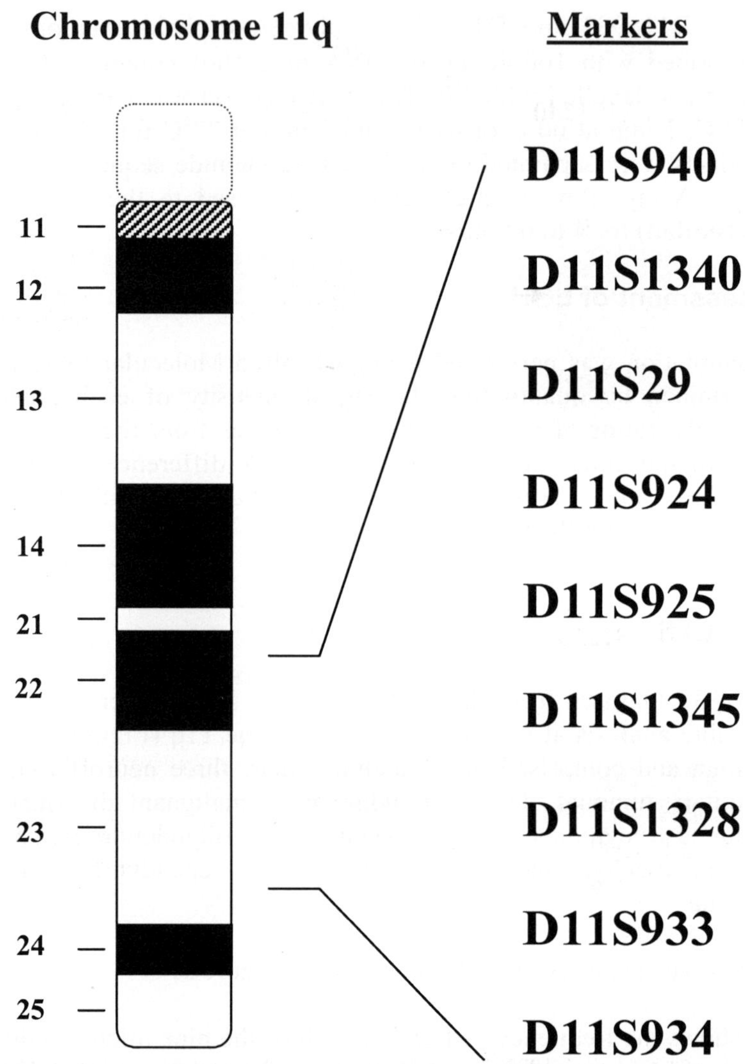

Fig. 1 LOH on chromosome 11.

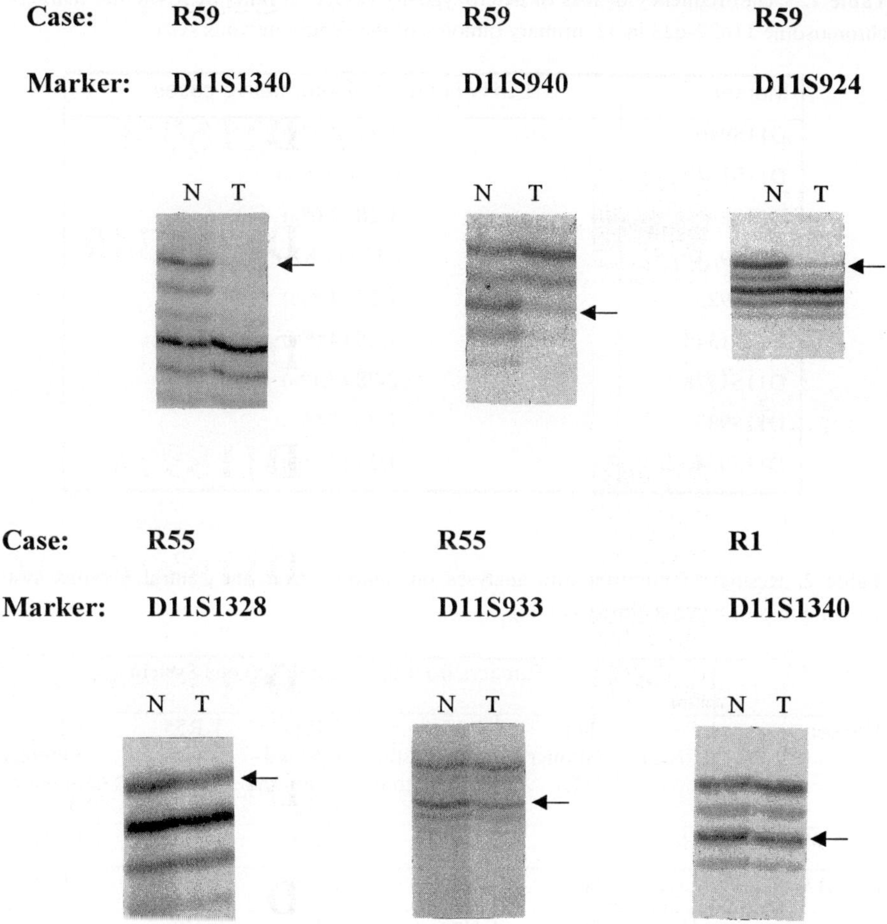

Fig. 2 Representative examples of LOH on chromosome 11q in primary tumours of the central nervous system. Normal DNA (N) and tumour DNA (T) from the same patient are enclosed within the same box. Arrows indicate the alleles showing loss.

Table 1. The frequency of loss of heterozygosity (LOH) at nine microsatellite markers on chromosome 11q22-q23 in 32 primary tumours of the central nervous system.

Marker	LOH / Number of informative cases
D11S940	1/23 (4.3%)
D11S1340	3/27 (11.1%)
D11S29	1/28 (3.6%)
D11S924	2/17 (11.8%)
D11S925	1/21 (4.8%)
D11S1345	1/25 (4.0%)
D11S1328	2/28 (7.1%)
D11S933	1/20 (5.0%)
D11S934	1/23 (4.3%)

Table 2. Results of microsatellite analyses on tumours from the central nervous system showing loss of heterozygosity (LOH).

	Tumours from the Central Nervous System					
Marker	R1 PNET (M)	R11 Glioma (M)	R15 Anaplastic astrocytoma (M)	R49 Neuro-fibroma (B)	R55 Glioma (M)	R59 Pituitary tumour (B)
D11S940	He	He	He	Ho	Ho	LOH
D11S1340	LOH	He	He	LOH	Ho	LOH
D11S29	He	He	He	He	Ho	LOH
D11S924	He	Ho	LOH	He	Ho	LOH
D11S925	He	LOH	Ho	He	Ho	Ho
D11S1345	He	Ho	LOH	He	He	Ho
D11S1328	He	He	Ho	He	LOH	LOH
D11S933	He	Ho	He	Ho	LOH	Ho
D11S934	He	He	Ho	He	Ho	LOH

M = malignant; B = benign; He = heterozygous; Ho = homozygous

4. DISCUSSION

We have studied the chromosomal region 11q22-q23 for LOH in tumours of the central nervous system. Although LOH at this chromosomal region is common in other tumours, the frequency of LOH in tumours of the CNS is low. In breast cancer, allelic loss occurred at a high frequency (59%) at D11S29[1], but was only 3.6% in this present study on tumours of the central nervous system. In this present study, we have found an overall frequency of LOH in 6 of 32 (18.8%) of the tumours from the CNS. In contrast, high overall frequencies of between 30% and 63% have been reported in studies on breast cancer, cervical cancer, ovarian cancer, stomach adenocarcinoma, lung adenocarcinoma and nasopharyngeal carcinoma.[1–7]

The low frequencies found in this present study suggest that a different process of tumorigenesis occurs in tumours of the CNS. One key difference between these tumours and those with high frequencies of LOH is the origin of the cell type. There are no epithelial cells in the CNS, which instead has three main structural elements: the neurones, specialised supporting cells (astrocytes, oligodendrocytes, ependymal cells and microglia) and connective tissue.

Since LOH was detected in both benign and malignant tumors and only in a small proportion of these tumors, we conclude that LOH at the chromosomal region 11q22-q23 in primary tumors of the central nervous system is a random event and may not play an important role in the pathogenesis of these tumors.

References

1. Tomlinson IPM, Stickland JE, Lee ASG, Bromley L, Evans MF, Morton J and McGee JO'D, Loss of heterozygosity on chromosome 11q in breast cancer, *J Clin Pathol* 1995; **48:** 424–428.
2. Hampton GM, Penny LA, Baergen RN, Larson A, Brewer C, Liao S, Busby-Earle RM C, Williams AWR, Steel CM, Bird CC, Stanbridge EJ and Evans GA, Loss of heterozygosity in cervical carcinoma: Subchromosomal localization of a putative tumor-suppressor gene to chromosome 11q22-q24, *Proc Natl Acad Sci USA* 1994; **91:** 6953–6957.

3. Gabra H, Watson JEV, Taylor KJ, Mackay J, Leonard RCF, Steel CM, Porteous DJ, and Smyth JF, Definition and refinement of a region of loss of heterozygosity at 11q23.3-q24.3 in epithelial ovarian cancer associated with poor prognosis, *Cancer Res* 1996; **56:** 950–954.
4. Robertson G, Coleman A and Lugo TG, A malignant melanoma tumor suppressor on human chromosome 11, *Cancer Res* 1996; **56:** 4487–4492.
5. Baffa R, Negrini M, Mandes B, Rugge M, Ranzani GN, Hirohashi S and Croce CM, Loss of heterozygosity of chromosome 11 in adenocarcinoma of the stomach, *Cancer Res* 1996; **56:** 268–272.
6. Rasio D, Negrini N, Manenti G, Dragani TA and Croce CM, Loss of heterozygosity at chromosome 11q in lung adenocarcinoma: Identification of three independent regions, *Cancer Res* 1995; **55:** 3988–3991.
7. Hui ABY, Lo KW, Leung SF, Choi PHK, Fong Y, Lee JCK and Huang DP, Loss of heterozygosity on the long arm of chromosome 11 in nasopharyngeal carcinoma, *Cancer Res* 1996; **56:** 3225–3229.
8. Miller SA, Dykes DD and Polesky HF, A simple salting out procedure for extracting DNA from human nucleated cells, *Nucl Acids Res* 1988; **16:** 1215.

FRONTIERS IN HUMAN GENETICS
Diseases and Technologies
© 2001 by World Scientific Publishing Co. Pte. Ltd.

IDENTIFICATION OF *RB1* GENE MUTATIONS WITH CONSTITUTIONAL ORIGIN

R RONG*, YFL LING †, AEJ YEOH*, TC QUAH*, PS LOW* and PS LAI*,‡
Department of Paediatrics, National University of Singapore
†*Singapore National Eye Centre, Singapore*

Retinoblastoma is an embryonic neoplasm of retinal origin. Tumour development in retinoblastoma is initiated by the somatic inactivation of both alleles of the *RB1* gene. It is predicted that 40% of all cases of retinoblastoma are hereditary with all bilateral and 10-15% of unilateral cases being constitutional in origin. In this study, the incidence of somatic loss of heterozygosity (LOH) as a mutational event in retinoblastoma was studied using four intragenic DNA markers. These include a microsatellite polymorphism, RB1.20 and four restriction fragment length polymorphisms (RFLPs) namely, intron 1/*Bam*HI, intron 17/*Xba*I, intron 24/*Tth*111I and intron 25/*Dra*I. LOH was found in 17 out of 27 (63%) tumours analysed. A higher frequency of maternal loss compared to paternal loss was observed in bilateral cases (75%) as compared to unilateral cases (54%). Direct exon-by-exon sequencing revealed mutations in 15 out of 17 (88%) tumours with LOH. In three out of ten tumours without LOH, one mutation was detected and two mutations were found in five tumours without LOH. Seven of the mutations were found to be constitutional (germline) in origin with three of these mutations occurring in unilateral cases of retinoblastoma. The presence of constitutional mutations did not correlate with early age of tumour development. Our results show that there was no association between the types of mutation and constitutional origin of the mutation.

Keywords: retinoblastoma, *RB1*, tumour, LOH, mutation, constitutional

‡ Corresponding author

1 INTRODUCTION

Retinoblastoma is an embryonic neoplasm of retinal origin with an incidence of 1/15,000-20,000 births.[1,2,3] Two-thirds of these tumours are unilateral, of which 10-15% are hereditary and the remaining one third are bilateral (hereditary) cases.

This disease is caused by the loss of function of both copies of the retinoblastoma susceptibility gene, *RB1*, which is located on chromosome 13q14.11. The *RB1* gene is 200 kb in size and comprises of 27 exons.[5] It encodes a cell cycle regulatory protein. The analysis of the gene structure and expression in retinoblastoma tumours has confirmed that *RB1* is an anti-oncogene.[6,7] In the tumorigenesis of retinoblastoma and a number of other cancers, two mutation events are required as proposed by Knudson's two-hit model.[8] Hereditary retinoblastoma occurs due to a combination of a constitutional anomaly of the *RB1* gene and a second mutation occurring in the retinal cells. In non-hereditary retinoblastoma, both mutations are somatic in origin. Hereditary retinoblastoma patients can transmit the constitutional mutation to their offspring and they also stand a higher chance of developing secondary tumours as compared to non-hereditary patients. Phenotypically, it is difficult to differentiate between hereditary and non-hereditary forms of retinoblastoma. Identification of constitutional mutations is important for genetic counselling in affected families.

Attempts to define the mutations resulting in the inactivation of the *RB1* gene have led to the identification of different types of alterations. These include constitutional cytogenetic rearrangements, gross molecular deletions and subtle mutations (single point mutations, short inversions or deletions).[9,10] It has been reported that 70% of second mutations occurring as somatic events involve the loss of a normal allele. These are detected using analysis of polymorphic intragenic markers as loss of heterozygosity (LOH) in the allele.[11,12] This study reports the results of the analysis of *RB1* gene mutations with constitutional origin in 27 retinoblastoma patients.

2 MATERIALS AND METHODS

2.1 Intragenic LOH Analysis

To determine the frequency and nature of *RB1* gene mutations in both unilateral and bilateral retinoblastoma, we analyzed DNA from peripheral blood and from tumor tissue samples. Out of the 27 patients in our study, 21 had unilateral retinoblastoma while six had bilateral retinoblastoma.

Amplification of DNA by polymerase chain reaction (PCR) was performed for four introns (intron 1/*Bam*HI, intron 17/*Xba*I, intron 24/*Tth*111I and intron 25/DraI) in 50 µl reaction volumes containing 200 ng DNA, 50 pmol of each of the forward and reverse primers, 1x *Taq* polymerase buffer (500 mM KCl, 100 mM Tris-HCl, 1% Triton X-100), 0.4 mM of each of the dNTPs and 1 unit of *Taq* polymerase, using standard amplification conditions.[9] The primer sequences and PCR conditions have been previously described. The PCR products were digested by the appropriate restriction endonucleases, using the conditions specified by the manufacturers and resolved by gel electrophoresis. The microsatellite marker, RB1.20 consists of variable number of [CTTT(T)] repeats. The 5' primer flanking this region was labelled with {γ-P^{33}} ATP before carrying out PCR using similar conditions as above. The products were resolved on 6% sequencing gel and detected by autoradiography at –70°C.

2.2 DNA Sequencing Analysis

PCR amplifications were performed for all 27 exons and the promoter region of *RB1* gene in 50 µl reaction volumes containing 400 ng DNA, 100 pmol of each of the forward and reverse primers, 1x *Taq* polymerase buffer (500 mM KCl, 100 mM Tris-HCl, 1% Triton X-100), 0.2 mM of each of the dNTPs and 1 unit of *Taq* polymerase, using the primer sequence and amplification conditions previously described.[10] The PCR products were purified by standard phenol chloroform method and used as templates for cycle sequencing reactions using Bigdye Ready Reaction Kit (ABI Prism).

The cycle sequencing reaction was performed on an oil-free temperature cycler (Hybaid PCRExpress) in 20 μl volume containing 30 to 90 ng of each purified PCR product, 8 μl of Bigdye Terminator Ready Mix (Applied Biosystems), 4.2 pmol of primer. The reaction conditions were 25 cycles of 30 seconds of denaturation at 96°C, 15 seconds of annealing at 50°C and 4 minutes of extension at 60°C.

Each reaction mix was precipitated by adding 80 μl of 75% isopropanol and centrifuged for 20 minutes at 14000rpm. After washing with 75% isopropanol, the pellet was dissolved in 6 μl of sequencing loading buffer (ABI Prism) and subjected to polyacrylamide gel electrophoresis for sequencing analysis on the ABI 377 sequencer.

3 RESULTS

3.1 LOH Study

In order to detect the loss of heterozygosity in *RB1* gene, a panel of five polymorphic markers, consisting of four restriction fragment length polymorphisms (RFLPs) and one microsatellite polymorphism was used to analyse 27 cases of retinoblastoma tumours. The first four affect restriction sites in introns 1, 17, 24 and 25, respectively, while the last is a four-to-five-base-repeat polymorphism occurring in intron 20. Presence of LOH was detected by the absence of the allele in the tumour DNA as compared to lymphocyte DNA from patients (Fig. 1).

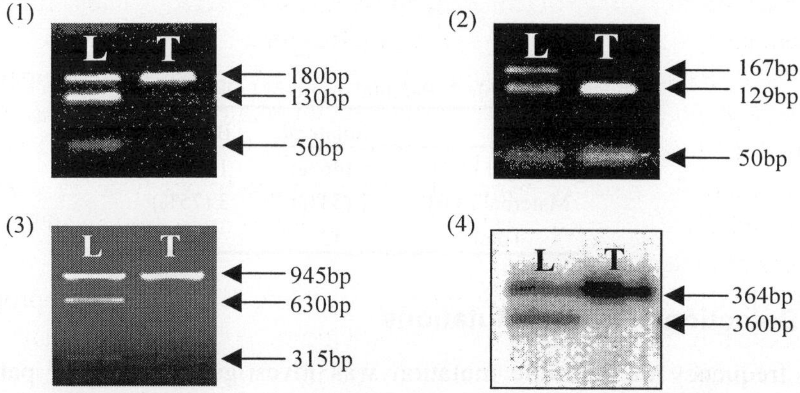

Fig. 1 Heterozygosity at the retinoblastoma (*RB1*) locus at polymorphic regions of (1) intron 1, (2) intron 17, (3) intron 25, (4) intron 20. Loss of one allele is seen in tumour DNA (T) as compared to lymphocyte DNA (L).

In all 27 cases, the blood lymphocyte DNA was informative for at least one marker. LOH was observed in a total of seventeen cases, representing an incidence of 63% (Table 1). In the determination of paternal origin of LOH, loss of maternal alleles was observed in seven unilateral tumours with the remaining six unilateral cases showing loss of paternal alleles (Table 2). There was no significant difference in LOH observed in 13 out of 21 (62%) unilateral cases and four out of six bilateral cases (66%). Thus, there was no significant difference in incidence of LOH between the two types of retinoblastoma. For the bilateral cases, maternal loss was found in three tumours and paternal loss in one tumour. The differences in maternal loss for both unilateral and bilateral also do not appear to be statistically significant ($P > 0.05$).

Table 1 Incidence of LOH in the 27 cases analysed

	Unilateral	Bilateral	
LOH present	13	4	17 (63%)
No LOH	8	2	10 (37%)
Total	21	6	27 (100%)

Table 2 Parental origin of LOH

	Unilateral	Bilateral
Paternal LOH	6 (46%)	1 (25%)
Maternal LOH	7 (54%)	3 (75%)
Total	13	4

3.2 Detection of Subtle Mutations

The frequency of germline mutation was investigated in all 27 patients. Mutations could be found in tumour and blood samples confirming their constitutional origin (Fig. 2). Of the 17 cases of retinoblastoma with LOH, the second mutation could be identified in 15 tumours (Table 3). In the remaining cases without LOH, there were five cases in which both mutations were detected, three cases in which only one mutation was detected and two cases in which no mutation could be detected. Nine out of the total of 23 subtle mutations identified were novel.

Table 3 Number of mutations identified in the 27 tumours

	LOH	No LOH
Two mutations	–*	5
One mutation	15	3
No mutation (unidentified)	2	2
Total	17	10

*For LOH cases, only one other mutation is required to trigger tumorigenesis while for non-LOH cases, two mutations should be identified. The mutations found were either point mutations, splice site errors or small base insertions or deletions.

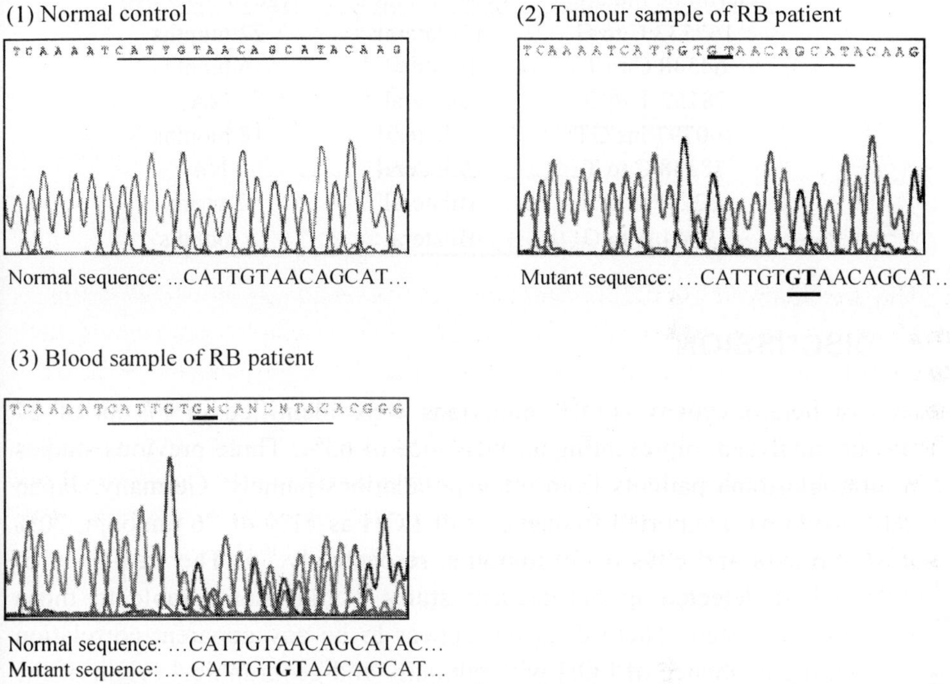

Fig. 2 Homozygous 160797insGT mutation was found in the tumour sample from a RB patient. Heterozygous 160797insGT mutation was found in the blood sample from the same patient.

Out of the total of 23 patients with subtle mutations identified by sequencing, seven had mutations which were constitutional in origin with three of these mutations originating from unilateral cases of retinoblastoma (Table 4).

Table 4 *RB1* gene mutations with constitutional origin that were identified and the age of diagnosis of RB in the patients.

Germline mutation	Type of tumour	Age of diagnosis
162333 C to G	Unilateral	22 months
64348 C to T	Bilateral	16 months
78152 T to G	Bilateral	NA
160797 ins GT	Unilateral	18 months
78238 C to T	Unilateral	NA
162237 C to T	Bilateral	2 years
39444 A to G	Bilateral	3 months

4 DISCUSSION

Loss of heterozygosity (LOH) mutations were identified in 17 out of 27 tumours analysed, representing an incidence of 63%. Three previous studies in retinoblastoma patients from other populations namely, Germany, Japan and Canada have reported frequencies of LOH as 71% of 76 tumours, 70% of 46 tumours and 63% of 30 tumours, respectively.[10,12] The frequency of LOH (63%) detected in our present study is thus comparable to those previously reported. There does not appear to be any apparent correlation between the incidence of LOH with ethnicity and geographical origin of the patients.

It has been previously reported that bilateral cases show a higher incidence of maternal loss as compared to unilateral cases.[13] The higher frequency of loss of maternal allele in bilateral cases would suggest that the first predisposing germline mutations occur more frequently on the paternal alleles. However, the frequency of maternal loss to paternal loss in our bilateral cases is not significantly different due to the small sample size. The mutations in the majority of the bilateral cases analysed in our study appear to be sporadic as most of the parents did not carry these mutations. Therefore, most of the germline mutations in these bilateral cases are due to new mutations occurring in germ cells.

A total of seven constitutional mutations were identified in this study. Four of these were from the 15 cases with LOH and the remaining three from cases with either no detectable LOH or in which no tumour sample was available for LOH study. Among the somatic and germline mutations observed, nine are novel mutations involving small deletions/insertions and point substitutions. No region of the *RB1* gene was found to be preferentially affected by these novel mutations. In the 17 tumours with LOH, the second mutation could be identified in 15 cases. In the ten tumours without LOH, both mutations could be identified in only five cases. In the remaining five cases, only one mutation could be detected in three tumours and no mutation could be identified in two tumours. Hence, in the latter, both mutational events triggering tumorigenesis could not be identified. This could be due to presence of mutations occurring in promoter or regulatory regions outside the sequences screened, or due to inactivation of the gene by other unknown mechanisms such as methylation, epigenic effects, transcriptional errors, interactions with other genes (developmental, regulatory, oncogenes), or aberrant mature mRNA processing, etc.

The information on ages at diagnosis was available from our medical records in five out of the seven patients with germline mutations. With the exception of one bilateral case, all of the five patients were diagnosed before 22 months of age, which is closely similar to those reported in other populations.[1,2] These results further confirm the observation that the presence of the constitutional mutations is associated with the early development of tumour in retinoblastoma patients. Patients carrying constitutional mutations are at higher risk of secondary tumours and they can transmit the mutations to their offsprings. Therefore, identification of *RB1* mutations with constitutional origin plays an important role in the genetic counselling of retinoblastoma, especially for unilateral cases. It was found in this study that three (42.8%) out of the seven cases with constitutional mutations were from patients with unilateral tumours. The high percentage of constitutional mutations in unilateral tumours further emphasizes the importance of identification of such mutations in the management of patients with unilateral retinoblastoma.

References

1. Zimmerman LE, Retinoblastoma and retinocytoma. In: Spencer WH, ed. Ophthalmic pathology; an atlas and textbook, Vol.2. Philadelphia: W.B. Saunders, 1985: 1292-351.
2. Abramson DH, Servodidio CA, Retinoblastoma, *Optom Clin* 1993; **3**(3): 49-61.
3. Vogel F. Genetics of retinoblastoma. In: Human Genetics: Problems and Approaches, New York: Springer-Verlag; 1979 52: 1-54.
4. Mateu E *et al.*, Genetics of retinoblastoma: a study, *Cancer Genet Cytogenet* 1997; **95**(1): 40-50.
5. Toguchida J *et al.*, Complete genomic sequence of the human retinoblastoma susceptibility gene, *Genomics* 1993; **17**(3): 535-43.
6. Dryja TP *et al.*, Molecular detection of deletions involving band q14 of chromosome 13 in retinoblastomas, *Proc Natl Acad Sci USA* 1986; **83**: 7391-4.
7. Knudson AG Jr, Mutation and cancer: statistical study of retinoblastoma, *Proc Natl Acad Sci U S A* 1971; **68**(4): 820-3.
8. Lai PS and Tay JSH, Wilm's tumour In: Molecular Biology in Cancer Medicine 2nd edition, 1999 Pg531-547 eds. Kurzrock R and Talpaz M. Martin Dunitz, London.
9. Lohmann DR *et al.*, The spectrum of RB1 germ-line mutations in hereditary retinoblastoma. *Am J Hum Genet* 1996; **58**(5): 940-9.
10. Lohmann DR *et al.*, Constitutional RB1-gene mutations in patients with isolated unilateral retinoblastoma. *Am J Hum Genet* 1997; **61**(2): 282-94.
11. Cavenee WK, Dryja TP, Phillips RA, Benedict WF, Godbout R, Gallie BL, Murphree AL, Strong LC and White RC, Expression of recessive alleles by chromosomal mechanisms in retinoblastoma, *Nature* (Lond) 1983; **305**: 779-784.

12. Kato VM, Ishizaki K, Shimizu T, Ejima Y, Tanooka H, Takayama J, Kaneko A, Togichida J, Sasaki M, Parental origin of germ-line and somatic mutations in the retinoblastoma gene, *Hum Genet* 1994; **94**: 31-38.
13. Zhu X, Dunn JM, Goddard AD, Squire JA, Becker A, Philips RA, Gallie BL, Mechanisms of loss of heterozygosity in retinoblasotma, *Cytogenet Cell Genet* 1992; **59**: 248-252.

FRONTIERS IN HUMAN GENETICS
Diseases and Technologies
© 2001 by World Scientific Publishing Co. Pte. Ltd.

MOLECULAR AND CLINICAL PROFILES OF SINGAPORE FAMILIAL ADENOMATOUS POLYPOSIS PATIENTS

XIA CAO, KONG WENG EU, FRANCIS SEOW-CHOEN and PEH YEAN CHEAH *
Department of Colorectal Surgery, Singapore General Hospital, Singapore 169608
**E-mail*: *gcscpy@sgh.com.sg*

Familial adenomatous polyposis (FAP) is a familial form of colon cancer caused by mutation of the adenomatous polyposis coli (*APC*) gene. We investigated the *APC* mutation and phenotypic spectrum in 172 members of 36 Singapore FAP families. The protein truncation test (PTT) and DNA sequencing were used to screen the entire *APC* coding region for germline mutations. *APC* mutations were found in 28 families (78 %). 65 patients tested positive while 63 non-affected members tested negative. The correlation of PTT to clinical diagnosis is therefore 100%, suggesting that PTT is a highly reliable presymptomatic test for FAP. Twenty different *APC* mutations were identified, eleven of those were novel. All mutations, except one, resulted in the classical colonic phenotype. Interestingly, mutation at codon 332 resulted in attenuated FAP with left-sided predominance of polyps rather than the right. For the eight families without *APC* mutations, we screened for *β-catenin* mutation which was shown to be able to substitute for *APC* mutation in sporadic colorectal cancer. No germline *β-catenin* mutation was found. Further analysis reveals atypical clinical features such as the co-existence of adenomatous and hyperplastic polyps and other non-FAP associated cancers in these patients. Our results suggest the involvement of other genes and possibly new variants for the polyposis syndrome.
Keywords: *APC*, *β-catenin*, FAP, PTT, germline mutation

1. INTRODUCTION

Familial adenomatous polyposis (FAP) is an autosomal dominantly inherited form of colorectal cancer (CRC) characterized by the appearance of hundreds to thousands of adenomatous polyps in the colon and rectum of an affected individual by the second or third decade of his life. The risk that one

* Corresponding author

of these polyps will progress to cancer in a person's lifetime is virtually 100% if the polyps are not detected and removed in time. Further, FAP patients often develop extracolonic manifestations such as osteomas, desmoids, retinal lesions, brain and thyroid tumors.[1]

The adenomatous polyposis coli (*APC*) gene at chromosome band 5q21 has been implicated in the pathogenesis of FAP.[2–4] The *APC* gene spans about 100 kilobases of genomic region. It contains 16 exons, one noncoding and 15 coding exons. The coding sequence is 8.5 kilobases long and codes for a large protein that comprises several functional domains including the binding domains for β-catenin, axin, EB1, human discs large protein (DLG) and microtubule.[5] More than 98% of the *APC* mutations reported to date are translation terminating mutations resulting in truncated proteins.[6] Direct genetic test for FAP became feasible with the development of the protein truncation test (PTT)[7–8] which combines polymerase chain reaction (PCR) and *in vitro* transcription and translation techniques to detect translation terminating mutations. Unlike linkage study, no detailed family history is required to carry out the test.

Several studies have demonstrated that the inactivation of APC could lead to the transcription of genes mediated by the β-catenin/T-cell transcription factor (Tcf) pathway. A majority of the *APC* mutations lack the β-catenin binding domain.[9] The ability of APC to form a multimolecular complex with β-catenin and glycogen synthase kinase 3β (GSK-3β) is crucial to its ability to degrade β-catenin, probably as a consequence of the phosphorylation of the amino terminal of β-catenin by GSK-3β.[10] Colon carcinoma cells lacking the wildtype APC protein were shown to contain constitutively active β-catenin-human/Tcf complex, which was abrogated upon introduction of APC.[11] A recent study has shown that about half of the sporadic CRC lacking *APC* mutation has *β-catenin* mutations, involving either the GSK-3β phosphorylation sites or interstitial deletion of exon 3, suggesting that *β-catenin* mutation can substitute for *APC* mutation in the initiation of sporadic colorectal tumorigenesis.[12] The role of *β-catenin* mutation in the initiation of FAP, however, is unclear.

Although the *APC* gene has been extensively studied in the Caucasian population, it has not been previously described in the Chinese population. In this study, we used the PTT to screen the entire coding region of the *APC* gene for germline mutation in 36 unrelated, predominantly Chinese, FAP families from the Singapore Polyposis Registry and found mutation in 28 families. For the remaining families, we screened for mutation in the GSK-

3β phosphorylation sites and interstitial deletion of exon 3 of the *β-catenin* gene. No germline *β-catenin* mutation was found in the 8 families.

2. MATERIALS AND METHODS

2.1 Sample Selection

Blood was collected over the past ten years from 172 members of 36 unrelated FAP families. Pedigrees and clinical data for these families were retrieved from the Singapore Polyposis Registry in the Department of Colorectal Surgery, Singapore General Hospital, Singapore 169608.

2.2 DNA/RNA Extraction

DNA was obtained from peripheral blood lymphocytes by a simple salting out procedure.[12] RNA extraction was performed from fresh blood lysates using the Promega Total RNA extraction kit according to the manufacturer's protocol.

2.3 PTT Assay

The *APC* coding region of the human gene (Genebank accession number M73548) was amplified by 6 overlapping PCR/RT-PCR segments corresponding to nucleotides 1-1800 (Segment 1.1); 1041-2514 (Segment 1.2); 2056-3831 (Segment 2); 3297-5410 (Segment 3); 4783-7011(Segment 4) and 6364-8543 (Segment 5) as previously described.[14] The unpurified PCR product was *in vitro* transcribed and translated into protein with the TNT/T7 coupled reticulocyte lysate system (Promega).

2.4 Sequence Analysis

The size of the truncated protein was used to estimate where the mutation was. The relevant PCR fragment was identified and primers made for PCR-based sequencing using Perkin-Elmer automated ABI™ sequencer.

2.5 *β-catenin* Mutation Screening

β-catenin exons 2 to 4 from genomic DNA was amplified using the primers C-F (5'-CCAGCGTGGACAATGGCTAG-3') and C-R (5'-TGAGCTCG AGTCATTGCATAC-3')[15] corresponding to sequences in exons 2 and 4, respectively, to detect interstitial deletion involving exon 3. The products were separated on 1% agarose gel. Similarly amplified RT-PCR products were sequenced for missense mutations at the serine/threonine residues in exon 3.

3. RESULTS AND DISCUSSION

3.1 Correlation of PTT with Clinical Diagnosis

We used the PTT to screen the entire *APC* gene for germline mutation in 36 unrelated FAP families and found mutation in 28 families.

The detailed analysis of 128 members of these 28 FAP families whose blood samples were available show that the mutated *APC* gene was found in all 65 FAP patients. In fact, truncated proteins were found in some young at-risk individuals and subsequent colonoscopy confirmed the presence of polyps. None of the 63 individuals clinically diagnosed as unaffected members have the truncated protein. There was thus no false positive or false negative case in the PTT analysis of the 128 members. In other words, there is a 100% correlation between clinical diagnosis by colonoscopy and molecular diagnosis by PTT. Fig. 1 shows representative PTT results for two of the FAP families.

In contrast to clinical diagnosis which can only determine post-symptomatically who amongst the members carry the defective gene, PTT can detect FAP carriers presymptomatically. Thus, patients can be better counseled and prophylactic colectomy timed to ensure compliance. For unaffected members of the family, the relief from emotional turmoil and stress cannot be measured in monetary terms while the cost saved from unnecessary repeated colonoscopy is highly significant.[16]

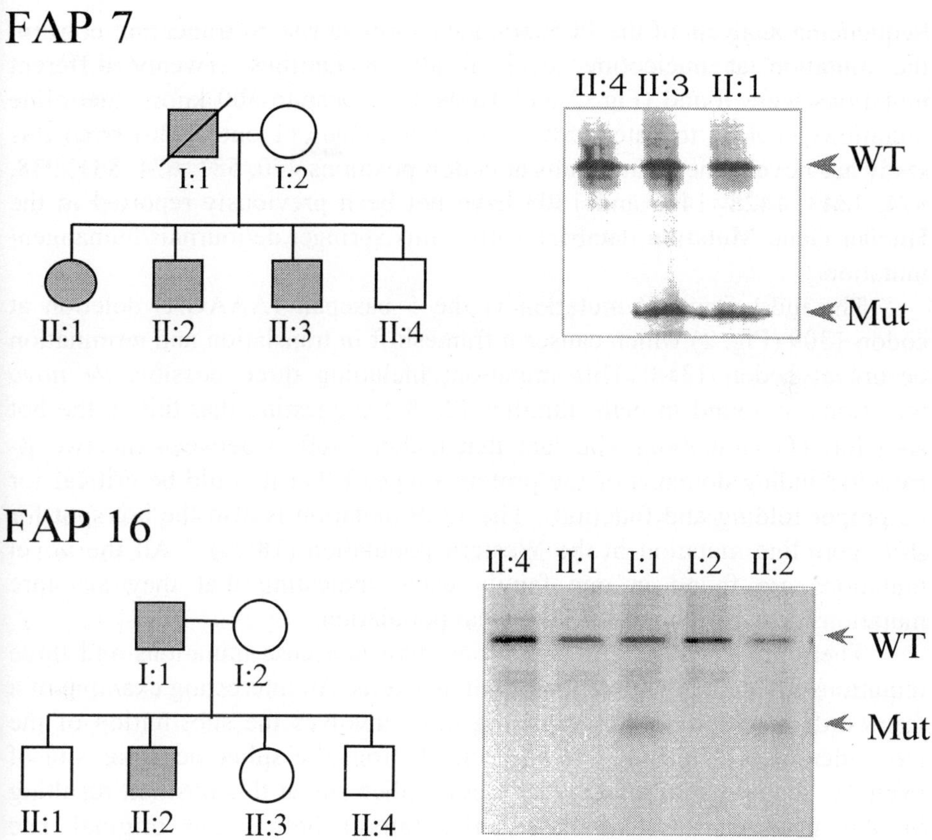

Fig. 1 PTT results from FAP families 7 and 16. PTT was carried out whenever blood was available. The full-length protein (WT) is found in all members of the family. FAP patients from the same family have truncated proteins (Mut) of the same size and unaffected members have no truncated protein. In the pedigrees, FAP patients and unaffected members of the family as diagnosed by colonoscopy are represented by filled and unfilled symbols respectively.

3.2 *APC* Germline Mutations

Sequencing analysis of the PCR fragments giving rise to truncation confirm the mutation at nucleotide level in all 28 families. Twenty different mutations were found (Fig. 2 and Table 1).[17] Despite 300 known germline mutations in *APC* to date, 55% of the mutations (11 out of 20) from this study are novel. These mutations at codon positions 470, 581, 674, 831, 938, 974, 1343, 1428, 1465 and 1503 have not been previously reported in the Human Gene Mutation database (http://link.springer.de/journals/humangen/mutation/).

The most frequent mutation is the 5 basepair (AAAGA) deletion at codon 1309 (Fig. 2) which causes a frameshift in translation and termination occurs at codon 1313. This mutation, including three possible *de novo* mutations, is found in eight families (22 %) suggesting that this is the hot spot for *APC* mutation. The fact that codon 1309 is between the two β-catenin binding domains of the protein suggests that it could be critical for its proper folding and function. The 1309 mutation is also the hot spot for *APC* germline mutation in the Western population (18 %).[18] All the novel mutations are found in one family each, indicating that they are rare mutations possibly endemic to the local population.

There are 12 frameshift mutations, five nonsense mutations and three mutations involving the deletion of whole exons. An interesting example of a frameshift mutation creating splicing error involves the substitution of the base adenine with guanine at position –6 from the splice acceptor site of exon 11. This substitution created a new splice site at this position resulting in the insertion of five base pairs (attag) before the original site (tttaaattag/GGG → tttag/attagGGG) and premature termination of the reading frame at codon 470. Two deletions of the whole of exons 11 and 14 were found to be large genomic rearrangements involving introns 10–11 and introns 13–14 respectively.

3.3 Genotype-Phenotype Correlation

The severity and extracolonic manifestations of FAP have been correlated with the mutation sites. Mutations between codons 1250 and 1464 of APC are associated with increased number of polyps.[19] Mutation in the

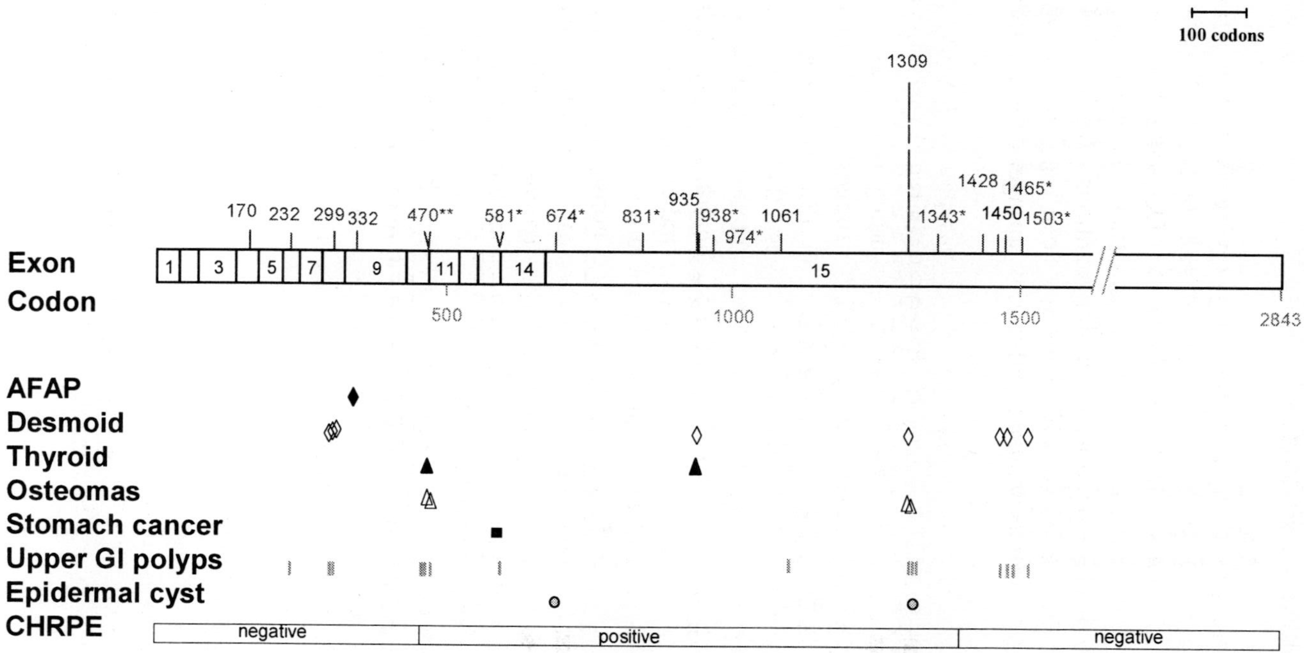

Fig. 2: Distribution of *APC* germline mutations and phenotypic spectrum in 28 Singapore FAP families. The length of the bar indicates the number of mutations at the indicated position. Novel mutation is marked by an asterisk (*). Extracolonic manifestations associated with each mutation is indicated under the mutation site. Positive and negative CHRPE domains are indicated by the horizontal boxes. Figure adapted from Fig. 2 of Ref. 17.

Table 1: APC germline mutations and phenotypic spectrum of FAP patients (adapted from Ref. 17)

Exon	Position	Mutation	FAP Patient	APC Mutation	Age at Diagnosis	Colonic Manifestation Polyps*	Colonic Manifestation Cancer**	Extracolonic Manifestation CHRPE[@]	Extracolonic Manifestation Others
4	170	ATAGATAGTC→ ATAGTC	45 II:1	N.D.	49	N.A.	+(D)	N.D.	
			45 II:2	+	39	+		N.D.	
6	232	CGA→TGA	9 I:1	N.D.	60	N.A.	+ (N.A.)	N.D.	
			9 II:1	+	44	+	+ (D)	N.D.	
			9 II:2	N.D.	29	++		–	
			9 II:3	N.D.	27	+	+(A)	–	duodenum fundic glandular polyps
8	299	ACTCTG→ACTG	14 II:1	+	32	++	+(C)	–	desmoid, fundic glandular polyps
			14 II:2	+	31	++	+(A)	–	desmoid
			14 II:4	+	26	++		–	desmoid, fundic glandular polyps
			14 III:3	+	14	+		–	
			14 III:6	+	10	+		–	
			14 III:7	+	12	N.A.		–	
9	332	CGA→TGA	28 I:1	N.D.	66	<100	+ (C)	N.D.	
			28 II:1	+	49	<100	+ (C)	–	
11	I10-I11[#]	Delete E11 (2 kb)	4 I:1	N.D.	39	N.A.	+ (N.A.)	N.D.	
			4 II:1	N.D.	47	N.A.	+ (C)	N.D.	
			4 II:3	N.D.	36	N.A.	+	N.D.	
			4 II:4	N.D.	41	m		N.D.	
			4 II:6	N.D.	34	N.A.	+ (D)	N.D.	
			4 II:7	N.D.	37	+		N.D.	
			4 II:8	N.D.	29	N.A.	+ (D)	N.D.	
			4 III:6	+	21	N.A.		N.D.	
			4 III:7	N.D.	20	N.A.		N.D.	duodenum fundic glandular polyps
			4 III:8	+	16	++		+	fundic glandular polyps
			4 III:9	N.D.	21	++		+	fundic glandular polyps
			4 III:10	N.D.	17	N.A.		+	
			4 III:12	+	20	m		+	
			4 III:15	+	20	+	+ (A)	+	
			4 III:16	N.D.	17	N.A.		+	
			4 III:17	+	15	N.A.		+	thyroid cancer

Table 1 *(continued)*

Exon	Position	Mutation	FAP Patient	APC Mutation	Age at Diagnosis	Colonic Manifestation Polyps*	Colonic Manifestation Cancer**	Extracolonic Manifestation CHRPE[@]	Extracolonic Manifestation Others
11	470[#]	tttaaattag/GGG→ tttag/attagGGG (ATTAG insertion)	5 I 1	+	49	+	+ (C)	+	osteomas
			5 II:1	+	24	+		+	
			5 II:2	+	23	N.A.		+	
			5 II:4	+	20	+		+	left mandible osteomas
14	I13-I14[#]	Delete E14 (6 kb)	35 I:1	N.D.	50	N.A.	+ (N.A.)	N.D.	
			35 II:5	+	43	m	+ (A)	N.D.	
14	581	ctag/GAA→ctgg/GAA (exon 14 deletion)	11 I:1	+	47	NA		+	stomach cancer
			11 II:2	+	30	++		+	fundic glandular polyps
			11 II:4	+	22	m		+	
			11 III:9	+	13	+		+	
15	674[#]	ATAGTTTGACAATAGT CAG→ATAGTCAG	34 II:3	+	35	+	+ (C)	N.D.	
			34 II:6	+	30	+		N.D.	sebacious cyst
			34 III:1	+	16	N.A.		N.D.	
15	831[#]	TTA→TTAA	44 I:1	N.D.	31	N.A.	+ (D)	N.D.	
			44 II:1	+	23	++	+ (C)	N.D.	
15	935	TAC→TAA	13 I:1	N.D.	32	N.A.	+ (N.A.)	N.D.	
			13 II:1	+	29	+	+	+	thyroid, desmoid cancer
			51 I:1	+	53	++	+ (C)		
			51 II:2	N.D.	24	N.A.		+	
			51 II:3	N.D.	16	N.A.		+	
15	938[#]	ACTAAG→AG	7 I:1	N.D.	32	+	+ (C)		
			7 II:1	+	23	+		N.D.	
			7 II:3	+	21	N.A.		N.D.	
15	974[#]	GGT→GT	48 I:1	N.D.	33	N.A.	+ (D)	N.D.	
			48 II:3	+	15	N.A.			
15	1061	AAACAAAGT→AAGT	23 I:1	+	37	+	+(C)	–	fundic glandular polyps, duodenum adenoma

Table 1 *(continued)*

Exon	Position	Mutation	FAP Patient	APC Mutation	Age at Diagnosis	Colonic Manifestation Polyps*	Colonic Manifestation Cancer**	Extracolonic Manifestation CHRPE@	Extracolonic Manifestation Others
15	1309	AAAGAAAAGATT→ AAAGATT	1 I:1	N.D.	34	+	+ (D)	N.D.	
			1 II:2	+	14	+		+	osteomas, duodenal adenomas
			1 II:3	+	17	+		+	
			1 II:4	+	14	++		+	osteomas of jaw
			10 II:5	+	30	+		+	
			18 I:1	N.D.	34	N.A.	+ (N.A.)	N.D.	
			18 II:2	N.D.	28	N.A.	+ (N.A.)	N.D.	
			18 II:6	+	21	++	+ (D)	+	
			19 III:4	+	25	++	+ (C)	+	fundic glandular polyps, desmoid
			27 I:1	N.D.	30	N.A.	+ (N.A.)	N.D.	fundic glandular polyps
			27 II:1	N.D.	19	++		+	
			27 II:2	N.D.	26	+		+	
			27 II:3	+	23	+		N.D.	
			27 II:4	+	18	++		N.D.	
			32 II:1	N.D.	30	N.A.	+ (N.A.)	N.D.	
			32 II:2	+	29	++		N.D.	fundic glandular polyps
			39 II:1	+	21	N.A		N.D.	
			49 II:6	+	33	++	+ (D)	+	epidermal cyst
15	1343#	TCT→TT	33 II:4	+	33	+	+	N.D.	
15	1428#	GGA→GG	16 I:1	+	50	+	+ (C)	–	
			16 II:2	+	21	++		–	
15	1450	CGA→TGA	52 I:1	+	15	+		–	osteomas, desmoid, fundic glandular polyps
15	1465#	$(AG)_5\rightarrow(AG)_6$	15 I:1	N.D.	58	N.A.	+	N.D.	
			15 II:1	N.D.	23	++		+	fundic glandular polyps, duodenal adenomas
			15 II:2	+	22	++		+	duodenal adenomas, desmoid
15	1503#	TCA→TAA	50 II:2	+	17	+		N.D.	desmoid, fundic glandular polyps

Novel mutation; polyps* m = multiple; + = 100-1000 polyps; ++ = more than 1000 polyps; cancer ** (X) =Dukes stage; CHRPE@: Congenital hypertrophy of the retinal pigment epithelium, – negative; + positive; N.A. Not available; N.D. Not done

very 5′ and 3′ ends of the protein as well as the alternatively spliced sites in exon 9 have been associated with a more attenuated form of FAP[20–22].

Truncating mutations between codons 463 and 1387 are associated with congenital hypertrophy of the retinal pigment epithelium (CHRPE)[23] while truncating mutations between codons 1403 and 1578 are associated with Gardner's syndrome[24] such as increased desmoids and mandibular lesions but not CHRPE.

The phenotypic spectrum of the *APC* germline mutations is documented in Table 1 and represented schematically in Figure 2.

All the mutations in this study have resulted in the loss of either one or both of the β-catenin binding sites which is crucial for APC to down-regulate β-catenin. Patients with these germline mutations exhibit the classical FAP phenotype with more than 100 polyps in the colon except for mutation at codon 332 (exon 9) which is a case of attenuated FAP (AFAP) with less than 100 polyps. Interestingly though, both patients with mutation 332 in the present study manifested with more polyps on the left rather than the right side of the colon, as has been reported for Caucasian patients. This indicates that right-sided predominance is not an absolute requisite for AFAP. It is also intriguing that the patients with mutation at 170 (exon 4) exhibit classical colonic phenotype rather than AFAP as has been suggested by an earlier study[22].

Further, in this study, we found new positions associated with desmoid (299 and 935); thyroid cancer (470 and 935) and stomach cancer (581) that have not been previously reported.

The mutations with CHRPE manifestation wherever it is determined are at positions 470, 581, 935, 1309 and 1465. The mutations with no CHRPE manifestations are at 299, 332, 1061 and 1428 respectively (Table 1 and Figure 2). Thus, with the exception of mutations at codon 1061 and 1465, the demarcation appears to be consistent with the findings of an earlier study[23] delineating the CHRPE limits to the region between codons 463 and 1387. CHRPE is also the only phenotype with no intra-family variation. This enables clinicians to predict affected members and provide useful lead for genetic testing. In contrast, although there are 16 FAP patients from family 4, only one patient has thyroid carcinoma, indicating the large intra-family variability (Table 1). It is as yet unclear why CHRPE is the only phenotype that is not influenced by varying modifying genes and/or environment (dietary) factors.

3.4 FAP Families Without Detectable *APC* Germline Mutation

In order not to miss truncating mutations at the very ends of the *APC* coding region, we sequenced the 5′ and 3′ ends (covering the first 400 and last 300 nucleotides respectively) of the PCR fragments for the eight families without PTT detectable mutations. No further mutation was found. The probability of finding a missense mutation amongst the thirty six families is very low indeed as less than 2% of *APC* germline mutations are missense mutations[18]. Furthermore, there are doubts as to whether missense mutations are disease causing as observations from others' and ourselves suggest that the APC protein has to be truncated in order to be inactivated.

Since the function of APC is to regulate β-catenin degradation, activating *β-catenin* mutation was suggested to be responsible for the initiation of sporadic colorectal cancers lacking *APC* mutation.[12] To our knowledge, all the *β-catenin* mutations reported to date[12,15] involve either the GSK-3β phosphorylation sites (codons 33, 37, 41 and 45), the codons next to these sites (for example, codons 32 and 34) or the interstitial deletion of exon 3. We have thus focused on exon 3 in our search for *β-catenin* germline mutations. No *β-catenin* germline mutation was found in these eight families indicating that *β-catenin* mutation cannot substitute for *APC* mutation in the initiation of the FAP syndrome.[14] The molecular data thus suggest that these families may have germline mutation in other genes that are involved possibly in an APC-independent pathway.

Moreover, phenotypic analysis revealed that these PTT negative families have atypical FAP features such as mixture of adenomatous and hyperplastic polyps and no documentation of extracolonic manifestation in some of these families. The rest of the families differs from classical FAP families by the manifestation of other cancers (which hitherto have not been associated with FAP) such as breast, ovarian, and nasopharyngeal carcinomas. The atypical FAP features for these PTT negative and *β-catenin* negative families are suggestive of new polyposis variants.

4. CONCLUSIONS

We have identified the *APC* germline mutation in 28 out of 36 Singapore FAP kindreds. The specificity and sensitivity of the PTT test is 100% and 78% respectively. DNA sequencing analyses revealed that eleven of the twenty mutations found are novel, indicating that the mutational spectrum of

APC is highly heterogeneous. We have also shown that *β-catenin* germline mutation cannot substitute for *APC* mutation in FAP and that other genes are possibly involved in these patients with atypical FAP features.

Acknowledgments

The authors would like to thank Dr. Zhao Yi from the Department of Clinical Research, SGH for help with the automated sequencer; Ms. Jocelyn Yao from the laboratory for help with RT-PCR, and Ms. Carol Loi and Koh Yin Fung from the Singapore Polyposis Registry for help in retrieving the pedigrees and clinical data. This work is funded in part by a grant from the National Medical Research Council (NMRC/0239/1997) of Singapore.

References

1. Kinzler KW, Vogelstein B, Lessons from hereditary colorectal cancer. *Cell* 1996; **87:** 159–170.
2. Groden J, Thliveris A, Samowitz W *et al.,* Identification and characterization of the familial adenomatous polyposis coli gene. *Cell* 1991; **66:** 589–600.
3. Joslyn G, Carlson N, Thliveris A *et al.,* Identification of deletion mutations and three new genes at the familial polyposis locus. *Cell* 1991; **66:** 600–613.
4. Kinzler KW, Nilbert MC, Su LK *et al.,* Identification of FAP locus genes from chromosome 5q21. *Science* 1991; **253:** 661–665.
5. Goss KH, Groden J, Biology of the Adenomatous Polyposis Coli Tumor Suppressor. J Clin Oncol 2000; **18:** 1967-1979.
6. Beroud C, Soussi T, *APC* gene: database of germline and somatic mutations in human tumors and cell lines. *Nucleic Acids Research* 1996; **24:** 121–124.
7. Roest PAM, Roberts RG, Sugino S, van Ommen GB, den Dunnen JT, Protein truncation test (PTT) for rapid detection of translation-terminating mutations. *Hum Mol Genet* 1993; **2:** 1719–1721.

8. Powel SM, Peterson GM, Krush AJ *et al,* Molecular diagnosis of familial adenomatous polyposis. *New Eng J Med* 1993; **329:** 1982–1987.
9. Nakamura Y. The role of the adenomatous polyposis coli (APC) gene in human cancers. *Adv Cancer Res* 1993; **62**: 65–87.
10. Rubinfeld B, Albert I, Porfiri E, Fiol C, Munemitsu S, Polakis P. Binding of GSK-3β to the APC-β-catenin complex and regulation of complex assembly. *Science* 1996; **272**: 1023–1026.
11. Korinek V, Barker N, Morin PJ, van Wichen D, de Weger R, Kinzler K, Vogelstein B, Clevers H. Constitutive transcriptional activation by a β-catenin-Tdf complex in $APC^{-/-}$ colon carcinoma. *Science* 1997; **275**: 1784–1786.
12. Sparks AB, Morin PJ, Vogelstein B, Kinzler KW, Mutational analysis of the APC/β-catenin/Tcf pathway in colorectal cancer. *Cancer Res* 1998; **58**: 1130–1134.
13. Miller SA, Dykes DD, Polesky HF, A simple salting out procedure for extracting DNA from human nucleated cells. *Nucl Acids Res* 1988; **16:** 739.
14. Cao X, Eu KW, Seow CF, Cheah PY, Germline mutations are frequent in the APC gene but absent in β-catenin gene in Familial Adenomatous Polyposis patients. *Genes Chromosomes & Cancer* 1999; **25:** 396–398.
15. Iwao K, Nakamori S, Kameyama M, Imaoka S, Kinoshita M, Fukui T, Ishiguro S, Nakamura Y, Miyoshi Y, Activation of the β-catenin gene by interstitial deletions involving exon 3 in primary colorectal carcinomas without Adenomatous Polyposis coli mutations. *Cancer Res* 1998; **58**: 1021–1026.
16. Cromwell DM, Moore RD, Brensinger JD, Petersen GM, Bass EB, Giardiello FM, Cost analysis of alternative approaches to colorectal screening in familial adenomatous polyposis. *Gastroenterology* 1998; **114:** 893-901.
17. Cao X, Eu WK, Seow FC, Cheah PY, *APC* mutation and phenotypic spectrum of Singapore familial adenomatous polyposis patients. *Eur J Hum Genet* 2000; **8:** 42–48.
18. Laurent-Puig, Beroud C, Soussi T, *APC* gene: database of germline and somatic mutations in human tumors and cell lines. *Nucl Acids Res* 1998; **26**: 270–273.

19. Nagase H, Nakamura Y, Mutations of the *APC* (adenomatous polyposis coli) gene. *Hum Mutat* 1993; **2:** 425–434.
20. Spirio L, Olschwang S, Groden J *et al.*, Alleles of the *APC* gene: an attenuated form of familial polyposis. *Cell* 1993; **75**: 951–957.
21. Friedl W, Meuschel S, Caspari R *et al.,* Attenuated familial adenomatous polyposis due to a mutation in the 3′ part of the APC gene. A clue for understanding the function of the APC protein. *Hum Genet* 1996; **97:** 579–584.
22. Soravia C, Berk T, Madlensky L *et al.,* Genotype-phenotype correlations in Attenuated adenomatous polyposis coli. *Am J Hum Genet* 1998; **62:** 1290–1301.
23. Olswang S, Tiret A, Laurent-Purg P, Muleris M, Parc R and Thomas G, Restriction of occular fundus lesions to a specific subgroup of APC mutations in Adenomatous Polyposis Coli patients. *Cell* 1993; **75:** 959–968.
24. Davies DR, Armstrong JG, Thakker N *et al.*, Severe Gardner Syndrome in families with mutations restricted to a specific region of the APC gene. *Am J Hum Genet* 1995; **57:** 1151–1158.

FRONTIERS IN HUMAN GENETICS
Diseases and Technologies
© 2001 by World Scientific Publishing Co. Pte. Ltd.

LINKAGE OF DOPAMINE RECEPTOR D2 (*DRD2*) MARKERS WITH ESSENTIAL HYPERTENSION IN SINGAPOREAN CHINESE SUBJECTS

VMS OH[*,†], HM WU[‡], EPH YAP[‡], X ZHOU[*] and EA TAYLOR[*]
**Division of Clinical Pharmacology & Therapeutics*
Department of Medicine, National University Hospital
5 Lower Kent Ridge Road, Singapore 119074
‡Defence Medical Research Institute
Defence Science & Technology Agency
10 Medical Drive #02-04, Singapore 117597
†E-mail: mdcohms@nus.edu.sg

In the National University of Singapore Genetics of Hypertension (NUSIGHT) Study, we compared allele and genotype frequencies of three dopamine receptor D2 (*DRD2*) gene markers in Chinese hypertensive sibling-pairs and unrelated Chinese servicemen to test the hypothesis that the markers are associated with essential hypertension. Essential hypertension is a complex multifactorial disease caused by the interaction of genetic and behavioural factors. Several candidate predisposing genes have been proposed, including those in the dopaminergic neurotransmission pathway. We selected 62 ethnic Chinese non-obese, non-diabetic sibling-pairs on blood pressure > 150/90 mmHg or 24-hour ambulatory blood pressure > 140/90 mmHg, and disease onset < 60 years. We enrolled 176 unrelated ethnic Chinese men as population controls. Two polymorphisms in the coding region (Ser311Cys and *Nco*I) and one in the 3'UTR (*Taq*IA) of *DRD2* were genotyped. Sib-pair linkage analysis of alleles identical by state and case-control linkage dysequilibrium analysis of allele, genotype and haplotype frequencies were performed. Significantly increased sharing of the marker *Taq*IA (pZ < 0.004), but not *Nco*I and Ser311Cys, was observed. However, no association between the three polymorphisms and hypertension was detected. The results suggest that linkage exists between *DRD2* and essential hypertension in Singaporean Chinese, hence justifying further and larger genetic studies of this locus.

† Corresponding author

1 INTRODUCTION

Essential hypertension in humans is a complex multifactorial disease that arises from the interaction of both genetic and environmental (including behavioural) factors. The specific genes involved have not been confirmed, but studies of rare Mendelian hypertension syndromes and rat models have suggested certain candidate genes. These candidates include components of the renin-angiotensinogen-aldosterone axis and the dopaminergic neurotransmission system[1].

Several lines of evidence implicate the dopaminergic neurotransmission system in the pathogenesis of hypertension. The diverse physiological actions of dopamine are mediated by at least five distinct G protein-coupled receptor subtypes[2,3]. Dopamine receptors D1 and D5 activate, whereas D2, D3 and D4 inhibit adenylyl cyclase. In arteriolar smooth muscle, the inhibition of adenylyl cyclase might decrease the intracellular concentration of cyclic adenosine monophosphate, a mediator of vasomotor tone, thus producing vasodilatation.

We selected dopamine receptors (DRs) for study for the following reasons. In animal models, the Dahl salt sensitive rat and the spontaneously hypertensive rat (SHR) have defective D1 receptor signal transduction, resulting in lower salt excretion. This defect could produce a net sodium retention and thus promote hypertension. The SHR also shows decreased levels of dopamine transporter protein[4] and *DRD2* mRNA.[5] Secondly, *DRD1* and *DRD3* knock-out mice have systolic and diastolic hypertension.[6] Thirdly, dopamine receptors in the brain and pituitary mediate neuroendocrine secretion (e.g. of vasopressin), and function as a pressor mechanism.[7] Finally, in the peripheral nervous system, dopamine receptors are found in the kidney, blood vessel musculature, adrenal glands and sympathetic nerves, and function as a depressor mechanism. [8] For a review of the role of dopamine in human hypertension, we refer to an excellent review on the topic[9].

In this study we focused on the 8-exon human *DRD2* gene, and we investigated a genetic linkage of three *DRD2* markers with essential hypertension.

2 MATERIALS AND METHODS

2.1 Subjects

As part of the Singapore NUSIGHT (National University of Singapore Genetics of Hypertension) study, volunteers of Chinese descent were ascertained and selected based on these criteria:- resting blood pressure > 150/95 mmHg or 24-h ambulatory BP > 140/90 mmHg, and disease onset < 60 years. Siblings who did not meet these criteria were regarded as unknown rather than unaffected. Sixty-two concordant affected sib-pairs were recruited. As population controls, 176 unrelated Chinese subjects were recruited from a cohort enlisting for military service. The criterion for ethnicity was that all paternal and maternal grandparents professed to be Chinese. Venous blood (40-80 ml) was drawn from each participant after informed consent. From each subject's blood, we processed one aliquot into plasma or serum for clinical chemistry tests. Genomic DNA was extracted from 5 ml blood using resin columns (Qiagen), and quantitated by UV spectrophotometry. The study was approved by the Research and Ethics Committee of the National University Hospital, Singapore.

2.2 Genotyping of *DRD2* Markers

Three single nucleotide polymorphisms (SNP) in or near the *DRD2* gene were selected as DNA markers. Genotyping was performed by PCR-RFLP and agarose gel separation of restricted products.

2.2.1 TaqIA polymorphism

Amplification of the 310 bp region of the *Taq*IA polymorphic site was performed using the primers described by Grandy *et al.*[10]: 5'–CCGTCGACGGCTGGCCAAGTTGTCTA–3'(Taq1F) and 5'–CCGTCGACCCTTCCTGAGTGTCATCA–3'(Taq1B). PCR was carried out on 96-well microtitre plates using a MJ PTC200 thermocycler with following conditions: 94°C, 5 min; 30 x (94°C, 30 s; 60°C, 30 s; 72°C, 30 s); 72°C, 5 min, in a 20 μl of reaction mixture that contained 40 ng sample DNA, 200 μM of each dNTP, 1.5 mM $MgCl_2$, 1x PCR buffer, 1 μM of each primer, and 0.5 U Taq DNA polymerase (Promega). The PCR product was then incubated with 5 units of *Taq*I overnight at 65°C, and subjected to electrophoresis in 2% agarose gel. The intact 310 bp allele was designated A1, whereas the A2 allele was digested into two fragments of 130 and 180 bp.

2.2.2 Nco*I polymorphism*

The fragment surrounding the *Nco*I polymorphism site was amplified using primers NcoIF (5'–ATCCTGCAGCCATGG–3') and NcoIB (5'–ATTGTCCGGCTTTACC–3').[11] PCR conditions and reaction mixture were similar to that of TaqIA except that annealing was performed at 55°C. The PCR product was digested with 3 units of *Nco*I and then fractionated on 2% agarose gel. The N1 allele remained intact, whereas the N2 allele was cut into two fragments of 194 and 252 bp.

2.2.3 Ser311Cys polymorphism

According to Arinami *et al.*,[12] primer A (5'–ACCAGCTG ACTCTCCCCGACCGCT–3') and primer B (5'–GGAAGGACAT GGCAGGGAATGGGAC–3') were used to amplify a 274 bp region at Ser311Cys polymorphism site. Primer A contained a modified 3' end, which created an artificial *Sau*96 restriction site within the Ser allele. PCR conditions and reaction mixture were similar to that of *Taq*IA except that annealing temperature was performed at 65°C. When the PCR product was incubated with *Sau*96, the Ser (S1) allele gave a 148 bp fragment, while the Cys (S2) allele was digested to a 126 bp.

2.3 Statistical Analysis

Allele and genotype frequencies of the three *DRD2* markers from one randomly selected hypertensive sibling from each family was compared against those of the general Singaporean Chinese population, using Fisher's exact test of significance. The data were determined to be in Hardy-Weinberg equilibrium. Haplotype analysis was performed using the EH program (J. Ott, Rockefeller University, New York, USA). Sib-pair linkage analysis was performed to assess the alleles shared identical by state (IBS) using the Genetic Analysis System (GAS) package, version 2.0 (Young, University of Oxford, UK). Lange's Z statistic was applied to calculate the p value for IBS, with adjustment for multiple sibships.

Association study was carried out in a second data set of cases and controls. The allele, genotype and haplotype frequencies among cases were compared with the control group using the CLUMP program (Curtis, Institute of Psychiatry, London, UK). Significance was assessed using a Monte Carlo approach, by performimg repeated simulations to generate unbiased chi-square value. [13]

3 RESULTS

3.1 Phenotypic Features

The 124 individual hypertensive patients (68 men and 56 women) were aged between 24 and 78 years at the onset of clinical hypertension. Table 1 shows the mean age, mean blood pressure, and mean body mass index of the patients at enrolment into the study.

Table 1 Phenotypic data of 124 hypertensive subjects (68 men and 56 women)

	Mean	SD
Age at enrolment (years)	50.8	8.8
Disease onset age (years)	43.0	8.2
Body mass index (kg/m^2)	24.7	3.9
Onset systolic blood pressure (mmHg)	168	19
Onset diastolic blood pressure (mmHg)	107	10

3.2 *DRD2* Polymorphisms in the Chinese Population

Three single nucleotide polymorphisms, two in the coding sequence (Ser311Cys, *Nco*I) and one in the 3' flanking region (*Taq*IA), were genotyped in 176 unrelated Singaporean Chinese (Table 2). All three markers were polymorphic in the local Chinese population studied. The allelic and genotype frequencies were determined, and did not deviate significantly from Hardy Weinberg equilibrium (Table 2). The heterozygosities for markers *Nco*I and *Taq*IA in the population were 0.54 and 0.49, respectively. The Cys311 allele is rare, with a frequency of 0.01, and heterozygosity was 0.03 in this study population. Haplotype analysis for all three markers was carried out in EH program. The markers *Nco*I and *Taq*IA were in linkage dysequilibrium with each other, but not with Ser311Cys.

Table 2. Polymorphisms of *DRD2* gene and their frequencies in 176 Singaporean Chinese.

Polymorphism of *DRD2*	Location	Restriction Enzyme	Alleles	Allelic Frequencies	Het*
Ser311Cys	Exon 6, Codon 311	*Sau*96	C/G (Ser/Cys)	Ser(S1): 0.99 Cys(S2): 0.01	0.03
*Nco*I	Exon 6, Codon 313	*Nco*I	C/T (silent)	N1: 0.43 N2: 0.57	0.54
*Taq*IA	10542 bp 3' downstream of termination codon	*Taq*I	C/T	A1: 0.37 A2: 0.63	0.49

* Heterozygosity

3.3 Sibling-Pair Linkage Analysis

Sibling-pair linkage analysis (identity by state, IBS) of the three markers described was carried out separately. The number of alleles shared IBS at each marker was determined for the sib-pair data set. Observed IBS sharing for affected sib pairs was compared with expected IBS sharing, which was calculated using the allele frequencies. There was a significant increased sharing for the marker *Taq*IA among the affected siblings (pZ = 0.0035) (Table 3). However, no evidence of increased sharing was found for marker *Nco*I and Ser311Cys.

'able 3. Sib-pair linkage analysis identical by state for *DRD2* markers and hypertension

Polymorphism	Sib-pairs	Affected			Expected			$p(\chi^2)$	pZ
Alleles shared IBS		2	1	0	2	1	0		
Ser311Cys	47	46	1	0	45.2	1.8	0	0.83	0.27
*Nco*I	42	28	14	0	25.2	15.5	1.3	0.42	0.13
*Taq*IA	60	46	14	0	36.2	22.0	1.8	0.026	0.0035

.4 Case-Control Association Study

. case-control comparison between the *Nco*I, Ser311Cys, and *Taq*IA olymorphisms and essential hypertension was carried out in a data set of 2 unrelated cases and 176 controls. No significant difference in allele and enotype frequencies was found between the case and control group for all markers (Table 4). The observed haplotype frequencies were compared to ose expected if they were independent variables. Haplotypes N1-A1, N2-2 were more frequent, and haplotypes N1-A2 and N2-A1 were rarer, than xpected if association were absent (Table 5). The haplotype frequencies r the three markers in the case group were not significantly different with e controls.

Table 4. *DRD2* polymorphisms and hypertension

		Allele Frequency			Genotype Frequency			
NcoI	**No**	**N1**	**N2**	X^2	**N1\N1**	**N1\N2**	**N2\N2**	X^2
Cases	45	0.42	0.58	0.02	0.18	0.49	0.33	0.05
Controls	163	0.43	0.57		0.19	0.49	0.32	
Ser311Cys	**No**	**S1**	**S2**	X^2	**S1\S1**	**S1\S2**	**S2\S2**	X^2
Cases	49	0.98	0.02	0	0.96	0.04	0	0.19
Controls	176	0.98	0.02		0.97	0.03	0	
TaqIA	**No**	**A1**	**A2**	X^2	**A1\A1**	**A1\A2**	**A2\A2**	X^2
Cases	62	0.37	0.63	0.03	0.13	0.48	0.39	0.44
Controls	162	0.38	0.62		0.16	0.48	0.36	

Table 5. Haplotype analysis for DRD2 markers and hypertension

Polymorphism	Haplotype	Control Frequency	*n*	Hypertensive Frequency	*n*	Expected* Frequency
Ser-*Nco*I	S1-N1	0.432	176	0.433	45	0.424
	S1-N2	0.549		0.556		0.557
	S2-N1	0.000		0.000		0.008
	S2-N2	0.018		0.011		0.011
Ser-*Taq*IA	S1-A1	0.397	176	0.359	45	0.390
	S1-A2	0.584		0.625		0.591
	S2-A1	0.000		0.016		0.007
	S2-A2	0.019		0.000		0.011
*Nco*I-*Taq*IA	N1-A1	0.331	176	0.337	45	0.168
	N1-A2	0.102		0.097		0.266
	N2-A1	0.056		0.019		0.219
	N2-A2	0.510		0.548		0.347

*Nco*I-Ser-*Taq*IA	N1-S1-A1	0.327	176	0.332	45	0.162
	N1-S1-A2	0.098		0.097		0.260
	N1-S2-A1	0.000		0.004		0.003
	N1-S2-A2	0.004		0.000		0.005
	N2-S1-A1	0.054		0.013		0.214
	N2-S1-A2	0.500		0.547		0.344
	N2-S2-A1	0.002		0.007		0.004
	N2-S2-A2	0.014		0.000		0.007

*Expected assuming that polymorphisms independently assort, with no linkage dysequilibrium.

4 DISCUSSION

Within the National University of Singapore Genetics of Hypertension (NUSIGHT) study, we investigated the contribution of the *DRD2* gene to essential hypertension. Initially, using affected sib-pair identity-by-state (IBS) analysis, we found a significantly increased sharing for marker TaqIA in 62 affected sib-pairs (pZ = 0.0035). This suggested a linkage between marker *Taq*IA and hypertension. However, we did not detect a significantly increased sharing among 45 affected sibling pairs for the markers *Nco*I and Ser311Cys. IBS analysis scores allele sharing of affected siblings identical by state, rather than by descent, and does not use parental genotypes. Therefore, it is less likely to detect small gene contributions. Also, the lack of increased sharing for marker *Nco*I and Ser311Cys might also be due to the relatively small sample size. Furthermore, the Ser311Cys is poorly informative, which limits its usefulness in linkage analysis.

In a separate case-control comparison between the *Nco*I, Ser311Cys, and *Taq*IA polymorphisms and essential hypertension, we did not find a significant difference in allelic and genotype frequencies for all 3 markers between the two subject groups. The lack of detectable association between *DRD2* and hypertension, in the presence of positive linkage, may be attributed to several reasons. Linkage can be detected over a relatively wide genetic distance of several centiMorgans (equivalent to several megabases), whereas linkage dysequilibrium occurs over a smaller distance, depending on the nature of the population studied. Hence the hypertension susceptibility gene at this locus may not be *DRD2* itself, but another gene nearby. Alternatively, there could be multiple disease mutations in the DRD2 gene, which collectively are not associated with any of the three selected markers.

Direct observation of haplotypes was not possible, as the PCR assays genotyped one marker at a time. Phase information was not available, as the sample comprised unrelated individuals or sib-pairs without parental genotypes. Haplotype frequencies were therefore estimated computationally. However, both the genetic history of the Chinese population (their origins, and the presence and age of population bottlenecks) and the heterogeneity of the population are not known. It was therefore useful to determine whether these markers were in linkage dysequilibrium with each other, and therefore likely to be in association with novel or functional mutations in the *DRD2* coding sequence. As expected, certain haplotypes (Table 7) appeared more frequently than if the three markers were independently assorted, as the three markers lie within 10 kb of each other. This result suggests that the Singaporean population of ethnic Chinese descent would be useful for linkage dysequilibrium studies for disease susceptibility genes.

This study found a linkage between the *DRD2* marker *Taq*IA and essential hypertension. We consider that the results of this preliminary investigation justify studying larger numbers of hypertensive patients and population controls, and other candidate genes at this locus.

Acknowledgements

We thank Ms Selen Tan, Defence Medical Research Institute, for expert technical assistance, and Professor Mark J Caulfield, Department of Clinical Pharmacology, St Bartholomew's and The Royal London Hospital School of Medicine & Dentistry, London UK, for expert advice. We are very grateful to the following for permission to study patients at their respective centres:- Professor Chia Boon Lock & Associate Professor Lim Yean Teng, Cardiac Department, National University Hospital; Clinical Associate Professor Chee Yam Cheng, Head, Department of General Medicine, Tan Tock Seng Hospital, Singapore; Clinical Professor Lim Yean Leng, Director, National Heart Centre, Singapore; Dr Paul Ho, Ho Medical Centre, Singapore. The study was supported by the National Medical Research Council (Singapore) through grants NUMI 66000009 and R-172-000-071-213.

References

1. Iimura O, Ura N, Nakagawa M. Comparative hypertensionology - renal dopaminergic activity in experimental hypertensive rats. *Clin Exp Hypertens* 1999; **19**: 117-130
2. Lachowicz J, Sibley DR. Molecular characteristics of mammalian dopamine receptors. *Pharmacol & Toxicol.* 1997; **81**: 105-113.
3. Missale C, Nash SR, Robinson SW, Jaber M, Caron MG. Dopamine receptors: From structure to function. *Physiol Rev.* 1998;**78**: 189-225
4. Watanabe Y, Fujita M, Ito Y, Okada T, Kusuoka H, Nishimura T. Brain dopamine transporter in spontaneously hypertensive rats. *J Nucl Med* 1997; **38**: 470-474.
5. Jose PA, Eisner GM, Drago J, Garey RM, Felder RA. Dopamine receptor signalling defects in spontaneous hypertension. *Am J Hypertension* 1996;**4** :400-405.
6. Jose PA, Drago J, Accili D, Eisner GM, Felder RA. Transgenic mice to study the role of dopamine receptors in cardiovascular function. *Clin Exp Hypertens* 1997; **19**:15-25.
7. Autelitano DJ, van den Buuse M. Concomitant up-regulation of proopiomelanocortin and dopamine D2-receptor gene expression in

the pituitary intermediate lobe of the spontaneously hypertensive rat. *J Neuroendocrinol* 1997; **9**: 255 –262.

8. Kuchel O. Peripheral dopamine in hypertension and associated conditions. *J Hum Hypertens* 1999; **13**:605-615.
9. Jose PA, Eisner GM, Felder RA. Renal dopamine receptors in Health and hypertension. *Pharmacol. Ther* 1998; **80**:149-182.
10. Grandy DK, Litt M, Allen L, Bunzow JR, Marchionni M, Makam H, Reed L, Magenis RE, Civelli O. The human dopamine D2 receptor gene is located on chromosome 11 at q22-q23 and identifies a TaqI RFLP. *Am J Hum Genet* 1989; **45**:778-785.
11. Sarkar G, Kapelner S, Grandy DK, Marchionni M, Civelli O, Sobell J, Heston L, Sommer SS. Direct sequencing of the dopamine D2 receptor (*DRD2*) in schizophrenics reveals three polymorphisms but no structural change in the receptor. *Genomics* 1999; **11**:8-14.
12. Arinami T, Itokawa M, Enguchi H, Tagaya H, Yano S, Shimizu H, Hamaguchi H, Toru M. Association of dopamine D2 receptor molecular variant with schizophrenia. *Lancet* 1994; **343**:703-4
13. Sham PC, Curtis D. Monte Carlo tests for associations between disease and alleles at highly polymorphic loci. *Ann Hum Genet*. 1995; **59**: 97-105.

III GENE THERAPY

FRONTIERS IN HUMAN GENETICS
Diseases and Technologies
© 2001 by World Scientific Publishing Co. Pte. Ltd.

MOLECULAR MEDICINE– POTENTIAL THERAPIES FOR GENETIC DISEASES

STEPHEN D WILTON
Australian Neuromuscular Research Institute, 4th Floor "A" Block, QE II Medical Centre, Nedlands, WA 6009
E-mail: swilton@cyllene.uwa.edu.au.

After identification of the dystrophin gene in 1987, there were confident claims that a cure for Duchenne muscular dystrophy would soon be available. It was thought that introducing a new or functional dystrophin gene into dystrophic tissue would compensate for the defective copy. Sadly, this approach has not turned out to be as simple as was anticipated.

The introduction of a functional dystrophin gene through cell therapy was encouraging in mice but failed dismally in human trials. A range of viral vectors have been evaluated for their potential to introduce a replacement dystrophin gene into muscle but the immune response and long-term transgene expression is an on-going problem.

Some alternative therapies have been considered, with intervention at one of the basic steps in the gene expression: (1) Transcription: re-activation of homologous genes. (2) Splicing: specific removal of an exon carrying a nonsense mutation. (3) Translation: suppression of nonsense mutations.

Gene therapy is a very young science. To date, it has promised much but delivered little. Nevertheless, there is still great optimism that therapies for some disorders, either gene replacement or genetically based, will prove effective in the near future.
Keywords: genetic therapies, up-regulation, induced splicing, nonsense suppression, Duchenne muscular dystrophy

Our interest in human heredity and the transmission of various traits from generation to generation can be traced back many thousands of years. More than 60 birth defects were found listed on Babylonian clay tablets written around 3000 BC. The Jewish Talmud contains an accurate description of the inheritance of hemophilia, an X-linked recessive trait. Heritable human traits have played a very significant role in shaping social customs through the ages. For example, sacred Hindu Scriptures gave instructions for selecting a wife, emphasizing that no heritable illness should be present. In addition, her family should show evidence of "good character" for several preceding generations.

Despite these general observations, it has only been in the last few decades that we have come to understand the basics of gene expression and some of the genes involved in "simple" serious genetic disorders such as cystic fibrosis, thalassemia and Duchenne muscular dystrophy. The term "simple" here only refers to those conditions which show clear recessive or dominant genetic inheritance, it is not meant to imply a simple molecular basis to any of these conditions. Indeed the precise role of skeletal muscle dystrophin, or its numerous muscle and non-muscle isoforms, has yet to be elucidated.

More complex disorders (Alzheimer's dementia, asthma, Parkinson's, diabetes and so on) are assumed to involve several genes or susceptibility alleles in addition to environmental factors.[1] It is worth noting that while it is going to be much harder to identify all the factors (including both genetic AND environmental) implicated in these complex traits, it is almost assured that these conditions will be even harder to treat.

When the genetic basis of a serious disorder has been identified (how does one define "serious"?), we are in the position of being able to do something more than just offering general advice to assist some of those family members at risk. This help could range from suggesting changes to diet or life style to prenatal screening and genetic counselling so that those families can make informed decisions as to their reproductive options. However, even in the presence of sophisticated genetic screening, individuals affected with one (or more) of the many common disorders will continue to be appear.

For example, there are many hundreds of different mutations in the CFTR gene and it is not possible to routinely screen for all possible cystic fibrosis gene defects at the moment. The huge dystrophin gene poses another screening challenge due to the size of the gene where the 14 kb muscle specific mRNA consists of 79 exons spanning almost 0.1% of our total genome. Possibly because of this enormous size, the dystrophin gene has a very high *de novo* mutation rate where one in three DMD cases arise in families with no prior history of that disease. It is not yet feasible or practical to offer screening of all major genes implicated in genetic diseases although the development of new sequencing and scanning technologies (chips, mass spectroscopy, etc) may facilitate this task in the future.[2]

The aim of this chapter is to briefly mention conventional gene therapy, that is the introduction of a functional gene to compensate for a defective one and then consider some more lateral approaches.

The concept of introducing a functional gene to compensate for an inherited defective copy is both straightforward and logical so this approach must be pursued with vigour[3]. In some metabolic disorders where a functional enzyme is missing, levels of replacement would generally not have to be 100%. Many clinical studies have shown that generally 5–10 % of normal enzyme activity is sufficient for adequate cellular/tissue function.

However, there are many different genetic disorders, each with its own subtle nuances, influenced to various extents by genetic and environmental backgrounds, and different patterns of inheritance (dominant vs recessive). It is unlikely that there will be one all-powerful, all applicable, universal gene therapy technique. Although gene replacement seems an obvious approach in conditions where the defective gene product is inactive or non-functional (typically in recessive disorders), such an approach is unlikely to be applicable in most genetic diseases with a dominant mode of transmission. For example, missense mutations in the superoxide dismutase gene (*SOD1*) have been associated with familial amyotrophic lateral sclerosis (F-ALS) or motor neuron disease. The normal SOD1 enzyme exists as a homodimer which is responsible for the dismutation of superoxide radicals to hydrogen peroxide and oxygen. The mutated *SOD1* allele directs production of an altered protein which can also dimerize with the normal gene product. Consequently, these F-ALS patients would have 25% normal SOD1 homodimer, 50% normal:mutated subunits and 25% mutant SOD1. The mutant enzyme still has considerable dismutase activity, yet somehow, even in the presence of some normal SOD1, the altered enzyme is involved in the onset of motor neuron disease, presumably though some gain of function. Direct gene replacement is unlikely to be an effective treatment for F-ALS as the normal gene product is still present and it is the mutant protein, with the as yet unknown gain of function, that is responsible for the disease. One of the more immediate goals in developing a therapy for amyotrophic lateral sclerosis must be identifying the activity of the mutant enzyme so that rational avenues of therapy can be developed. Perhaps some of the experiences from those researchers studying viral infections or cancer control through down-regulation of oncogenic transcripts may be applicable to F-ALS. Two approaches for the specific silencing or correction of dominant mutant alleles will be considered later.

In DMD, it is generally assumed that the missing/defective dystrophin protein cannot fulfil its normal structural role and this leads to more fragile muscle fibres which tear or rupture upon contraction. This structural hypothesis has not been confirmed and an alternate theory, the channel aggregation model suggests the primary role of dystrophin is in the organization of the membrane cytoskeleton and a dystrophic cytoskeleton disrupts ion channel function in muscle fibres. In either model, it has been assumed that introducing a normal or functional dystrophin gene into a DMD patient will cure the disorder or, at least minimize the clinical progression of the disease.

Developing a therapy for DMD is considered to be one of the greatest challenges as this is the largest known gene with a very complex organization. At least seven promoters are expressed in a tissue specific manner and the major muscle specific isoform is expressed in terminally differentiated non-dividing cells.[4] I will consider several potential therapies being considered for DMD as a general example for disorders where the normal gene product is defective or absent.

Approaches to introduce a functional dystrophin gene through cell therapy were encouraging in mice but failed dismally in human trials, highlighting the importance of relevant animal models. Other experiments demonstrated that a functional dystrophin cDNA (naked DNA) could be directly injected into dystrophic muscle tissue with subsequent and persistent dystrophin expression (albeit at very low levels). A range of viral delivery systems are currently being evaluated for their potential to introduce a replacement dystrophin cDNA (both full length and the minigene) into muscle. There is no viral delivery system capable of delivering the entire dystrophin gene of 2400 kb. Some of the newly developed "gutless" adenoviral vectors can carry the entire dystrophin coding region while the adenoviral associated virus only has the capacity for the dystrophin minigene.

One method that may be capable of delivering the entire dystrophin gene into dystrophic tissue could be through the development and application of artificial chromosomes.[6] These vectors may be assembled through combining 3 kinds of DNA elements: a centromere, several DNA replication origins and two telomeric repeats. The resulting arrangement could carry and express one or more selected genes. The minimum length is unknown but is thought to be up to several Mb, more than enough to handle the entire dystrophin gene.

The problem of immune response to "foreign" proteins (both viral and dystrophin) remains to be clarified but it should be noted that one of the most promising aspects of gene therapy is the application of DNA vaccines. The very successes of DNA vaccines must cause some concern about simple gene replacement approaches. Plasmid DNA, carrying coding sequences for bacterial or viral antigens, have been injected into muscle to elicit cell-mediated immune responses to the "foreign peptides".[5] A DMD patient has never been exposed to normal dystrophin and thus his immune system will recognize it as a foreign protein.

While there can be no doubt as to the logical nature of the gene replacement approach, some researchers have begun examining therapies where intervention may occur at one of the three basic steps in the expression of the gene product.

1.(Transcription) 2.(Splicing/processing) 3.(Translation)

Gene ——〉 pre-mRNA ——〉 mRNA ——〉 Protein

(1) <u>Transcription intervention</u> could take the form of re-activation of redundant or back-up copies of the defective gene. These back-up copies are normally switched off at birth or are restricted to limited expression in certain tissues or sites, even though the normal adult gene is defective. The "back-up" copy for dystrophin appears to be the utrophin gene, and it has been shown that an up-regulation of utrophin can compensate for a lack of dystrophin in mice. Work by Kay Davies and colleagues in Oxford has shown that a utrophin transgene expressed at high levels in an *mdx*

(dystrophin-negative) mouse seemed to reduce the dystrophic pathology in skeletal and diaphragm muscle.[7] Immunostaining of tissue sections from these animals demonstrated that utrophin was over-expressed and localized at the sarcolemma where dystrophin should be found. Most importantly, the utrophin associated with the sarcolemma glycoproteins to form a complex normally found with dystrophin.

There are other examples of back-up genes that, if induced, may compensate for another defective gene. Fetal globin expression, if it could be induced to persist, may compensate for β-globin gene defects. In McArdle's disease (an autosomal recessive disorder), there are mutations in the muscle specific isoform of glycogen myophosphorylase, an enzyme that breaks glycogen down to lactic acid. Patients with McArdle's disease have exercise intolerance, myalgia and cramps in exercising muscle, even though every cell involved carries the genes for two other glycogen phosphorylase isoforms (liver and brain). The myophosphorylase defect in McArdle's disease should be overcome if either of the other two glycogen phosphorylase genes could be switched back on and expressed in muscle.

A significant advantage of this approach is that any protein product of these induced genes would not be recognized as being foreign so that immunological problems should be minimized.

(2) Splicing intervention has the potential to specifically remove an exon carrying a disease-causing nonsense mutation or restore the reading frame around a genomic re-arrangement.

After, or in some cases during, the transcription of a gene, it is necessary to remove the intronic sequences from the protein coding exons. Intron removal from the pre-mRNA must be carried out with single base precision as anything less would disrupt the reading frame of the mature mRNA. Numerous RNA/protein factors have been identified in the spliceosome as well as consensus nucleotide sequences at the exon:intron and intron:exon junctions and other motifs within the exon (exon recognition sequence) and intron (branch point). To further complicate this mechanism, other factors can influence final splice site selection so that several isoforms can be generated from one gene in a tissue specific fashion.

Ryszard Kole and colleagues demonstrated that RNA oligonucleotides directed to an activated cryptic splice site in a β-globin intron could block the abnormal splicing.[8] These activated splice sites arose from intronic mutations that made the cryptic splice site "stronger" than the normal one. The antisense oligonucleotide annealed to the cryptic splice site and blocked assembly of the splicing machinery to re-direct the normal selection of exons. If antisense oligonucleotides could suppress strong aberrant splicing, it should be possible to influence spliceosome assembly during normal splicing so that specific exons could be skipped.

This approach may have only limited application to the correction of cryptic splice mutations and some structural proteins (such as dystrophin). The CFTR gene, with its many subtle missense mutations or microdeletions (Δ508) is unlikely to be a candidate for the antisense approach. The dystrophin gene consists of 79 exons and may be amenable to manipulation, especially since it has been shown that shorter versions of dystrophin are functionally or biologically active as seen in the Becker muscular dystrophy patients[9]. In-frame genomic deletions in the dystrophin gene can still allow the production of a shorter but partially functional dystrophin protein. This will be discussed in more detail in a subsequent chapter.

(3) <u>Translation intervention</u> may be applicable to those disorders which have arisen from nonsense mutations (about 25% of DMD and 5% of CF mutations). Aminoglycoside antibiotics have been shown to reduce the fidelity of translation so that nonsense mutations may be suppressed. This phenomena was observed in 1964 by Gorini and colleagues where streptomycin was used to reduce the fidelity of translation by inhibiting ribosomal proof-reading. This proof-reading activity excludes poorly or mismatched amino acyl-tRNAs from incorporation into the polypeptide chain. David Bedwell and co-workers have used this method to suppress nonsense mutations in the CFTR gene and have shown restoration of cAMP activated chloride channel activity.[1]

Although many aminoglycoside antibiotics have been approved for human usage, there are still many limitations to this approach. Aminoglycosides must enter the cell to influence the eukaryotic proof-reading activity during translation. Some aminoglycosides are ototoxic and nephrotoxic, thereby limiting continual delivery. Cell culture experiments have shown that not all nonsense mutations will be influenced at the same rate as some are easier to suppress than others. The nonsense mutation in the dystrophin gene of the *mdx* mouse (UAA) is supposed to be one othe least susceptible termination codons for aminoglycoside suppression. However, Lee Sweeney's group have recently shown that a daily administration of gentamicin over a two week period resulted in dystrophin expression in the *mdx* mouse.[11] Not only was this dystrophin correctly localized but there were other significant improvements including a reduction in serum creatine kinase levels.

Clinical trials involving CF patients being treated with gentamicin have already been undertaken in the US. There has been anecdotal evidence that some CF patients treated with aminoglycoside antibiotics for respiratory infections responded far better to the treatment than others. It is possible that the positively responding CF patients may have had a nonsense mutation in their CFTR gene that was suppressed by the antibiotic treatment. Trials involving appropriate DMD patients are already planned.

I have briefly mentioned those disorders which may respond to a treatment that by-passes or suppresses the primary defect. The genetic lesion will still be present and continued treatment would be necessary. These paarticular approaches would not be applicable to those disorders where there is a "gain-of-function" associated with the mutant allele. In these dominant cases, it would be necessary to inactivate or silence the mutant allele while leaving the normal transcript untouched. This could be a difficult task requiring very high specificity as there is often only a single nucleotide difference between the normal and mutant transcripts. The majority of *SOD1* mutations (responsible for F-ALS) are single nucleotide missense mutations so any targeting of the mutated transcript for degradation must be very specific.

Antisense oligonucleotides with a different backbone chemistry and deoxyribonucleotide residues to those used to induce exon skipping have been applied to suppress gene expression.[12] These antisense oligodeoxyribonucleotides can bind to a specific mRNA so the resultant RNA:DNA hybrid is targeted for degradation by RNase H. Such an approach has been used in abolishing/reducing oncogenic transcripts and as antibacterial or antiviral compounds. There is extensive research in examining new chemistries for improved specificity and stability.

Ribozymes are self-catalytic RNAs that bind to and then cleave specific RNA sequences.[13] Like the antisense oligodeoxyribonucleotide approach, these have been used to target specific gene transcripts with the aim of blocking expression. Recently, DNA has been found to be able to act in a similar fashion and has the advantages of easier synthesis and greater stability.

Finally, one approach that has generated considerable interest and controversy is that of chimeric oligonucleotide mediated gene repair[14]. These chimeric oligonucleotides are large, chemically synthesized oligonucleotides (typically 68-mers) comprised of RNA and DNA nucleotides. The sequence is such that the a single oligo will fold back onto itself to form a duplex structure where one strand is comprised of only DNA bases while the complementary strand contains RNA:DNA blocks. This arrangement is thought to anneal specifically to a target gene sequence and induce correction of a point mutation to the normal sequence during DNA repair.

Although this approach has not been successfully applied in many laboratories and hence is still somewhat controversial, the potential of permanent gene correction (at least of subtle DNA mistakes) must be thoroughly explored. Since oligonucleotides are not generally immunogenic, repeated systemic delivery is possible so that a gradual correction may be achieved. Chimeric oligos could be used to treat missense mutations as found in the CF gene, allowing the repaired gene to remain under the control of its own promoters. While there do not appear to be many missense mutations in the dystrophin gene, it is estimated that up to 25% of mutations are nonsense or point mutations that inactivate splice sites. Finally, gene repair could be used in selected dominant disorders such as *SOD1* missense mutations responsible for F-ALS. Rather than trying to inactivate a mutant allele, it may be possible to permanently correct the defect at the genomic level.

Gene therapy is a very young field with the first trials only commencing a decade ago. To date, it has promised much but delivered little although there is now a much greater appreciation of some of the biological complexities. Nevertheless there is still great optimism that effective genetically based therapies are close at hand for some disorders.

Acknowledgments

This work was supported by the Muscular Dystrophy Association of Western Australia, the Neuromuscular Foundation of Western Australia and by grants from the Muscular Dystrophy Association of USA. I wish to thank Dr Sue Fletcher for the helpful discussion, suggestions and for critically reviewing this manuscript.

References

1. Ellsworth DL., Manolio TA, The emerging importance of genetics in epidemiologic research II. Issues in study design and gene mapping, *Annals of Epidemiology* 1999; **9**:75–90.

2. Wilgenbus KK, Lichter P, DNA chip technology ante portas, *Journal of Molecular Medicine* 1999; **77**:761–768.

3. Karpati G, Lochmuller H, The scope of gene therapy in humans: scientific, safety and ethical considerations, *Neuromuscular Disorders* 1997; **7**:273–276.

4. Sadoulet-Puccio HM, Kunkel LM, Dystrophin and its isoforms, *Brain Pathology* 1996; **6**:25–35.

5. Ramsay AJ, Ramshaw IA, Ada GL, DNA immunization, *Immunology & Cell Biology* 1997; **75**:360–363.

6. Grimes B, Cooke H, Engineering mammalian chromosomes, *Human Molecular Genetics* 1998; **7**:1635–1640.

7. Tinsley JM, Potter AC, Phelps SR, Fisher R, Trickett JI, Davies KE, Amelioration of the dystrophic phenotype of *mdx* mice using a truncated utrophin transgene, *Nature* 1996; **384**:349–353.

8. Sierakowska H, Sambade MJ, Agrawal S, Kole R, Repair of thalassemic human beta-globin mRNA in mammalian cells by antisense oligonucleotides, *Proceedings of the National Academy of Sciences of the United States of America* 1996; **93**:12840–12844.

9. Wilton SD, Lloyd F, Carville K, Fletcher S, Honeyman K, Agrawal S, Kole R, Specific removal of the nonsense mutation from the *mdx* dystrophin mRNA using antisense oligonucleotides, *Neuromuscular Disorders* 1999; **9**:330–338.

10. Howard M, Frizzell RA, Bedwell DM, Aminoglycoside antibiotics restore CFTR function by overcoming premature stop mutations, *Nature Medicine* 1996; **2**:467–469.

11. Barton-Davis ER, Cordier L, Shoturma DI, Leland SE, Sweeney HL, Aminoglycoside antibiotics restore dystrophin function to skeletal muscles of *mdx* mice, *Journal of Clinical Investigation* 1999; **104**: 375–381.

12. Agrawal S, Importance of nucleotide sequence and chemical modifications of antisense oligonucleotides, *Biochimica et Biophysica Acta* 1999; **1489**: 53–68.

13. Tanner NK. Ribozymes: the characteristics and properties of catalytic RNAs, *FEMS Microbiology Reviews* 1999; **23**:257–275.

14. Kren BT, Metz R, Kumar R, Steer CJ, Gene repair using chimeric RNA/DNA oligonucleotides, *Seminars in Liver Disease* 1999; **19**:93–104.

FRONTIERS IN HUMAN GENETICS
Diseases and Technologies
© 2001 by World Scientific Publishing Co. Pte. Ltd.

DEVELOPMENT OF HVJ-LIPOSOMES AND CANCER GENE THERAPY

YASUFUMI KANEDA
Division of Gene Therapy Science, Graduate School of Medicine, Osaka University, Suita, Osaka 565-0871, Japan
E-mail: kaneday@gts.med.osaka-u.ac.jp

We have developed HVJ (hemagglutinating virus of Japan; Sendai virus)-liposomes that are efficient *in vitro* and *in vivo* gene delivery vehicles using the fusion-mediated gene delivery. The HVJ-liposome was highly efficient for the introduction of oligonucleotides into cells *in vivo* as well as the transfer of genes less than 100 kbp without damaging cells. By coupling the Epstein-Barr (EB) virus replicon apparatus with HJV-liposomes, transgene expression was sustained *in vitro* and *in vivo*. Most animal organs were found to be suitable targets for the fusigenic-viral liposomes, and numerous gene therapy strategies using this system were successful in animals. Here we demonstrate cancer gene therapy using HVJ-liposomes.
Keywords: gene transfer, viral-liposome, HVJ (Sendai virus), EBV replicon, synthetic vector, gene therapy

1 DEVELOPMENT OF HVJ-LIPOSOMES

To promote human gene therapy, development of more effective gene transfer vector systems is desired.[1] There are several barriers to be overcome in human gene therapy. First is improvement of gene transfer efficiency. Numerous viral and non-viral (synthetic) methods for gene transfer have been developed for human gene therapy.[2,3] Generally, viral methods are more efficient than non-viral methods for gene delivery to cells. However, viral vectors are safety hazards because of the co-introduction of essential genetic elements from the parent viruses, leaky expression of viral genes, immunogenicity, and alterations of host genomic structure. In general, non-viral vectors are less toxic and less immunogenic. However, most non-viral methods are less efficient for gene transfer, especially *in vivo* compared to some of the viral vectors. Thus, both viral and non-viral vectors have limitations as well as advantages. Therefore, to develop an *in vivo* gene transfer vector with high efficiency and low toxicity, the limitations of one type of vector system should be compensated for by introducing the strengths of another type of system. With this idea of compensation in mind, we developed a novel hybrid gene transfer vector by combining viral and non-viral vectors.

1.1 Fusion-Mediated Gene Delivery System

Liposomes attach to the cell membrane and are taken up into cells by phagocytosis or endocytosis. However, most of the macromolecules in liposomes are degraded in lysosomes before reaching the cytoplasm. Consequently, introduction of molecules directly into the cytoplasm has been attempted. To address this problem, we have constructed liposomes with a fusigenic envelope derived from hemagglutinating virus of Japan (HVJ; Sendai virus).[4-6] As shown in Fig. 1, DNA-loaded liposomes were fused with viral proteins or viral envelope to form HVJ-liposomes. Thus liposomes acquire viral functions, and can then fuse with cell membrane, as would HVJ itself, to directly introduce DNA into the cytoplasm. HVJ is a paramyxovirus 300 nm in diameter that contains two distinct glycoprotein on the envelope that are involved in cell fusion.[7] Hemagglutinating (HN) protein is required for binding to a receptor consisting of sialoglycoprotein or sialolipids and removal of sugars by neuraminidase activity, and Fusion (F) protein interacts with the lipid bilayer of the cell membrane to induce cell fusion. HVJ-liposomes as well as HVJ alone can fuse with almost all cells except peripheral lymphocytes. Fusion is complete within 10 to 30 minutes at 37°C. Transfer of molecules by cationic liposomes requires a much longer time, generally 5 to 20 hr. The short incubation time for delivery by HVJ-liposomes is beneficial for *in vivo* use.

Another advantage of fusion-mediated delivery is the protection of molecules from degradation by endosomes and lysosomes. When fluorescence isothiocianate (FITC) - oligodeoxynucleotides (ODN) wa

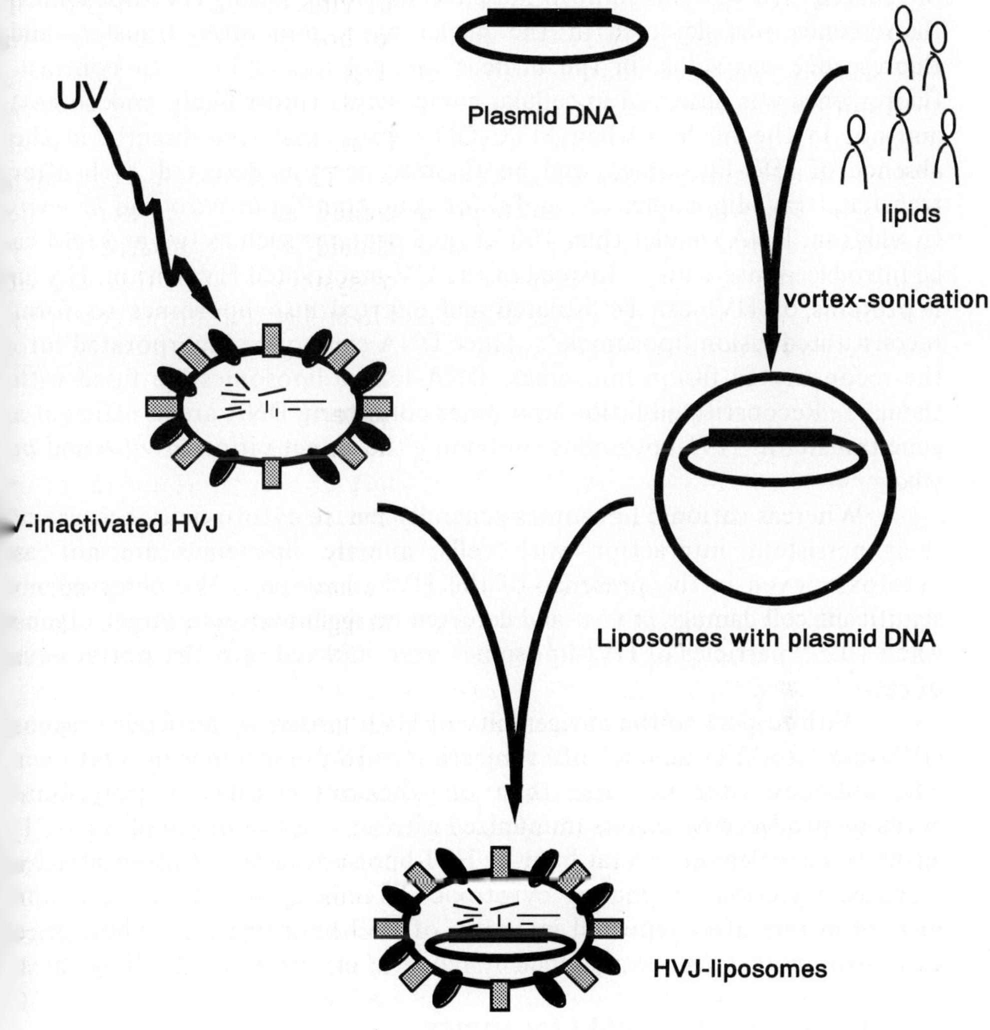

Fig.1 Delivery of the transgene into the target cell using HVJ-liposomes. Plasmid DNA is incorporated into liposomes by vortex-sonication, and the liposomes are fused with UV-inactivated HVJ. The resulting fusigenic viral liposomes can fuse with the plasma membrane of target cells to introduce the DNA directly into the cytoplasm.

introduced into vascular smooth muscle cells (VSMC) using HVJ-liposomes, fluorescence was detected in the nuclei by 5 min after transfer, and fluorescence was stable in the nucleus for at least 72 hr.[8] In contrast, fluorescence was observed in cellular components (most likely endosomes) and not in the nucleus when FITC-ODN was transferred directly in the absence of HVJ-liposomes, and no fluorescence was detected 24 h after transfer. HVJ-liposomes are useful for gene transfer *in vitro* and *in vivo*. In addition, DNA smaller than 100 kb, and proteins such as IgG and IgM ca be introduced into cells. Instead of the UV-inactivated HVJ virion, HN an F proteins of HVJ can be isolated and inserted into liposomes to form reconstituted fusion liposomes.[9] Since DNA can not be incorporated into the reconstituted fusion liposomes, DNA-loaded liposomes are fused with them. Reconstituted fusion liposomes containing DNA are as efficient a gene transfer as HVJ-liposomes containing the intact virion *in vitro* and *in vivo*.

Whereas cationic liposomes generally induce cytotoxicity because of their persistent interaction with cells, anionic liposomes are not as cytotoxic, even in the presence of the HVJ envelope. We observed no significant cell damage *in vivo* and detected no dysfunction in target organs when $10^{10\text{-}11}$ particles of HVJ-liposomes were injected into the portal vein of rats.[10]

With respect to the antigenicity of HVJ-liposomes, antibodies against HVJ were detected one week after injection of HVJ-liposomes into rat liver The antibody titer was less than one-thousandth that of polyclonal antibody produced by rabbits immunized with adjuvant-conjugated HVJ. In addition, gene delivery to rat liver by HVJ-liposomes was not attenuated by repeated injection. Finally, cytotoxic T cells against HVJ were not induced in rats after repeated injection of HVJ-liposomes.[10] Thus, gene expression *in vivo* can be sustained by repeated injection of HVJ-liposomes.

1.2 Improvement of HVJ-Liposomes

To examine whether the use of cationic liposomes can augment the transfection efficiency of the HVJ-liposomes method, we prepared two ifferent dried lipid mixtures, a conventional anionic mixture containing hosphatidylserine (PS) and a new mixture containing positively-charged)C-cholesterol.[11] HVJ-cationic liposomes were 100 times more effective n luciferase gene expression than the anionic liposomes. We then xamined the optimum lipid composition of HVJ-DC liposomes.)ptimized DC-Chol (opDC) liposomes consisting of PtdCho, DOPE, phingomyelin (Sph), Chol and DC-Chol at the molar ratio of 5:5:5:12:3 howed the highest gene expression in cultured cells with the aid of the HVJ nvelope. Thus, HVJ-cationic liposomes are useful for in vitro delivery of nacromolecules. We also found that the delivery of ribozyme to cells was to 5 times more efficient by HVJ-cationic liposomes than by the anionic ype.[12]

Ve next investigated the transfection efficiency of these new gene transfer ehicles *in vivo* in the muscle and liver of adult mice.[11] Total luciferase xpression in organs transfected with HVJ-cationic liposomes was 10 to 150 imes lower than that with conventional HVJ-PS liposomes, which were less fficient for *in vitro* transfection. It is difficult to optimize *in vivo* ransfection strategies with *in vitro* experiments.

We then developed a new type of liposomes[11], the artificial viral nvelope (AVE) liposomes, with a lipid composition similar to that of uman immunodeficiency virus (HIV) envelopes.[13] AVE liposomes ontain PtdChol, Sph, DOPE, PS and Chol at molar ratio of 3.3:13.3:13.3:10:50, respectively. AVE liposomes were modified further o create AVE+DC10 (contains 10% PS and 10% DC-Chol), AVE+DC20 contains 10% PS and 20% DC-Chol), and AVE-PS (contains neither PS or DC-Chol) liposomes. We examined *in vivo* gene transfection fficiency with these liposomes after conjugation with the HVJ envelope. VE yielded the highest luciferase expression in liver (6 to 7 times higher han PS, 60 to 300 times higher than opDC), AVE-PS and AVE+DC10 iposomes, which have a net neutral charge, showed intermediate luciferase ctivities. AVE+DC20 liposomes, which have an excessive amount of ationic lipid, were similar to opDC liposomes. Similar results were btained with luciferase gene expression in mouse skeletal muscle (data not hown). However, we recently found that HVJ-cationic liposomes were nore effective for *in vivo* gene transfer in some cases. High expression of he *LacZ* gene was obtained in restricted regions of chick embryos after injection of HVJ-cationic liposomes[14], whereas HVJ-anionic liposomes wer ineffective. In addition, when HVJ-cationic liposomes containing the *LacZ* gene were administered to rat lung by jet nebulizer, more efficient gene expression in the epithelium of the trachea and bronchus was observe compared to that via HVJ-anionic liposomes.[15] HVJ-cationic liposome were also very effective for antisense (AS)-ODN transfer to a restricte region of rat brain, whereas AS-ODN was broadly distributed in the brain by HVJ-anionic liposomes.

Therefore, we conclude that anionic HVJ-liposomes should be used f gene transfer to broad regions of tissues, and that HVJ-cationic liposome should be used for gene transfer to restricted regions of tissues. We believ that the variable gene transfer efficiency of these liposomes is influenced b the net charge and the size of the liposomes.

1.3 Sustained Gene Expression *In vitro* and *In vivo* by the EBV Replicon Vector Plasmid

Another barrier for human gene therapy is long-term gene expression The transient expression of genes in tissues and inability for targete delivery are the main limitations of current HVJ-liposomes. The mai reason for transient gene expression *in vivo* appears to be instability of th transgene in the cell nucleus. Southern blotting of the transgene in th liver revealed that the gene transferred to liver nucleus by HVJ-liposom existed extrachromosomally without integration into host genome and th episomal form of the transgene was degraded in the liver nucleus after tw weeks.[4] To achieve sustained gene expression, there are two distinc approaches. One is the integration of a transgene into the host genome Insertion of a transgene into the genome of non-dividing cells has bee problematic, and site-specific integration is desirable. Another approach the stable retention of an episomal piece of DNA. Epstein-Barr viru (EBV) has been analyzed thoroughly in terms of its latent infection. Th *cis*-acting oriP (the latent viral DNA replication origin) sequence and th *trans*-acting EBNA-1 (EBV nuclear antigen-1) are required for EBV latent infection, which is characterized by autonomous replication and nucle retention of the EBV genome in host cells.[16-19] However, this EBV vect can transfer genes dominantly to B lymphocytes, but hardly to othe somatic cells. Then, to augment long-term transgene expression in various animal organs, we constructed an EBV replicon-based plasmid, pEB, that contains oriP and EBNA-1, and coupled the replicon vector with HVJ-liposomes. Using this sytem,we succeeded in sustained and enhanced luciferase expression in several cell lines and in mouse liver.[20] When luciferase activity was measured in HEK 293 and BHK-21 cells at various time-points after gene transfer, luciferase gene expression by the introduction of pActLuc was highest on day 1 after the transfer, but then rapidly decreased. However, luciferase activity by the introduction of pEBActLuc was almost equal to that by pActLuc on day 1 in HEK 293 cells, and then increased and it was maintained for at least 10 days at 5 to10 times higher level compared with that by pActLuc. Similar sustained gene expression was observed in human cervical cancer HeLa-S3 and human diploid fibroblast FS3 cells.[20] Southern blot analysis revealed that in human cells transfected with the EBV replicon vector, the plasmid was episomally retained in the nucleus and that the amplification of the plasmid was detected probably due to the autonomous replication of the DNA.[20] In BHK-21 cells, luciferase gene expression was highest on day 1 and then decreased for both plasmids, but it was maintained on and after day 3 post-transfection for pEBActLuc. Southern blot analysis of luciferase gene after gene transfer showed that the plasmid containing EBV sequence was retained extrachromosomally in the nucleus of BHK-21 cells, while the plasmid DNA without EBV sequence was lost between days 3 and 9 after transfer. About 100-150 copies of the pEBActLuc were retained in the nucleus on day 9, decreasing between days 9 and 14.[20] The EBV replicon vector was also very useful for obtaining stable transformants. Approximately 410 colonies resistant to G418 resistant were obtained by the transfer of pcEB vector, while only 10 colonies were resistant to G418 by the transfer of pcDNA3.[20]

Although we have succeeded in gene expression in many organs using HVJ-liposomes, the main limitation of the vector is transient gene expression *in vivo*. Then, we attempted to prolong *in vivo* gene expression by combining EBV replicon vector system with HVJ-AVE liposome. HVJ-AVE liposomes incorporating pEBActLuc or pActLuc were directly injected under perisplanchnic membrane of Balb/c mouse liver. Luciferase activity of the liver was assayed at various time points after the transfer (Fig. 2). When pActLuc without EBV sequence was introduced, luciferase gene

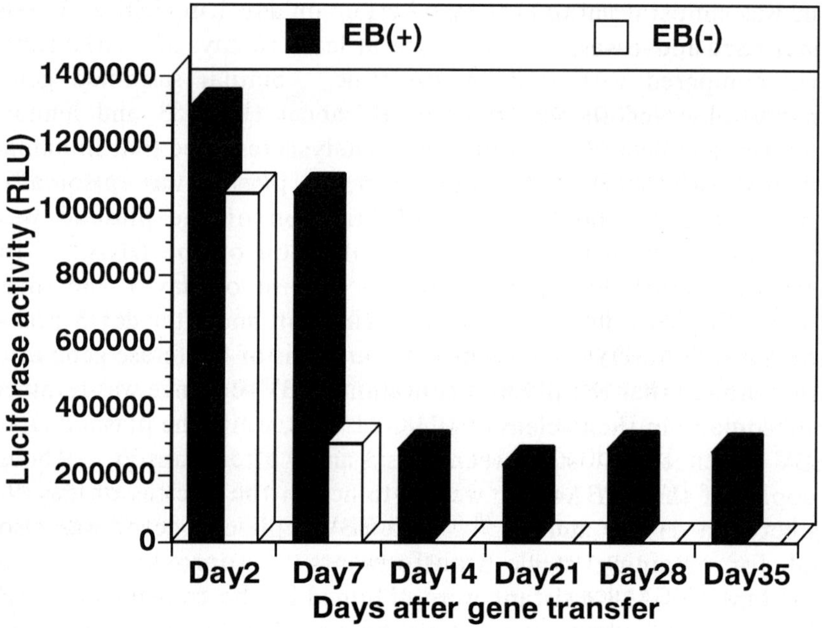

Fig. 2 Sustained gene expression in mouse liver by the Epstein-Barr virus (EBV) replicon vector coupled with HVJ-AVE liposomes. With EBV replicon vector, luciferase gene expression was sustained for at least 35 days in mouse liver after transfer using HVJ-AVE liposome. Without EB virus replicon apparatus, gene expression in liver ceased by 14 days.

expression was detected on day 7, but rapidly decreased to insignificant levels by day 14. However, with pEBActLuc, luciferase gene expression also decreased between days 7 and 14, but the expression remained constant after day 14 for at least 35 days.[6, 20]

2 APPLICATION OF HVJ-LIPOSOMES TO CANCER GENE THERAPY

The potential of HVJ-liposome gene delivery for clinical trials has been evaluated in several animal models.[21] Here, we describe cancer gene therapy using HVJ-liposomes.

2.1 Suicide Gene Therapy of Hepatocellular Carcinoma

Among various strategies of gene therapy of cancer, suicide gene therapy seems to be one of the most promising anti-cancer strategies because cancer cells lacking the transgene are also killed.[22, 23] Clinical trials are under way to treat hepatoma with suicide gene therapy. The success of suicide gene therapy depends on efficient gene transfer to cancer cells to induce a bystander effect.[24] Another important consideration is specific expression of herpes simplex virus-thymidine kinase (HSV-TK) in cancer cells but not in normal tissues. Therefore, to drive the HSV-TK gene, we constructed a system in which HSV-TK gene expression is driven by the early growth response-1 (EGR-1) promoter,[25] which is specifically induced by ionizing radiation. With 10 gray gamma-irradiation, luciferase gene expression driven by EGR-1 promoter was 30-fold enhanced in human hepatoma cell line, HepG2, but less than two-fold in primary pig hepatocytes. HepG2 cells were transplanted under the skin of nude mice. By direct injection of *LacZ* gene by HVJ-liposomes into the tumor mass, about 40% of HepG2 cells expressed *LacZ*.[26] Injection of HVJ- liposomes containing the HSV-TK gene driven by the EGR-1 promoter, followed by 20-gray irradiation and intraperitoneal injection of gancyclovir, led to complete eradication of tumors in mice.[26] When one of the three tools (HSV-tk, irradiation and GCV) was omitted, tumors significantly grew in nude mice.

2.2 Prevention of Melanoma

Interest in cancer prevention, technological advances, and improved understanding of the biology of the immune response have resulted in a resurgence of interest in the use of cancer vaccination as a means o controlling cancer. DNA vaccine is a novel and promising strategy, whicl has been shown to be effective in eliciting both humoral and cell-mediatec immunity that are protective against numerous infectious diseases[27, 28] anc potentially some cancers.[29, 30] (29, 30). Experiments showed tha immunization of mice with liposome-entrapped mRNA encoding th influenza virus nucleoprotein induced anti-influenza CTL *in vivo* .[31] We were interested in the development of RNA-based cancer vaccines becaus RNA has some advantages over DNA, such as safety factor and less potent to integrate into host chromosomes after transfection. RNA expression i transient, lasting for only a few days, making it relatively easier to evaluat and control gene expression *in vivo*. This may be important in some type of gene therapy when a limited expression of the foreign gene is desired However, mRNA is less stable in the host`s body fluids than DNA. A efficient method for delivery of RNA directly into host cells is required fo effective RNA-based vaccines. HVJ-liposome delivery is a liposome-base gene transfer method that has been proven to efficiently deliver th contents such as nucleic acids and proteins directly into host cells *in vivo* b means of virus-cell fusion mechanism.[5, 6, 32] We employed this gen transfer system to deliver RNA to hosts *in vivo* to study mRNA-based tum antigen vaccination.

Human melanoma-associated antigen *gp*100 is a melanocyt differentiation antigen recognized by HLA-A restricted cytotoxic T cells a well as antibodies in patients with melanoma.[33, 34] The human *gp*10 antigen gene which has 75.5% identity to mouse melanoma *gp*100 gene ha been shown to induce some level of protective anti-melanoma immunity i mice after immunization by adenovirus-mediated gene transfer.[35] Ou hypothesis was to test whether vaccination with human *gp*100 mRNA cou induce anti-*gp*100 immune responses and protective anti-tumor immunit in mice challenged with syngeneic melanoma cells. To prepare mRNA fo full-length *gp*100 protein, *in vitro* translation was performed using pSFV vector. The pSFV3 eukaryotic expression vector is a novel gen expression system based on the SFV replicon. The gene of interest can b cloned directly into the pSFV3 plasmid vector that serves as a template fo *in vitro* synthesis of recombinant RNA. When introduced into cells, th recombinant RNA leads to high expression of the heterologous protein wh competing out the host protein synthesis.[36] Initially ***Lac*Z** mRNA delivered by HVJ-liposomes was used as a model system to evaluate mRNA delivery and expression *in vivo*. The pSFV3 expression vector was used for *in vitro* preparation of both ***Lac*Z** mRNA and human *gp*100 mRNA. ***Lac*Z** mRNA was *in vitro* synthesized by pSFV3 expression vector and introduced into mouse spleen using HVJ-liposomes.[37] High level of beta-galactosidase activity was detected for 10 days in mouse spleen. Then the human melanoma-associated antigen *gp*100 mRNA was *in vitro* synthesized by pSFV3 vector and encapsulated in HVJ-liposomes. Immunization by direct injection of the *gp*100 mRNA HVJ-liposomes into mouse spleen induced both anti-*gp*100 antibody and cytotoxic T cell responses against B16 melanoma. Immunizations with *gp*100 mRNA into the spleen delayed tumor growth and significantly prolonged survival compared to control treated mice.[37] These pre-clinical studies demonstrate that an RNA tumor antigen vaccine strategy has potential application for human cancer treatment and prevention.

3 PERSPECTIVE

HVJ-liposomes combine some of the advantages of viral- (high efficiency gene transfer) and non-viral (lack of immunogenicity) vectors. Safety and efficacy tests of HVJ-liposomes have been already confirmed in primates. We are now under way of manufacturing clinical grade HVJ-liposomes. If this obstacle is overcome, HVJ-liposomes could represent a means of treating human cancers.

Acknowledgments This work was supported by grants from the Ministry of Public Welfare in Japan.

References

1. Marshall E, Gene therapy`s growing pains. *Science*, 1995; **269:** 1052 - 1055.

2. Mulligan RC, The basic science of gene therapy. *Science*, 1993; **260:** 926 - 932.

3. Ledley FD, Non-viral gene therapy; the promise of genes as pharmaceutical products. *Hum. Gene Ther.*, 1995; **6:** 1129 - 1144.

4. Kaneda Y, *et al.*, Increased expression of DNA cointroduced with nuclear protein in adult rat liver. Science, 1989; **243:** 375 - 378.

5. Kaneda Y, Fusigenic Sendai-virus liposomes: a novel hybrid type liposome for gene therapy. *Biogenic Amines*, 1998; **14:** 553 - 572.

6. Kaneda Y, *et al.*, Gene therapy using HVJ-liposomes; the best of both worlds. *Mol. Med. Today*, 1999; **5:** 298 - 303.

7. OkadaY, Sendai-virus induced cell fusion, in N. Duzugnes (ed.), *Methods in Enzymology*, Academic Press, Inc., San Diego, 1993; **221:** 18 - 41.

8. Morishita R,*et al.*, Pharmakokinetics of antisense oligonucleotides (cyclin B1 and cdc 2 kinase) in the vessel wall; enhanced therapeutic utility for restenosis by HVJ-liposome method. *Gene*, 1994; **149:** 3 - 9.

9. Suzuki K, *et al.*, Reconstitute fusion liposomes for gene transfer *in vitro* and *in vivo*. *Gene Ther. Regulation.*, 2000; **1:** 65 - 77.

10. Hirano T, *et al.*, Persistent gene expression in rat liver *in vivo* by repetitive transfections using HVJ-liposome. *Gene Ther.*, 1998; **5:** 459 - 464.

11. Saeki Y, *et al.*, Development and characterization of cationic liposomes conjugated with HVJ (Sendai virus): Reciprocal effect of cationic lipid for *in vitro* and *in vivo* gene transfer. *Hum. Gene Ther.*, 1997; **8:** 1965 - 1972.

12. Kitajima I, *et al.*, Efficient transfer of synthetic ribozymes into cells using hemagglutinating virus of Japan (HVJ)-cationic liposomes; application for ribozymes that target HTLV-1 tax/rex mRNA. *J. Biol. Chem.* 1997 **272:** 27099 - 27106.

13. Chander R, Schreier H, Artificial viral envelopes containing recombinant human immunodeficiency virus (HIV) *gp*160. *Life Sci.*, 1992 **50:** 481 - 489.

14. Yamada G, *et al.*, An efficient liposome-mediated gene transfer into the branchial arch, neural tube and the heart of chick embryos; A strategy to elucidate the organogenesis. *Cell. Moll. Biol.* 1997; **43:** 1165 - 1169.

15. Yonemitsu Y, *et al.*, HVJ(Sendai virus)-cationic liposomes; a novel and potentially effective liposome-mediated gene transfer technique to the delivery to the airway epithelium. *Gene Ther.* 1997; **4:** 631 - 638.

16. Reisman D, *et al.*, A putative origin of replication of plasmids derive

from Epstein-Barr virus is composed of two cis-acting components. *Mol. Cell. Biol.*, 1985; **5:** 1822 - 1832.

17. Sugden B, Warren N, A promotor of Epstein-Barr virus that can function during latent infection can be transactivated by EBNA-1, a viral protein required for viral DNA replication during latent infection. *J. Virol.*, 1989; **63:** 2644 - 2649.

18. Wisokenski DA, Yates J L, Multiple EBNA-1 binding sites are required to form an EBNA-1-dependent enhancer and to activate a minimal replication origin within oriP of Epstein-Barr virus. *J. Virol.*, 1989; **63:** 2657 - 2666.

19. Yates J L, *et al.*, Stable replication of plasmids derived from Epstein-Barr virus in various mammalian cells. *Nature*, 1985; **313:** 812 - 815.

20. Saeki Y, *et al.*, Sustained transgene expression *in vitro* and *in vivo* using an Epstein-Barr virus replicon vector system combined with HVJ-liposomes. *Gene Ther.* 1998; **5:** 1031 - 1037.

21. Dzau VJ, *et al.*, Fusigenic viral liposome for gene therapy in cardiovascular diseases. *Proc. Natl. Acad. Sci.(USA)*, 1996; **93:** 11421 - 11425.

22. Oldfield EH, *et al.*, gene therapy for the treatment of brain tumors using Intra-tumoral transduction with thymidine kinase gene and Intravenous gancyclovir. *Hum. Gene Ther.*, 1993; **4:** 39 - 69.

23. Ram Z, *et al.*, Therapy of malignant brain tumors by intratumoral implantation of retroviral vector-producing cells. *Nat. Med.*, 1997; **3:** 1354 - 1361.

24. Freeman SM, *et al.*, The bystander effect: tumor regression when a fraction of the tumor mass is gentically modified. *Cancer Res.*, 1993; **53:** 5274 - 5283.

25. Seung LP, *et al.*, Genetic radiotherapy overcomes tumor resistance to cytotoxic agents. Cancer Res., 1995; **55:** 5561 - 5565.

26. Kawashita Y, *et al.*, Regression of hepatocellular carcinoma *In vitro* and *In vivo* by radiosensitizing suicide gene therapy under the Inducible and spatial control of radiation. 1999; **10:** 1509 - 1519.

27. Ho TY, *et al.*, DNA vaccination induces a long-term antibody response

and protective immunity against pseudorabies virus in mice. *Arch Virol.* 1998; **143:** 115 - 125.

28. Larsen DL, *et al.*, Coadminstration of DNA encoding interleukin-6 and hemagglutinin confers protection from influenza virus challenge in mice. *J. Virol.* 1998; **72:** 1704 - 1708.

29. Iwasaki A, Barber BH, Induction by DNA immunization of a protective antitumor cytotoxic T lymphocyte response against a minimal-epitope-expressing tumor. *Cancer Immunol Immunother.* 1998; **45:** 273 - 279.

30. Chen Y, *et al.*, DNA vaccines encoding full-length or truncated Neu induce protective immunity against Neu-expressing mammary tumors. *Cancer Res.* 1998; **58:** 1965 - 1971.

31. Martinon F, *et al.*, Induction of virus-specific cytotoxic T lymphocytes *in vivo* by liposome-entrapped mRNA. *Eur. J. Immunol.* 1993; **23:** 1719 - 1722.

32. Okamoto T, *et al.*, Induction of antibody response to human tumor antigens by gene therapy using a fusigenic viral liposome vaccine. *Gene The.* 1997; **4:** 969 - 976.

33. Kawakami Y, *et al.*, Identification of a human melanoma antigen recognized by tumor-infiltrating lymphocytes associated with *in vivo* tumor rejection. *Proc Natl Acad Sci (USA)* 1994; **91:** 6458 - 6462.

34. Huang S K S, *et al.*, Antibody responses to melanoma/melanocyte anto-antigens in melanoma patients. *J. Invest. Dermatol.* 1998; **111:** 662-667.

35. Zhai Y, *et al.*, Antigen-specific tumor vaccines: development and characterization of recombinant adenoviruses encoding MART1 or *gp*100 for cancer therapy. *J Immunol.* 1996; **156:** 700 - 710.

36. Liljestrom P, Garoff H, A new generation of animal cell expression vectors based on the Semliki Forest virus replicon. *Bio/Technology* 1991; **9:** 1356 - 1361.

37. Zhou W-Z, *et al.*, RNA melanoma vaccine: Induction of antitumor Immunity by human glycoprotein 100 mRNA Immunization. *Hum. Gene Ther.* 1999; **10:** 2719 - 2724.

FRONTIERS IN HUMAN GENETICS
Diseases and Technologies

© 2001 by World Scientific Publishing Co. Pte. Ltd.

CURRENT STATUS OF DEVELOPMENT TO IMPROVE THE EFFICIENCY AND TARGETING SPECIFICITY OF LIPOSOMES FOR GENE THERAPY

KAM M. HUI* and HUI GAO†
Gene Vector Laboratory
Division of Cellular and Molecular Research
National Cancer Centre
11 Hospital Drive
Singapore 169610
**E-mail: cmrhkm@nccs.com.sg*
†*E-mail: cmrdgh@nccs.com.sg*

1 INTRODUCTION

Gene therapy plays an increasingly important role in providing an alternative for alleviating symptoms or preventing the occurrence of human diseases. The first concern for gene therapy is the ability to deliver large and heavily charged complexes containing the desired genes into cells. Currently, the vectors available for the delivery of DNA complexes fall into two categories: viral and non-viral. Non-viral vectors or so-called synthetic vectors have some advantages over the viral vectors, including their non immunogenicity, less restriction on the size of the DNA that can be incorporated, low acute toxicity, simplicity, and feasibility to be produced on a large scale. Among non-viral vectors, liposomes are widely used to deliver pharmacological agents and biological macromolecules to the target cells. Recently, intense research activity has been directed at the design,

* Corresponding author

synthesis, and evaluation of a number of purely chemical gene transfer agents as well as new formulations of currently used liposomes. This is also reflected in the numbers of gene therapy clinical trial protocol utilizing liposomes as gene delivery vehicles for genetic and acquired diseases.

This review will address the various current attempts to develop novel liposomal gene delivery systems for clinical applications. An overview on the development of liposomes as gene delivery vehicles for gene therapy to treat a variety of genetic and acquired diseases will be discussed.

2 DEVELOPMENT OF NOVEL LIPOSOMAL VECTORS

Historically, liposomes have been employed as delivery vehicles to encapsulate pharmacological agents and biological macromolecules to modify mammalian cells.[1] Liposomes were firstly used as internalization reagents of nucleic acids in the late 1970s.[2,3] Initially, the liposomes employed were mostly phosphatidylserine, large unilamellar vesicles that were prepared by the reverse phase evaporation method.[4] The employment of cationic liposomes for DNA transfer was first described by Papahadjopoulos *et al.* in 1982.[4] The stearylamine-doped liposomes employed, however, were later found to be rather toxic and not more efficient than large negatively charged liposomes with reference to their properties of delivery.[5,6] In the 1980s, Behr and colleagues[7] studied the interactions of DNA with cationic micelles and liposomes in details and it was only in 1987, however, that the potential of cationic liposomes for DNA transfection was fully realized.[8] Some of the early applications on employing liposomes for gene delivery were described by Nicolau and Papahadjopoulos.[9] Nicolau and colleagues[1] generated liposomes consisting of phosphatidylcholine, phosphatidylserine and cholesterol, and complexed them with a plasmid designed to express rat insulin. Following intravenous injection of the insulin-gene-containing liposomes into rats, a small increase in serum insulin was observed. The expression, however, lasted only for less than a day. Although this early result demonstrated the feasibility of employing liposomes for gene delivery, a lot of difficulties were associated with the application especially with the reproducible preparation of conventional liposomes. These early problems led to the development of cationic lipid formulations for gene transfer.

Among the liposomal systems described, cationic liposomes seem to be the most widely used DNA delivery system. Since the first report on manufacturing cationic liposomes, their interactions with DNA, and the use of such complexes for gene expression, hundreds of novel cationic lipids have been synthesized.[10–17] Numerous papers have been reported to illustrate the ability to obtain gene expression *in vitro* and *in vivo* by employing different cationic and neutral lipids, at different charge ratios, and with different transfection protocols.[18,19] Currently, gene delivery employing non-viral delivery vehicles *in vivo* relies mostly on employing DNA complexes with polycations, cationic liposomes, cationic polyelectrolytes, and combinations of these various components. Liposomes have been used effectively to introduce drugs,[20–22] radiotherapeutic agents,[23] enzymes,[24,25] viruses,[26] transcription factors[27] and allosteric effectors[28] into a variety of cultured cell lines and animals.

In an attempt to improve the efficiencies of gene transfection, researchers are working on improving the promoters within the DNA constructs,[29] synthesizing novel lipids and polymers, as well as the addition of functional ligands into the delivery systems. In addition, several clinical trials examining the effectiveness of liposome-mediated gene delivery have been successfully completed. In general, it has been recognized that the ideal liposome-based vector should offer the added advantages of : (i) be rapidly prepared in large quantities, (ii) be able to take up DNA inserts of chromosomal length into the host genome, (iii) absence of viral DNA sequence, (iv) inability to self-replicate *in vivo* and (v) potential for cell-specific targeting.

The first generation of liposomes used in drug delivery mainly consisted of phospholipids that were of relatively low efficiency of gene transfer and without the presence of targeting ligands on their surfaces.

The recent development of cationic liposomes reacting with DNA has been focused on improving the efficiencies of DNA-mediated gene transfer and the targeting specificity of the liposome-DNA complexes. A majority of the cationic lipids that have been described recently can be broadly divided into two classes: those that harbor cholesterol as the lipid anchor and those that utilize double fatty acids of varying lengths and different extent of saturation of the hydrocarbon chains. Both classes of liposomes bear positively charged short aliphatic chains as their head groups that allow them to interact potentially with both negatively charged DNA and biomembrane. Examples of these lipids include DC-chol, DOTAP, DOTMA and DMRIE. In comparison to the first generation lipids, the most notable difference in the structures of the more recent lipids is that they bear multivalent-chained head group[30,31] or employ heterocyclic rings with one or more heteroatoms as their headgroups.[32] The inclusion of multivalent-chained amines and heteroatoms might contribute to the interactions between liposome and DNA as well as liposome/DNA complexes and cell membrane. Liposomes prepared from these groups of novel lipids and neutral lipids showed more efficient gene delivery both *in vitro* and *in vivo*. Most importantly, some of them demonstrated certain degree of serum resistance so that they can be applied via systemic administration.[33,34]

Studies addressing the novel formulation of liposomes with improved gene delivery property have also been reported. One of these examples is the encapsulation of proteins,[35,36] small organic molecules[37] or polymers in the liposome-DNA complexes. These molecules substantially enhanced the efficiency of DNA encapsulation into liposome vectors by altering the physical state of the DNA. Histones were the first group of proteins employed for this purpose.[35] Lysozyme is a basic protein that could associate with DNA at physiological pH. Jay and Gilbert[35] reported that encapsulation of DNA in the presence of lysozyme resulted in 59% of the DNA being encapsulated. The DNA within the liposomes was very condensed with 0.5% aqueous volume. The condensation of DNA led to the increase of the efficiency of DNA transfection being 100-times greater than expected from random entrapment.

Considering the mechanism of transfection mediated by cationic liposomal systems and difference in the ability to condense DNA between monovalent cationic lipids and polyamines or multivalent cationic lipids, cationic polymers were introduced at appropriate ratios to the mixture of the cationic liposomes to generate novel formulations of liposomes.[38,39] Among the several polymers tested, polylysine with a molecular weight of 25Kd was found to be the most efficient in enhancing transfection.[38] The enhancing effect, of between 2 to 28 folds, was observed for a number of cell lines *in vitro*. It was observed that cells that were difficult to be transfected with liposome/DNA complexes could be transfected when a cationic polymer was included in the formulation of the liposomes (LPDI). The improved activity observed in transfecting the LPD-1 particles derived by the addition of polylysine might primarily be due to the condensed DNA structures. In comparison to the DC-chol liposome/DNA complexes that are large and heterogeneous in size, LPD-1 particles are highly compact with a size of less than or close to 100nm. These particles would be more favorable to enter cells via the endocytosis pathway, which is the major mechanism responsible for the cellular uptake of liposome/DNA complexes. Thus, LPD-1 is particularly suitable for transfecting cells that do not possess other DNA-uptake mechanism. Further reoptimization of LPD-1 led to the development of a lipidic vector that is suitable for intravenous gene delivery. The formulation of such a LPD complex composes of protamine, DNA and DOTAP/cholesterol liposomes.[39,40] Upon intravenous administration, this formulation gives a high level of systemic gene expression of the delivered DNA.

The development of novel liposomes as gene delivery vectors also addresses the targeting of liposomes to specific tissues or specific cell types. Three complementary strategies are used to increase the bioavailability of liposomal preparations to specific populations of cells: (1) changing the route of administration, (2) addition of targeting molecules to the surface of the liposomes, and (3) changing the composition of the liposomes. The composition of certain lipids could enhance the targeting of liposomes. For example, hepatocytes and splenic lymphocytes have asialoglycoprotein receptors.[41,42] The incorporation of lactosylceramide into liposomes for injection intravenously into mice increased the uptake of liposomes 3.5-fold into hepatocytes, and 1.5-fold into splenic lymphocytes.[43,44]

Another way to deliver molecules to specific target cells is to construct liposomes that would become structurally unstable under conditions that occur only near the desired target. This type of liposome has been commonly used for drug delivery.[45,46] Liposomes have been developed that destabilize in response to low pH.[47–49] Liposomes constructed with PE and palmitoyl-homocysteine, oleic acid,[50] or diacylsuccinylglycerols fuse spontaneously with other lipid bilayers at low pH. Because liposomes enter cells through the endocytic pathway, this type of liposome has potential for the intracellular delivery of various biological molecules. In this system, DNA adsorbed to the outside of the liposomes could also transfects cells. This is a characteristic common to all destabilizing liposomes.

Although a certain degree of targeting can be achieved by varying the lipid composition of the liposomes, a greater degree of specificity of targeting can be achieved by the addition of specific targeting molecules to the lipid vesicles (Figure 1). These molecules can be antibodies or ligands to specific cell-surface receptors. Antibodies and other proteins can either be attached to the outside of existing liposomes[51] or be integrated into the membrane as the liposome is formed.[52,53] Such liposomes are referred to as immuno-liposomes. When these liposomes are used for *in vivo* gene delivery, they bind efficiently to their target cells only when they are composed of lipid formulations with low RES uptake.[54] The use of antibodies with pH-sensitive liposomes change the tissue distribution of the immuno-liposomes and allow them to be targeted to tumor cells expressing the original antigens used to generate the antibodies.

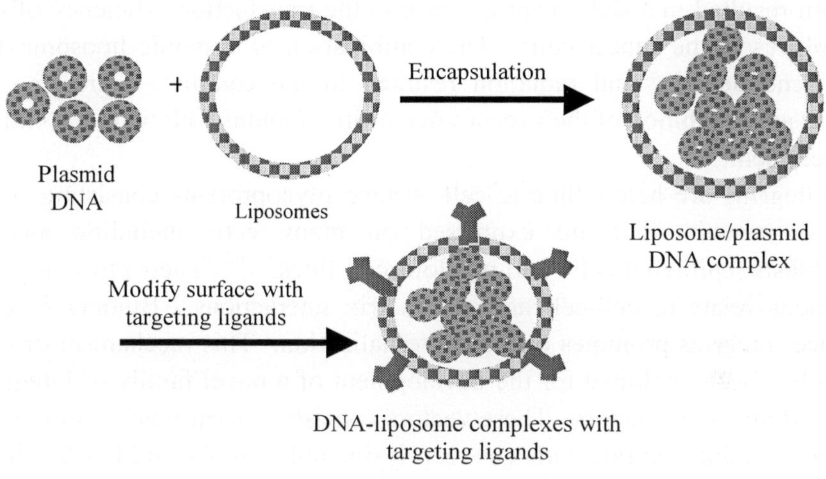

Fig. 1 Schematic representation showing plasmid DNA containing lipsomes with targeting ligands

Ligands can also be incorporated into the surface of liposomes by firstly modifying the ligand with a lipid group. The ligand will then be covalently inserted into the membrane during the formation of vesicles.[55,56] These ligands are intended to introduce biological functions into the complexes in order to make them more effective at delivering genes into the target cells. The peptides incorporated are usually recognized by a specific cell surface receptor and therefore as a result, the liposome-DNA complexes will be targeted to specific cells bearing this receptor. Potentially, various cell types may be transfected *in vivo* by this approach provided suitable receptor-ligand combinations could be identified. One example of such a process is the coating of liposomes with transferrin via complexing the transferrins to stearylamine present in the liposomal membrane. Transferrin is an important targeting ligand that could be incorporated in liposome/DNA complexes.[57] The level of transferrin receptors (TfR) have been reported to be elevated in various types of human cancer cells including breast,[58] prostate,[59–61] as well as squamous cell carcinomas.[62] The increase in the level of TfR correlated with the aggressive or proliferative ability of tumor cells.[63] Liang Xu and colleagues[64] demonstrated that the use of the transferrin ligand (Tf) for targeting a cationic liposome delivering system resulted in a significant increase in the transfection efficiency of the complexes to the cancer cells. The combination of systemic liposome-Tf-p53 gene delivery and radiation resulted in the complete regression of cancer and inhibition of their recurrence even 6 months following the end of all treatment.[65]

Integrins are heterodimeric cell surface glycoproteins consisting of α and β subunits that are expressed on many cells including airway fibroblasts, epithelial cells and myeloid cell lines.[66–68] Their physiological functions relate to cell-cell and cell-matrix interactions. Binding of cell surface integrins promotes cellular internalization. This mechanism of cell entry has been exploited for the development of a novel family of integrin-targeted non-viral vectors. These vectors consists of a cationic liposome, an integrin-binding peptide with a sixteen-lysine tail, and plasmid DNA, which combine electrostatically to form the LID vector complex. Such complexes demonstrated substantially higher expression of the reporter gene in animal model than vectors without the incorporation of integrin in the formulations.[69]

3 CLINICAL APPLICATIONS OF GENE THERAPY

As mentioned above, the identification of individual genes responsible for specific diseases has led to the prospect that these diseases might be "cured" through replacement of the defective gene with a normal gene. In addition to its logical role for the correction of monogenetic inherited diseases, gene therapy also holds promise for treatment of other acquired disorders of multifactorial origin, such as human cancers and cystic fibrosis. Three major approaches have been employed to design gene therapy strategies.[70] These include: (i) direct introduction of normal genes into the target cells to replace the defective ones, (ii) addition of genes for the stimulation of immune system to trigger immune self-attack and sensitization of cells to other treatments, (iii) genetic vaccination. Furthermore, the transfection of bone marrow cells with genes encoding the multi-drug resistance (MDR) gene have been adopted as a strategy to improve the clinical outcome of current chemotherapy protocols.

3.1 Gene Therapy for Cystic Fibrosis (CF)

One of the most exciting applications of non-viral gene therapy concerns the use of cationic-lipid-mediated gene transfer to treat cystic fibrosis (CF). Cystic fibrosis (CF) is a common lethal autosomal recessive disorder that affects approximately 1 in 2500 Caucasian births. The gene responsible for CF was isolated by positional cloning in 1989[71,72] and was shown to encode a membrane-associated glycoprotein referred to as the cystic fibrosis transmembrane conductance regulator (CFTR). Over 400 mutations in the CFTR gene have been identified[73] and the consequence of these mutations are varied but all lead to the generation of CFTR variants whose cAMP-stimulated Cl^- channel activity is diminished or absent from the cell surface. Conventional therapy for CF includes the use of physiotherapy, antibiotic, and pancreatic supplements. Although improvement in both the qualities of life and life expectancy are likely to occur from conventional therapy, these continual medical treatments are expensive. Attempts to improve the clinical outcome of this disease by gene therapy have been initiated. Present attempts to alleviate the disease by gene therapy have focused primarily on delivering a copy of the wild-type CF gene to the affected cells in the lungs. Both viral and non-viral vectors are currently being evaluated and several protocols have been reported to achieve efficient gene delivery to the airway epithelial cells.[74–77]

To facilitate the transfection of airway epithelial cells, the vectors need to harbor properties that allow them to traverse several biological barriers. An effective treatment of CF lung disease will likely require the uniform, low level transfection of the entire lung with the wild type CFTR gene. To achieve this, nebulization of the gene delivery vector represents the most simple and direct approach. One of the potential advantages of liposomal vectors to the treatment of CF is that DNA-liposome complexes can be produced as an aerosol, thus gaining access to pulmonary epithelial cells, which are present over a large surface area. In theory, the liposomal approach offers the potential for efficient gene transfer without the attendant risks of the adenovirus and adeno-associated virus gene-therapy approaches. A large number of preclinical *in vivo* studies have been performed with cationic lipids/pDNA complexes in the lung using either instillation or nebulization.[78–81] Hyde and colleagues utilized lipofectin (DOTMA/DOPE) to successfully delivered the CFTR gene to the trachea of transgenic mice that were homozygous for a null mutation in CFTR.[82] Alton and co-workers performed similar studies.[83] These results, taken in combination with the finding that repeated aerosol administration of DNA-lipid complexes to the lungs of rabbits are well tolerated, indicated that a variety of different cationic lipids and formulations are capable of mediating gene transfer to the airway epithelial cells. In some instances, these cationic lipid-mediated levels were sufficiently high to partially correct some of the electrophysiological defects shown to be associated with CF.

These preclinical results promoted the clinical studies of cationic liposomes gene therapy for CF. The first liposome-mediated CF gene therapy trial was carried out by Alton and colleagues in 1995.[84] Initial protocols (both non-viral and viral) to treat CF have focused on the administration of the complexes to the nasal epithelium, which shares some properties with the pulmonary epithelium and may minimize the risk to subjects involved in the Phase I clinical studies. Several groups are delivering normal CFTR genes in early clinical trials to patients with CF0.[85–88] Results from several intra-nasal instillation clinical studies using different cationic lipid formulations including DOTAP, GL67/DOPE, DMRIE/DOPE, EDMPC/ cholesterol have since been reported.[89]

3.2 Cancer Gene Therapy

The development of cancer is a highly complex process, of which some of the possible causes include genetic, environmental or dietary factors.[90] It has been demonstrated that the immune system plays an important role in the control of malignancies. Therefore, the introduction of genes encoding highly immunogenic novel antigens or immuno-potentiating agents into tumor cells both *in vivo* and *ex vivo* could result in the stimulation of the host's immune response against tumors. Based on this logic, cancer vaccines and cancer gene therapy protocols have been developed using both the viral and non-viral gene delivery systems. Clinical application of liposome/pDNA complexes to deliver therapeutic genes to cancer cells with the objective to achieve therapy has been extensively studied by many investigators and biotechnology companies. The current cationic liposomes being employed for clinical studies are listed in Figure 2.

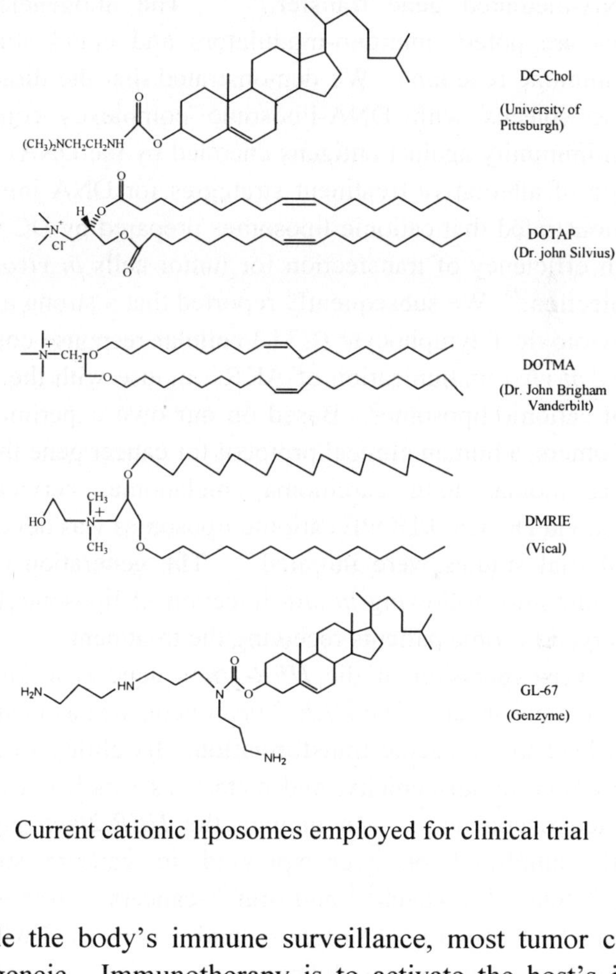

Fig. 2 Current cationic liposomes employed for clinical trial

To evade the body's immune surveillance, most tumor cells are poorly immunogeneic. Immunotherapy is to activate the host's immune system to combat tumor cells. To achieve this goal, tumor cells must be immunogenic and can be recognized as foreign to the host. Several approaches have been developed employing DNA-mediated gene transfer technology to modify antigens expressed on tumor cells with the ultimate aim of enhancing their immunogenicity.

In our earlier work, we expressed the allogeneic major histocompatibility complex (MHC) class I molecules on tumor cells via *ex vivo* DNA-mediated gene transfer.[91–94] The allogeneic MHC class I molecules are potent immuno-modulators and could stimulate a strong cellular immune reactions. We demonstrated that the direct immunization of mouse spleens with DNA-liposome complexes could induce cell-mediated immunity against antigens encoded by the DNA. This may allow the design of alternative treatment strategies for DNA immunization. We also demonstrated that cationic liposomes prepared by DC-Chol and DOPE give high efficiency of transfection for tumor cells *in vivo* following intra-tumor injection.[95] We subsequently reported that a strong allo-reactive anti-H-2K^b cytotoxic T lymphocyte (CTL) cellular response could be generated following *in vivo* immunization of AKR/J mouse with the *H-2K^b* gene and DC-Chol cationic liposomes. Based on our own experimental studies and those of others, a human clinical protocol for cancer gene therapy, including breast carcinoma, lung carcinoma, melanoma, cervical and ovarian carcinoma via DC-Chol/DOPE cationic liposomes was developed and Phase I clinical trial studies were initiated.[96] The generation of strong tumor-specific immunity following *in situ* injection of liposome/DNA complexes was observed in some patients receiving the treatment

The over-expression of the *HER-2/neu* gene is a frequent molecular event in human cancers. The *HER-2/neu* oncogene has many characteristics that will lead to oncogenic transformation. Its ability to promote cellular transformation, tumorigenicity, and metastasis was first demonstrated with the rat neu oncogene.[97,98] In human, the *HER-2/neu* proto-oncogene is frequently amplified or over-expressed in cancers such as breast,[99] ovarian,[100] lung,[101] stomach[102] and oral[103] cancers. Over-expression of the *HER-2/neu* indicates a poor prognosis and is correlated with a reduction in the survival rate of patients with breast and ovarian cancers. Since the over-expression of *HER-2/neu* proto-oncogene significantly contributes to the malignant development of many human cancers, strategies that aim to down-regulate *HER-2/neu* gene expression have become highly attractive molecular approaches for cancer gene therapy. Hung *et al.*,[104–106] conducted a series of experiments to evaluate the efficacy and safety of introducing E1A and T antigen encoding DNA into cancer cells of tumor-bearing mice to suppress *HER-2/neu*-mediated cell transformation. Based on their early studies, these authors have a Phase I gene therapy clinical trial protocol ongoing to introduce the E1A-containing liposomes to patients with metastatic breast or epithelial ovarian cancer that over-expresses the *HER-2/neu* oncogene. The result indicates that appropriate gene expression can be observed at tolerable doses. Although further clinical trials are required to evaluate the full potential of the therapeutic efficacy, the preliminary results are encouraging.

4 PERSPECTIVE

Since the over-expression of oncogenic genes is a common mechanism for oncogenic transformation and other infectious diseases such as AIDS, the strategy to down-regulate genes that are over-express offer a potential alternative therapeutic approach for the treatment of genetic and acquired diseases. Besides cancers, there are at least ten other vaccines that are in advanced pre-clinical development.[107] Human clinical trials to evaluate DNA vaccines for HIV, influenza, malaria, and hepatitis are currently under way. The concerted efforts in gene therapy have provided fruitful achievements towards a new era to cure human diseases. A number of obstacles, however, still must be overcome for successful clinical applications. Ideal vectors that combine all the necessary safety parameters of non-viral vectors and possess the transfection efficiency of viral vectors have yet to be designed.

References

1. Nicolau C *et al.*, *In vivo* expression of rat insulin after intravenous administration of the liposome-entrapped gene for rat insulin I, *Proc Natl Acad Sci USA* 1983; **80:** 1068–1072.
2. Dimitriadis D, Entrapment of plasmid DNA in liposomes, *Nucleic Acids Res* 1979; **6:** 2697–2705.
3. Fraley P *et al.*, Introduction of liposome encapsulated sv-40 DNA into cells, *J Biol Chem* 1980; **255:** 10431–10435.

4. Fraley P *et al.,* Liposomes: The development of a new carrier system for introducing nucleic acids into plant and animal cells, *Curr Top Microbiol Immunol* 1982; **96:** 171–187.
5. Hoffman RM *et al.,* Binding and entrapment of high molecular weight DNA by lecithin liposomes, *FEBS Lett* 1978; **93:** 365–368.
6. Budker VG *et al.*, Interaction of polynucleotides with natural and model membranes, N*ucleic Acids Res* 1980; **8:** 2499–2515.
7. Behr JP, DNA strongly binds to micelles and vesicles containing lipopolyamines or lipointercalants, *Tetrahedron Lett* 1986; **27:** 5861–5864.
8. Felgner PL *et al.*, Lipofection: A highly efficient lipid-mediated DNA transfection procedure, *Proc Natl Acad Sci USA* 1987; **84:** 7413–7417.
9. Nicolau C and Papahadjopoulos D, (eds.), *Nonmedical Applications of Liposomes*, pp. 341–352. Amsterdam: Elseviers, 1998.
10. Lasic DD *et al.,*,Recent advances in medical applications of liposomes: Sterically stabilized and cationic liposomes. In Book of Abstracts (pp. 61–65). Proceedings of Fourth European Symposium on controlled Drug Delivery, 1996, Nordwijk ann Zee.
11. Leventis R *et al.,* Interactions of mammalian cells with lipid dispersions containing novel metabolizable cationic amphiphiles, *Biochim Biophys Acta* 1990; **1023:** 124–132.
12. Gao X. *et al.,*,Potentiation of cationic liposome-mediated gene delivery by polycations, *Biochemistry* 1996; **35:** 1027–1036.
13. Felgner JH *et al.,* Enhanced gene delivery and mechanism studies with a novel series of cationic lipid formulations, *J Biol Chem* 1994; **269:** 2550–2561.
14. Farhood H *et al.*, Effect of cationic cholesterol derivatives on gene transfer and protein kinase C activity, *Biochim Biophys Acta* 1992; **1111:** 239–246.
15. Solodin I *et al.*, A novel series of amphiphilic imidazolinium compounds for *in vitro* and *in vivo* gene delivery, *Biochemistry* 1995; **34:** 13537–13544.
16. Wheeler CJ *et al.*, Converting an alcohol to an amine in a cationic lipid dramatically alters the co-lipid requirement, cellular transfection

activity and the ultrastructure of DNA-cytofectin complexes, *Biochim Biophys Acta* 1996; **1280:** 1–11.

17. Lee ER *et al.*, Detailed analysis of structures and formulations of cationic lipids for efficient gene transfer to the lung, *Human Gene Ther* 1996; **7:** 1701–1717.
18. Eastman SJ *et al.*, Aerosolization of cationic lipid/pDNA complexes-*in vitro* optimization of nebulizer parameters for human clinical studies, *Human Gene Ther* 1998; **9:** 43–52.
19. Fasbender A *et al.,* Effect of co-lipids in enhancing cationic lipid-mediated gene transfer *in vitro* and *in vivo*, *Gene Ther* 1997; **4:** 716–725.
20. Heath TD *et al.*, Antiproliferative and anticontractile effects of liposome encapsulated fluoroorotate, *Invest Ophthalmol Vis Sci* 1987; **28:** 1365–1372.
21. Storm G *et al.*, Potential pitfalls in *in vitro* antitumor activity testing of free and liposome-entrapped doxorubicin, *J Pharm Sci* 1988; **77:** 823–830.
22. Balazsovits JA *et al.*, Analysis of the effect of liposome encapsulation on the vesicant properties, acute and cardiac toxicities, and antitumor efficacy of doxorubicin, *Cancer Chemother Pharmacol* 1989; **23:** 81–86.
23. Pikul SS *et al.*, *In vitro* killing of melanoma by liposome-delivered intracellular irradiation, *Arch Surg* 1987; **122:** 1417–1420.
24. Gregoriadis G and Buckland RA, Enzyme-containing liposomes alleviate a model for storage disease, *Nature* 1973; **244:** 170–172.
25. Tanwell AK *et al.*, Liposome-mediated augmentation of antioxidant defenses in fetal rat pneumocytes, *Am J Physiol* 1990; **258:** L165–172.
26. Faller DV and Baltimore D, Liposome encapsulation of retrovirus allows efficient superinfection of resistant cell lines, *J Virol* 1984; **49:** 269–272.
27. Debs RJ *et al.*, Regulation of gene expression *in vivo* by liposome-mediated delivery of a purified transcription factor, *J Biol Chem* 1990; **265:** 10189–92.

28. Nicolau C and Gersode K, Incorporation of inositol hexaphosphate into intact red blood cells. I. Fusion of effector-containing lipid vesicles with erythrocytes, *Naturwissenschaften* 1979; **66:** 563–566.
29. Kriegler M, Assembly of enhancers, promoters, and splice signals to control expression of transferred genes, *Methods Enzymol* 1990; **185:** 512–27.
30. Byk G *et al.,* Synthesis, activity, and structure-activity relationship studies of novel cationic lipids for DNA transfer, *J Med Chem* 1998; **41:** 229–235.
31. Cooper RG *et al.*, Polyamine analogues of 3β-[N-(N', N'-dimethylamino ethane) carbamoyl]-cholesterol (DC-chol) as agents for gene delivery, *Chem Eur J* 1998; **4:** 137–151.
32. Gao H and Hui KM, Synthesis of a novel series of cationic lipids that can act as efficient gene delivery vehicles through systematic heterocyclic substitution of cholesterol derivatives, (manuscript in preparation).
33. Lewis JG *et al.*, A serum resistant cytofectin for cellular delivery of antisense oligonucleotides and plasmid DNA, *Proc Natl Acad Sci USA* 1996; **93:** 3176–3181.
34. Fasbender AJ *et al.,* Optimization of cationic lipid-mediated gene transfer to airway epithelia, *Am J Physiol* 1995; **269**: L 45–L51.
35. Jay DG *et al.*, Basic protein enhances the incorporation of DNA into lipid vesicles: model for the formation of primordial cells, *Proc Natl Acad Sci USA* 1987; **84:** 1978–1980.
36. Mannio RJ *et al.,* Encapsulation of high molecular weight DNA in large unilamellar phospholipid vesicles. Dependence on the size of the DNA, *FEBS Lett* 1979; **101:** 229–232.
37. Tikchonenko TI *et al.*, Transfer of condensed viral DNA into eukaryotic cells using proteoliposomes, *Gene* 1988; **63:** 321–330.
38. Li S *et al.*, *In vivo* gene transfer via intravenous administration of cationic lipid-protamine-DNA (LPD) complexes, *Gene Ther* 1997; **4:** 891–900.
39. Sorgi FL *et al.,* Protamine sulfate enhances lipid mediated gene transfer, *Gene Ther* 1997; **4:** 961–968.

40. Li S *et al.* Characterization of cationic lipid-protamine-DNA (LPD) complxes for intravenous gene delivery, *Gene Ther* 1998; **5:** 930–937.
41. Kieda C *et al.,* Lymphocyte membrane lectins. Direct visualization by the use of fluoresceinyl-glycosylated cytochemical markers, *FEBS Lett* 1979; **99:** 329–232.
42. Kolb H, A ganglioside receptor on lymphocytes mediates recognition of self, *Biochem Biophys Res Commun* 1982; **105:** 1488–1495.
43. Soriano P *et al.*, Targeted and nontargeted liposomes for *in vivo* transfer to rat liver cells of a plasmid containing the preproinsulin I gene, *Proc Natl Acad Sci USA* 1983; **80:** 7128–3711.
44. Grosse E *et al.*, Flow cytofluorometric investigation of the uptake by hepatocytes and spleen cells of targeted and untargeted liposomes injected intravenously into mice, *Biochim Biophys Acta* 1984; **805:** 354–361.
45. Yatvin MB *et al.*, Temperature- and pH-sensitive liposomes for drug targeting, *Methods Enzymol* 1987; **149:** 77–87.
46. Ho RJ *et al.,* Target-sensitive immunoliposomes as an efficient drug carrier for antiviral activity, *J Biol Chem* 1987; **262:** 13973–13978.
47. Yatvin MB *et al.*, pH-sensitive liposomes: possible clinical implications, *Science* 1980; **210**: 1253–1255.
48. Connor J *et al.*, pH-sensitive liposomes: acid-induced liposome fusion, *Proc Natl Acad Sci USA* 1984; **81:** 1715–1718.
49. Wang CY *et al.*, pH-sensitive immunoliposomes mediate target-cell-specific delivery and controlled expression of a foreign gene in mouse, *Proc Natl Acad Sci USA* 1987; **84:** 7851–7855.
50. Heath DT, Covalent attachment of proteins to liposomes, *Methods Enzymol* 1987; **149:** 111–119.
51. Huang L *et al.*, Coupling of antibodies with liposomes. In G. Gregoriadis (ed.*), Liposome technology* (Vol. III, pp. 51–62). Boca raton, FL: CRC Press.
52. Holmberg E *et al.*, Highly efficient immunoliposomes prepared with a method which is compatible with various lipid compositions, *Biochem Biophys Res Commun* 1989 29; **165:** 1272–1278.

53. Maruyama K *et al.,* Lipid composition is important for highly efficient target binding and retention of immunoliposomes, *Proc Natl Acad Sci USA* 1990; **87:** 5744–5748.
54. Stavridis JC *et al.*, Construction of transferrin-coated liposomes for *in vivo* transport of exogenous DNA to bone marrow erythroblasts in rabbits, *Exp Cell Res* 1986; **164:**568–572.
55. Stripp BR *et al.*, Strategies for analysis of gene expression: pulmonary surfactant proteins, *Am J Physiol* 1990; **259:** L185–197.
56. Hart SL *et al.*, Cell binding and internalization by filamentous phage displaying a cyclic Arg-Gly-Asp-containing peptide, *J Biol Chem* 1994; **269:** 12468–12474.
57. Inoue T *et al.*, Differences in transferrin response and numbers of transferrin receptors in rat and human mammary carcinoma lines of different metastatic potentials, *J Cell Physiol* 1993; **156:** 212–217.
58. Keer HN *et al.*, Elevated transferrin receptor content in human prostate cancer cell lines assessed *in vitro* and *in vivo*, *J Urol* 1990; **143:** 381–385.
59. Chackal-Roy M *et al.*, Stimulation of human prostatic carcinoma cell growth by factors present in human bone marrow, *J Clin Invest* 1989; **84:** 43–50.
60. Rossi MC *et al.*, Selective stimulation of prostatic carcinaoma cell proliferation by transferrin, *Proc Natl Acad Sci USA* 1992; **89:** 6197–6201.
61. Thorstensen K *et al.*, The transferrin receptor: Its diagnostic value and its potential as therapeutic target, Scand. *J Clin Lab Invest Suppl* 1993; **215:** 113–120.
62. Miyamoto T *et al.*, Transferrin receptor in oral tumors, *Int J Oral Maxillofac Durg* (Denmark) 1994; **23:** 430–433.
63. Elliott RL *et al.*, Breast carcinoma and the role of iron metabolism. A cytochemical, tissue culture, and ultrastructural study, *Ann N Y Acad Sci* 1993; **698:** 159–166.
64. Xu L *et al.*, Transferrin-liposome-mediated p53 sensitization of squamous cell carcinoma of the head and neck to radiation *in vitro*, *Human Gene Ther* 1997; **8:** 467–475.

65. Xu L *et al.,* Transferrin-liposome-mediated systemic p53 gene therapy in combination with radiation results in regression of human head and neck cancer xenograft, *Human Gene Ther* 1999; **10:** 2941–2952.
66. Hart SL *et al.*, Gene delivery and expression mediated by an intergrin-binding peptide. *Gene Ther* 1995; **2:** 552–554.
67. Newham P *et al.,* Integrin adhesion receptors: structure, function and implications for biomedicine, *Mol Med Today* 1996; **2:** 304–313.
68. Hart SL *et al.*, Lipid-mediated enhancement of transfection by a nonviral integrin-targeting vector, *Human Gene Ther* 1998; **9:** 575–585.
69. Jenkins RG *et al.*, An integrin-targeted non-viral vector for pulmonary gene therapy, *Gene Ther* 2000; **7:** 393–400.
70. Lasic DD *et al.*, Liposomes in gene therapy, *Advanced Drug Delivery Reviews* 1996; **20:** 221–266.
71. Rommens JM *et al.*, Identification of the cystic fibrosis gene: Chromosome walking and jumping, *Science* 1989; **245:** 1059–1065.
72. Riordan JR *et al.,* Identification of the cystic fibrosis gene: Cloning and characterization of complementary DNA, *Science* 1989; **245:** 1066–1073.
73. Zielenski J *et al.*, Cystic fibrosis: Genotypic and phenotypic variations, *Annu Rev Genet* 1995; **29:** 777–807.
74. Drumm ML *et al.,* Correction of the cystic fibrosis defect *in vitro* by retrovirus-mediated gene transfer, *Cell* 1990; **62:** 1227–1233.
75. Rich D *et al.,* Expression of cystic fibrosis transmembrane conductance regulator corrects defective chloride channel regulation in cystic fibrosis airway epithelial cells, *Nature* 1990; **347:** 358–363.
76. Rosenfeld MA *et al.,* Gene transfer to freshly isolated human respiratory epithelial cells *in vitro* using a replication-deficient adenovirus containing the human cystic fibrosis transmembrane conductance regulator cDNA, *Human Gene Ther* 1994; **5:** 331–342.
77. Yoshimura K *et al.,* Expression of the human cystic fibrosis transmembrane conductance regulator gene in the mouse lung after *in vivo* intra-tracheal plasmid-mediated gene transfer, *Nucleic Acids Res* 1992; **20:** 3233–3240.

78. Logan JJ *et al.,* Cationic lipids for reporter gene and CFTR transfer to rat pulmonary epithelium, *Gene Ther* 1995; **2:** 38–49.
79. Wheeler CJ *et al.,* A novel cationic lipid greatly enhances plasmid DNA delivery and expression in mouse lung, *Proc Natl Acad Sci USA* 1996; **93:** 11454–11459.
80. Gorman CM *et al.,* Efficient *in vivo* delivery of DNA to pilmonary cells using the novel lipid EDMPC, *Gene Ther* 1997; **4:** 983–992.
81. Jiang C *et al.,* Efficiency of cationic lipid-mediated transfection of polarized and differentiated airway epithelial cells *in vitro* and *in vivo*, *Human Gene Ther* 1998; **9:** 1531–1542.
82. Hyde SC *et al.,* Correction of the ion transport defect in cystic fibrosis transgenic mice by gene therapy, *Nature* 1993; **362:** 250–255.
83. Alton EWFW *et al.,* Non-invasive liposome-mediated gene delivery can correct the ion transport defect in cystic fibrosis mutant mice, *Nature Genet* 1993; **5:** 135–142.
84. Caplen NJ *et al.,* Liposome-mediated CFTR gene transfer to the nasal epithelium of patients with cystic fibrosis, *Nature Med* 1995; **1:** 39–46.
85. Porteous DJ *et al.,* Evidence for safety and efficacy of DOTAP cationic liposome mediated CFTR gene transfer to nasal epithelium of patients with cystic fibrosis, *Gene Ther* 1997; **4:** 210–218.
86. Gill DR *et al.,* A placebo controlled study of liposome-mediated gene transfer to the nasal epithelium of patients with cystic fibrosis, *Gene Ther* 1997; **4:** 199–209.
87. Zabner J *et al.,* Comparison of DNA-lipid complexes and DNA alone for gene transfer to cystic fibrosis airway epithelium *in vivo*, *J Clin Invest* 1997; **100:** 1529–1537.
88. Alton EWFW *et al.,* A double-blinded placebo-controlled trial of cationic lipid-mediated CFTR gene transfer to the lungs and nose of CF subjects, *Lancet* 1999; **353:** 947–954.
89. Marshall J *et al.,* Cationic lipid-mediated gene delivery to the airways. In L Huang et al. (eds.), *Nonviral vectors for gene therapy* (pp.40–68), 1996. California: Academic Press.
90. Vogelstein B *et al.,* Thę multistep nature of cancer, *Trends Genet* 1993; **9:** 138–141.

91. Hui KM *et al.,* Rejection of transplantable AKR leukaemia cells following MHC DNA-mediated cell transformation, *Nature* 1984; **311:** 750–752.
92. Hui KM, Introduction of tumor-specific immunity by manipulating the expression of major histocompatibility complex molecules on tumor cells, *FEMS Microbiol Immunol* 1990; **64:** 215–222.
93. Hui KM *et al.,* Tumor rejection mediated by transfection with allogeneic class I histocompatibility gene, *J Immunol* 1989; **143:** 3835–3843.
94. Hui KM *et al.,* Promotion of tumor growth by transfecting antisense DNA to suppress endogenous H-2Kk MHC gene expression in AKR mouse thymoma, *Cell Immunol* 1991; **136:** 80–94.
95. Hui KM *et al.,* Induction of alloreactive cytotoxic T lymphocytes by intra-spenic immunization with allogeneic class I major histocompatibility complex DNA and DC-chol cationic liposomes, *J Liposome Res* 1994; **4:** 1075–1790.
96. Hui KM *et al.,* Phase I study of immunotherapy of cutaneous metastases of human carcinoma using allogeneic and xenogeneic MHC DNA-liposome complexes, *Gene Ther* 1997; **4:** 783–90.
97. Hung M-C *et al.,* Molecular cloning of the neu gene: Absence of gross structural alteration in oncogenic alleles, *Proc Natl Acad Sci USA* 1986; **83:** 261–264.
98. Hung M-C *et al.,* Amplication of the proto-neu gene facilitates oncogenic activation by a single point mutation, *Proc Natl Acad Sci USA* 1989; **86:** 2545–2548.
99. Slamon DJ *et al.,* Human breast cancer: Correlation of relapse and survival with amplification of the HER-2/neu oncogene, *Science* 1987; **235:** 177–182.
100. Slamon DJ *et al.,* Studies of the HER-2/neu proto-oncogene in human breast and ovarian cancer, *Science* 1989; **244:** 707–712.
101. Schrier PI *et al.,* Differential expression of the c-erbB-2 gene in human small cell and non-small cell lung cancer, *Cancer Res* 1989; **49:** 4968–4971.
102. Yokota J *et al.,* Genetic alterations of the c-erbB-2 oncogene occur frequently in tubular adenocarcinoma of the stomach and are often

accompanied by amplification of the v-erbA homolohue, *Oncogene* 1988; **2:** 283–287.

103. Xia W *et al.,* Strong correlation between c-erbB-2 overexpression and overall survival of patients with oral squamous cell carcinoma, *Clin Cancer Res* 1997; **3:** 3–9.

104. Xing X *et al.,* Safety studies of the intraperitoneal injection of E1A-liposome complex in mice, *Gene Ther* 1997; **4:** 238–243.

105. Xing X *et al.,* Safety study and characterization of E1A-liposome complexes gene-delivery protocol in an ovarian cancer model, *Gene Ther* 1998; **5:** 1538–1544.

106. Yu D *et al.,* Mechanism of c-erbB2/neu oncogene-induced metastasis and repression of metastatic properties by adenovirus 5 E1A gene products, *Oncogene* 1992; **6:** 2263–2270.

107. Felgner PL, Progress in gene delivery research and development. In L Huang et al. (eds.), *Nonviral vectors for gene therapy* (pp.40–68), 1996. California: Academic Press.

FRONTIERS IN HUMAN GENETICS
Diseases and Technologies
© 2001 by World Scientific Publishing Co. Pte. Ltd.

NAKED PLASMIDS: MUSCLING INTO GENE TRANSFER

OI LIAN KON*,†, SERENE HWEE LOON LOK*
IAIN BEE HUAT TAN‡ and EUGENIE WEI TING POH‡
*Department of Biochemistry,
National University of Singapore
10 Kent Ridge Crescent, Singapore 119260, and
Division of Medical Sciences,
National Cancer Centre,
11 Hospital Drive, Singapore 169610
‡Faculty of Medicine
National University of Singapore
10 Kent Ridge Crescent, Singapore 119260
Republic of Singapore.

The development of DNA vaccines and plasmid vectors for gene therapy in recent years is based on the capacity of mammalian cells and tissues to take up exogenous DNA spontaneously. Naked DNA uptake is a universal property of mammalian cells, having been demonstrated in several different mammalian cell lines, primary cells in culture, solid tumours as well as in normal tissues (e.g. skeletal muscle and liver) *in vivo*. While mechanisms of uptake of exogenous DNA by bacteria are understood in some detail, almost nothing is known about this process in mammalian cells. It is unlikely that the extremely poor efficiency of naked plasmid entry into mammalian cells and tissues can be overcome until eukaryotic mechanisms are elucidated. We review the current state of knowledge of naked plasmid entry into mammalian cells, and present preliminary evidence for saturable plasmid binding sites on the plasma membrane of C2C12 myoblasts followed by time-dependent translocation of internalized plasmid into the nuclear compartment. Myoblasts can be engineered for insulin secretion via plasmid mediated gene transfer. Thus, muscle and myoblast implants may be a suitable platform for developing genetic treatment of systemic metabolic disorders.
Keywords: plasmid, muscle, myoblasts, gene therapy, DNA vaccines

The field of human gene therapy has recently become clouded by the occurrence of severe adverse outcomes from clinical trials that involved the use of adenoviral vectors. Although unfortunate, these events may

---

† Corresponding author

serve to re-focus attention on non-viral vectors because plasmids lack many of the pathogenic hazards of viral vectors. Nonetheless, the efficacy of plasmids as vectors for *in vivo* gene transfer has thus far been seriously hampered by efficiencies that are significantly poorer when compared with viral vectors. While widely recognized, this problem of inefficient gene transfer cannot be overcome effectively by reliance on empirical approaches alone until fundamental mechanisms of plasmid-mediated gene transfer are understood. This review summarizes the current status of the applications of naked plasmids for gene transfer, possible mechanisms of naked plasmid DNA entry into mammalian cells and presents reasons for considering skeletal muscle as a potentially useful platform for systemic gene therapy.

1 GENE TRANSFER: UPTAKE OF NAKED PLASMID BY MAMMALIAN CELLS

Introduction of exogenous DNA into mammalian cells in culture or *in vivo* is commonly performed in genetic research and for the treatment of diseases. Biosafety considerations aside, the inefficiency of all gene transfer methods currently in use is generally acknowledged to be a severe limitation in the quest for clinically useful gene therapy. Prevailing techniques for introducing foreign DNA into cells include chemically inducing a temporary state of "competence" for uptake of foreign DNA, complexing DNA with lipids and/or polymers, coupling DNA to ligands for receptor-mediated endocytosis, exploiting tissue tropism of viruses and employing physical forces, e.g. electroporation and particle bombardment. Among the present generation of gene transfer vectors, it is no surprise that viral vectors are substantially more efficient than non-viral vectors. A majority of human gene therapy trials consequently employ retroviral and adenoviral vectors, although plasmid vectors — usually packaged with lipids — are another common class of gene transfer vector.

Mammalian cells were first reported to take up naked plasmid DNA spontaneously in 1986.[1] Several tissues of newborn rats expressed a reporter gene, hepatitis B surface antigen, growth hormone and proinsulin genes after a single intraperitoneal injection of naked plasmid. Wolff *et al*. later reported that single intramuscular injections of naked plasmid alone resulted in the expression of reporter transgenes *in situ* that was comparable to reporter gene expression in fibroblasts transfected under optimal conditions *in vitro*.[2] The ability to take up naked plasmid DNA has since been shown to be a property of many mammalian tissues including cardiac muscle,[3–5] arteries,[6,7] gastric smooth muscle,[8] skin,[9,10] liver,[11–13] lung,[14,15] brain,[16] thyroid follicular cells,[17,18] solid tumours (melanoma and cervical carcinoma)[19] and even bone.[20] It is worth noting that this phenomenon is not only limited to small laboratory animals, but has been shown to occur also in the skeletal muscle of non-human primates.[21]

While many different tissue types possess the spontaneous ability to take up plasmid DNA, skeletal muscle, cardiac muscle and liver appear to be most proficient.[13,22] The fact that mammalian cells not only take up naked plasmids from injected sites but also from plasmid solutions infused into bile ducts, portal and hepatic veins[12, 13] makes it unlikely that plasmid entry into mammalian cells is merely mechanically induced.

2 VACCINES AND THERAPY

The unusual capacity of skeletal muscle for spontaneous uptake of plasmid DNA forms the basis of two biomedical applications, i.e. DNA vaccines and gene therapy. In DNA vaccines, a plasmid expression construct encoding an immunogenic protein of the appropriate pathogen is administered by intramuscular or intradermal injection. This evokes a complex repertoire of humoural and cell-mediated immune responses[23] that, in experimental animals, are sustained and highly protective against subsequent lethal challenge by viral,[24,25] mycoplasma[26] and parasitic pathogens.[27] Genetic vaccination may also confer benefits in cancer[28] and autoimmune diseases.[29] The increasing corpus of literature on DNA vaccines is available from conventional databases and a specialty website (www.dnavaccine.com).

The prospects for adapting intramuscular injections of naked plasmid DNA for gene therapy arise from data obtained in small animal models. Several laboratories have reported that plasmid-borne transgenes are not only expressed but that the protein products of transgenes can also be secreted from muscle. The dual processes of *in situ* tissue expression and systemic secretion have produced therapeutically meaningful circulating levels of several model proteins, e.g. human α_1-antitrypsin,[30] growth hormone releasing hormone,[31] erythropoietin[32] and insulin[33] in murine models.

While the applications of plasmid-mediated gene transfer to the development of DNA vaccines and gene therapy are conceptually identical, there is nonetheless a practical difference of crucial importance. DNA vaccines evoke useful immunity yet typically express only nanogram amounts of proteins.[19] Such low levels of expression are clearly sub-therapeutic for many systemic protein deficiency states such as insulinopenic diabetes mellitus, other endocrine and clotting factor deficiencies. Thus, while current utilization of plasmid vectors may suffice for genetic immunization, vastly superior efficiencies of plasmid-mediated gene transfer and expression must be achieved before non-viral vectors become clinically applicable.

3 MECHANISMS OF PLASMID UPTAKE BY MAMMALIAN CELLS

There is a striking paucity of information on routes and mechanisms of plasmid entry into mammalian cells. At a minimum, this process might be envisaged to occur in at least four sequential stages: (i) association of plasmid DNA with the plasma membrane; (ii) passage across the plasma membrane; (iii) trafficking within the cytoplasmic compartment and (iv) entry into the nucleus.

3.1 Association of Naked Plasmids with Plasma Membrane

Wolff *et al*. observed that intramuscularly injected plasmid DNA was widely distributed in the interstitial space of mouse quadriceps muscle *in vivo*.[34] Gold-labeled plasmid was found associated with T-tubules and caveolae by electron microscopy, while colloidal gold complexed with polyethylene glycol, poly-L-glutamate and poly-L-lysine was not. This suggested that T-tubules and caveolae could be portals of entry. There was no morphological evidence for endocytosis of plasmid DNA. The observation from this study of much higher levels of transgene expression in primary myoblasts exposed to naked plasmid DNA than in immortalized C2C12 myoblasts might serve as encouragement for considering skeletal muscle as a suitable target tissue for therapeutic gene transfer.

Using confocal microscopy and biotin-labeled plasmid DNA (pBluescript), data from our laboratory shows the presence of binding sites for naked plasmid on the plasma membrane of C2C12 myoblasts (Fig. 1). Quenching of phycoerythrin fluorescence by a modest (50-fold) excess of unlabeled plasmid further indicates that these plasmid-binding sites are saturable and thus unlikely to represent non-specific adsorption to cellular surfaces. This is consistent with more indirect evidence for saturable DNA uptake by muscle reported by other workers.[35]

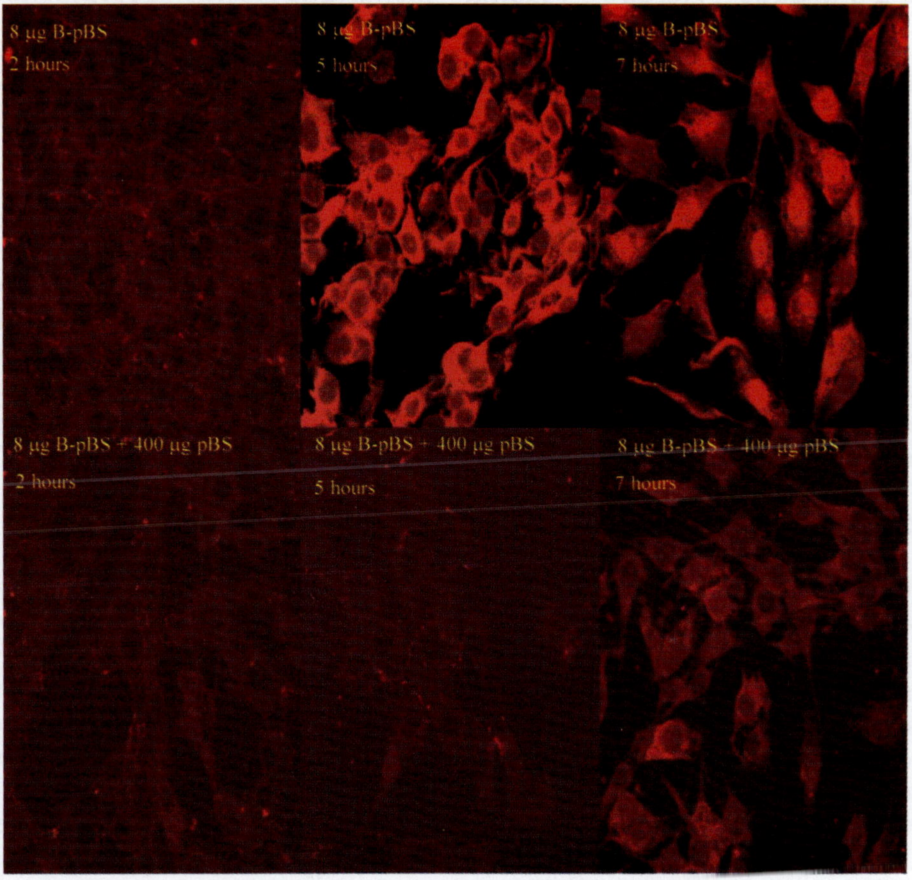

Fig. 1 C2C12 myoblasts were grown on 18 mm diameter glass coverslips in 12-well plates. Just before plasmid binding, the medium was changed to serum-free DMEM and 8 μg of biotinylated plasmid (B-pBS) was added to each well in the absence (top panel) or presence (lower panel) of 100 μg of the same unlabeled plasmid, and left for 2-7 hours. Unbound plasmid was removed by aspiration of the growth medium and 1.5 μg avidin-phycoerythrin was added. After 30 minutes, each glass coverslip was removed from the well, drained on absorbent paper, fixed with fluorescent mounting medium (Dako) and viewed on a Zeiss 410 LSM confocal microscope (excitation 514 nm, emission 576 nm). Selected images are shown above. Magnification = 630x.

3.2 Passage Across Plasma Membrane and Cytoplasmic Trafficking of Naked Plasmid

Figure 1 shows intracellular localization of biotinylated plasmid DNA five hours after exposure to C2C12 myoblasts, implying the operation of some transmembrane entry mechanism(s). Precisely what these mechanisms are is unclear on present evidence. Possiblities include potocytosis, receptor-mediated endocytosis or even a transport process. It is unlikely, however, that large membrane disruptions created by hydrostatic or ballistic forces mediate spontaneous plasmid entry into mammalian cells. Given that mechanisms of DNA uptake by bacteria are better understood and that cell surface DNA-binding sites have been identified in Gram-positive bacteria (*S. pneumoniae* and *B. subtilis*),[36] it is surprising that little effort has been expended in identifying eukaryotic homologs of membrane proteins that might function as "receptors" for naked DNA.

3.3 Naked Plasmid DNA in the Cytoplasmic Compartment

Using a combination of Southwestern analysis and DNA-cellulose chromatography, several triadin proteins from the sarcoplasmic reticulum of rabbit skeletal muscle have been reported to bind plasmid DNA specifically and with high affinity.[37] What function(s) such proteins serve (e.g. in cytoplasmic-nuclear transport) is yet uncertain. The highly-organized cytoskeleton of myotubes may be an important determinant of the nature of cytoplasmic-nuclear transport. When plasmid DNA was microinjected close to myotube nuclei, up to 56% of myotubes expressed the reporter transgene. However, this dropped precipitously to 8% when injections were placed 60 – 90 μm away from nuclei.[38] Such prolonged immobility of microinjected plasmid DNA for at least twenty-two hours from the time of injection could also be predicted to reduce eventual transgene expression because of rapid degradation of plasmid DNA by cytoplasmic nucleases.[39]

3.4 Nuclear Import of Plasmid DNA

Figure 1 demonstrates a distinct change in localization of biotinylated plasmid from the cytoplasmic to the nuclear compartment after seven hours. Emerging evidence supports the role of the nuclear pore complex and possibly active transport as pathways for entry of plasmid DNA into the nucleus. Thus, colloidal gold-labeled plasmid has been localized within or close to nuclear pore complexes,[38] although small DNA fragments may enter by passive diffusion.[40] There is some evidence that DNA entry into the nucleus may be sequence-dependent. A 72-bp repeat of the SV40 enhancer increases nuclear entry[41] of plasmid DNA as does the attachment of a single nuclear localization peptide.[42] Transfection with plasmids tagged with the peptide, PKKKRKVEDPYC, was 10 to 1000 times more efficient regardless of cell type or cationic vector used. Complete abolition of this effect by a single amino acid substitution (lysine to threonine at the third position of the nuclear localization peptide sequence) points to importin-mediated translocation. Evidence has also been presented to suggest that mitosis facilitates nuclear entry of plasmids.[43] Taken together, fragmentary evidence to date raises the possibility of a specific nuclear transport process that imports plasmid DNA from the cytoplasmic compartment.

4 SKELETAL MUSCLE: A PLATFORM FOR SYSTEMIC GENE THERAPY

Several features make skeletal muscle an attractive target tissue for therapeutic gene transfer to repair systemic protein deficiency states e.g. hormones and coagulation factors. It is an abundant and anatomically accessible tissue that is well vascularised and capable of protein secretion. It can also be transfected by viral[44] and non-viral vectors. Furthermore, skeletal muscle may have latent regenerative potential in the form of satellite cells, located between myofibres, which are believed to be myogenic precursors.[45] Stably modified satellite cells could thus serve as a permanent reservoir of myofibres that express a therapeutic transgene product, thereby ensuring sustained clinical effect with a single treatment. We have recently demonstrated transfection of murine skeletal muscle *in vivo* by injections of naked plasmid DNA constructs that contain dual proinsulin cDNA and furin cDNA expression cassettes. Using a radioimmunoassay that was highly specific for mature insulin with little cross-reactivity for proinsulin, the results showed that skeletal muscle was capable not only of expressing proinsulin but also of processing it to mature insulin and of insulin secretion into the systemic circulation.[33] This efficacy of this approach was tested in a severe model of type 1 diabetes mellitus, i.e. streptozotocin-induced murine diabetes with some amelioration of hyperglycemia.

While we and others have shown the possibility of using skeletal muscle as an *in vivo* platform for systemic gene therapy, the highly inefficient gene transfer nonetheless remains a problem that makes this simple approach impractical. We have therefore begun engineering myoblasts in culture for eventual implantation *in vivo*, using insulin secretion and diabetes mellitus as a model system.

Figure 2 shows the results of transfecting C2C12 myoblasts with two different plasmid expression constructs in which proinsulin cDNA was under the control of the CMV or β-actin promoter.

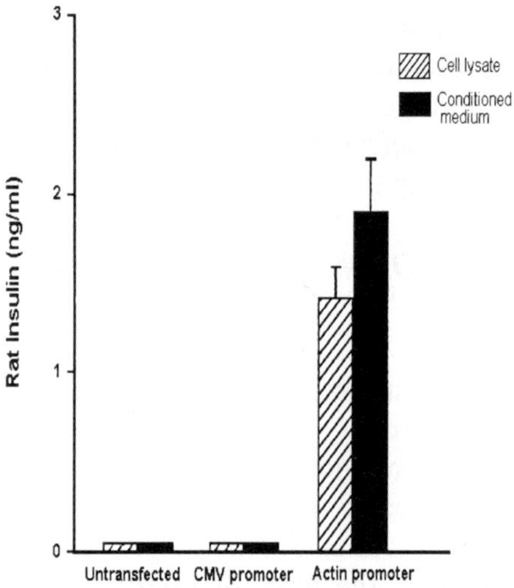

Fig. 2 C2C12 myoblasts, in Dulbecco's Modified Eagle's Medium (DMEM) with 10% fetal calf serum, were plated in replicate 75 cm^2 flasks (1.2 x 10^6 cells/flask) and transfected with dual expression plasmids bearing rat proinsulin· cDNA and furin cDNA using Lipofectamine PLUS (Life Technologies, USA) according to the supplier's protocol. Following transfection, myoblasts were placed in serum-free DMEM that was collected 24 hours later for rat insulin radioimmunoassay (Linco Research Inc., USA). Cell lysates from each flask were prepared by sonicating trypsinized myoblasts (0.5 x 10^6 cells) for 30 seconds in phosphate-buffered saline (0.1 ml) containing 1.5 μl N-ethylmaleimide (0.1M). The supernatants obtained by centrifugation (20000 x *g*, 5 minutes, 4°C) were also assayed to reflect intracellular concentrations of rat insulin. Rat insulin was undetectable in untransfected myoblasts and in myoblasts transfected with CMV promoter-rat proinsulin expression cassettes. Data are means and SEM of four replicate flasks.

Insulin was undetectable in the conditioned media of untransfected myoblasts and myoblasts transfected with the CMV promoter construct. In contrast, insulin was readily secreted from myoblasts transfected with the β-actin promoter construct. Myoglobin and heat shock promoters were also inactive in this myoblast cell line (data not presented). These studies indicate that, with appropriate promoter selection, myoblasts can be modified for insulin processing and secretion. Future directions for developing myoblast-mediated therapy of metabolic disorders could usefully focus on inducing a period of reversible immortalization[46, 47] in autologous cells during which stable modification for protein secretion could be achieved before re-implantation *in vivo*. By thus combining genetic and cellular engineering, safe and sustained metabolic reconstitution without the need for lifelong immunosuppression could, in theory, become a reality.

Acknowledgements

This work was supported by the National University of Singapore and the National Cancer Centre.

References

1. Benvenisty N, Reshef L, Direct introduction of genes into rats and expression of the genes, *Proc. Natl. Acad. Sci. USA* 1986; **83**: 9551–9555.
2. Wolff JA, Malone RW, Williams P, *et al.* Direct gene transfer into mouse muscle *in vivo*, *Science* 1990; **247**: 1465–1468.
3. Kistis RN, Buttrick PM, McNally EM, *et al.* Hormonal modulation of a gene injected into rat heart *in vivo*, *Proc. Natl. Acad. Sci. USA* 1991; **88**: 4138-4142.
4. Buttrick PM, Kass A, Kitsis RN, *et al.* Behavior of genes directly injected into the rat heart *in vivo* *Circ. Res.* 1992; **70**: 193–198.
5. Prentice H, Kloner RA, Prigozy T, *et al.* Tissue restricted gene expression assayed by direct DNA injection into cardiac and skeletal muscle, *J. Mol. Cell. Cardiol.* 1994; **26**: 1393–1401.
6. Chapman GD, Lim CS, Gammon RS, *et al.* Gene transfer into coronary arteries of intact animals with a percutaneous catheter, *Circ. Res.* 1992; **71**: 27–33.
7. Riessen R, Rahimizadeh H, Blessing E, *et al.* Arterial gene transfer using pure DNA applied directly to a hydrogel-coated angioplasty balloon, *Hum. Gene Ther.* 1993; **4**: 749–758.
8. Takehara T, Hayashi N, Yamamoto M, *et al.* *In vivo* gene transfer and expression in rat stomach by submucosal injection of plasmid DNA, *Hum. Gene Ther.* 1996; **7**: 589–593.
9. Raz E, Carson DA, Parker SE, *et al.* Intradermal gene immunization: the possible role of DNA uptake in the induction of cellular immunity to viruses, *Proc. Natl. Acad. Sci. USA* 1994; **91**: 9519–9523.
10. Hengge UR, Chan EF, Foster RA, *et al.* Cytokine gene expression in epidermis with biological effects following injection of naked DNA, *Nature Genet.* 1995; **10**: 161–166.
11. Malone RW, Hickman MA, Lehmann-Bruinsma K, *et al.* Dexamethasone enhancement of gene expression after direct hepatic DNA injection, *J. Biol. Chem.* 1994; **269**: 29903–29907.

12. Budker V, Zhang G, Knechtle S, *et al.* Naked DNA delivered intraportally expresses efficiently in hepatocytes, *Gene Ther.* 1996; **3**: 593–598.
13. Zhang G, Vargo D, Budker V, *et al.* Expression of naked plasmid DNA injected into the afferent and efferent vessels of rodent and dog livers, *Hum. Gene Ther.* 8: 1763–1772.
14. Tsan MF, White JE, Shepard B, Lung-specific direct *in vivo* gene transfer with recombinant plasmid DNA, *Am. J. Physiol.* 1995; **268**: L1052–L1056.
15. Sawa T, Miyazaki H, Pittet JF, *et al.* Intraluminal water increases expression of plasmid DNA in rat lung, *Hum. Gene Ther.* 1996; **7**: 933–941.
16. Schwartz B, Benoist C, Abdallah B, *et al.* Gene transfer by naked DNA into adult mouse brain, *Gene Ther.* 1996; **3**: 405– 411.
17. O'Malley BW Jr., Ledley FD, DNA- and viral-mediated gene transfer in follicular cells: progress toward therapy of the thyroid, *Laryngoscope* 1993; **103**: 1084–1092.
18. Sikes ML, O'Malley BW Jr., Finegold MJ, *et al.* *In vivo* gene transfer into rabbit thyroid follicular cells by direct DNA injection, *Hum. Gene Ther.* 1994; **5**: 837–844.
19. Yang JP, Huang L, Direct gene transfer to mouse melanoma by intratumor injection of free DNA, *Gene Ther.* 1996; **3**: 542–548.
20. Fang J, Zhu YY, Smiley E, *et al.* Stimulation of new bone formation by direct transfer of osteogenic plasmid, *Proc. Natl. Acad. Sci. USA* 1996; **93**: 5753–5758.
21. Jiao S, Williams P, Berg RK, *et al.* Direct gene transfer into nonhuman primate myofibers *in vivo*, *Hum. Gene Ther.* 1992; **3**: 21–33.
22. Acsadi G, Jiao SS, Jani A, *et al.* Direct gene transfer and expression into rat heart *in vivo*, *New Biol.* 1991; **3**: 71–81.
23. Hassett DE, Zhang J, Slifka M, *et al.* Immune responses following neonatal DNA vaccination are long-lived, abundant, and qualitatively similar to those induced by conventional immunization, *J. Virol..* 2000; **74**: 2620–2627.
24. Barouch DH, Craiu A, Kuroda MJ, *et al.* Augmentation of immune responses to HIV-1 and simian immunodeficiency virus DNA vaccines by IL-2/Ig plasmid administration in rhesus monkeys, *Proc. Natl. Acad. Sci. USA* 2000; **97**: 4192–4197.
25. Kodihalli S, Kobasa DL, Webster RG, Strategies for inducing protection against avian influenza A virus subtypes with DNA vaccines, *Vaccine* 2000; **18**: 2592–2599.

26. Lai WC, Bennett M, Johnston SA, *et al.* Protection against Mycoplasma pulmonis infection by genetic vaccination, *DNA Cell Biol.* 1995; **14**: 643– 651.
27. Le TP, Coonan KM, Hedstrom RC, *et al.* Safety, tolerability and humoral immune responses after intramuscular administration of a malaria DNA vaccine to healthy adult volunteers, *Vaccine* 2000; **18**: 1893–1901.
28. Bronte V, Apolloni E, Ronca R, *et al.* Genetic vaccination with "self" tyrosinase-related protein 2 causes melanoma eradication but not vitiligo, *Cancer Res.* 2000; **60**: 253–258.
29. Weissert R, Lobell A, de Graaf KL, *et al.* Protective DNA vaccination against organ-specific autoimmunity is highly specific and discriminates between single amino acid substitutions in the peptide autoantigen, *Proc. Natl. Acad. Sci. USA* 2000; **97**: 1689–1694.
30. Hickman MA, Malone RW, Lehmann-Bruinsma K, *et al.* Gene expression following direct injection of DNA into liver, *Hum. Gene Ther.* 1994; **5**: 1477–1483.
31. Draghia-Akli R, Li X, Schwartz RJ, Enhanced growth by ectopic expression of growth hormone releasing hormone using an injectable myogenic vector, *Nature Biotech.* 1997; **15**: 1285–1289.
32. Tripathy SK, Svensson EC, Black HB, *et al.* Long-term expression of erythropoietin in the systemic circulation of mice after intramuscular injection of a plasmid DNA vector, *Proc. Natl. Acad. Sci. USA* 1996; **93**: 10876–10880.
33. Kon OL, Sivakumar S, Teoh KL, *et al.* Naked plasmid-mediated gene transfer to skeletal muscle ameliorates diabetes mellitus, *J. Gene Med.* 1999; **1**: 186–194.
34. Wolff JA, Dowty ME, Jiao S, *et al.* Expression of naked plasmids by cultured myotubes and entry of plasmids into T tubules and caveolae of mammalian skeletal muscle, *J. Cell Sci.* 1992; **103**: 1249–1259.
35. Levy MY, Barron LG, Meyer KB, *et al.* Characterization of plasmid DNA transfer into mouse skeletal muscle: Evaluation of uptake mechanism, expression and secretion of gene products into blood, *Gene Ther.* 1996; **3**: 201–211.
36. Solomon JM, Grossman AD, . Who's competent and when: regulation of natural genetic competence in bacteria, *Trends Genet.* 1996; **12**: 150–155.
37. Hagstrom JE, Rybakova IN, Staeva T, *et al.* Nonnuclear DNA binding proteins in striated muscle, *Biochem. Mol. Med.* 1996; **58**: 113–121.

38. Dowty ME, Williams P, Zhang G, *et al.* Plasmid DNA entry into postmitotic nuclei of primary rat myotubes, *Proc. Natl. Acad. Sci. USA* 1995; **92**: 4572–4576.
39. Lechardeur D, Sohn KJ, Haardt M, *et al.* Metabolic instability of plasmid DNA in the cytosol: a potential barrier to gene transfer, *Gene Ther.* 1999; **6**: 482–497.
40. Ludtke JJ, Zhang GF, Sebestyen MG, *et al.* A nuclear localization signal can enhance both the nuclear transport and expression of 1 kb DNA, *J. Cell Sci.* 1999; **112**: 2033–2041.
41. Dean DA, Dean BS, Muller S, *et al.* Sequence requirements for plasmid nuclear import, *Exp. Cell Res.* 1999; **253**: 713–722.
42. Zanta MA, Belguise-Valladier P, Behr JP, Gene delivery: a single nuclear localization signal peptide is sufficient to carry DNA to the cell nucleus, *Proc. Natl. Acad. Sci. USA* 1999; **96**: 91–96.
43. Zelphati O, Liang XW, Hobart P, *et al.* Gene chemistry: functionally and conformationally intact fluorescent plasmid DNA, *Hum. Gene Ther.* 1999; **10**: 15– 24.
44. Kay MA, Manno CS, Ragni MV, *et al.* Evidence for gene transfer and expression of factor IX in haemophilia B patients treated with an AAV vector, *Nature Genet.* 2000; **24**: 257–261.
45. van Deutekom JC, Hoffman EP, Huard J, Muscle maturation: implications for gene therapy, *Mol. Med. Today* 1998; **4**: 214–220.
46. Rudolph KL, Chang S, Millard M, *et al.* Inhibition of experimental liver cirrhosis in mice by telomerase gene delivery, *Science* 2000; **287**: 1253–1258.
47. Kobayashi N, Fujiwara T, Westerman KA, *et al.* Prevention of acute liver failure in rats with reversibly immortablized human hepatocytes, *Science* 2000; **287**: 1258–1262.

FRONTIERS IN HUMAN GENETICS
Diseases and Technologies
© 2001 by World Scientific Publishing Co. Pte. Ltd.

ANTISENSE OLIGONUCLEOTIDES AS A THERAPY FOR DUCHENNE MUSCULAR DYSTROPHY

STEPHEN D WILTON
Australian Neuromuscular Research Institute, 4th Floor "A" Block, QE II Medical Centre, Nedlands, WA 6009
E-mail: swilton@cyllene.uwa.edu.au

Mutations in the dystrophin gene are responsible for the severe X-linked recessive Duchenne muscular dystrophy. Affected boys show signs of muscle weakness between the ages of 3 and 5 years and will be restricted to a wheelchair by 12 years. Becker muscular dystrophy also arises from mutations in the dystrophin gene but in these cases, the typical in-frame deletion allows a semi-functional dystrophin protein to be produced.

The *mdx* mouse has a nonsense mutation in exon 23 of the dystrophin gene. We wished to evaluate intervention at the stage of dystrophin pre-mRNA processing (splicing) where exon 23 could be deleted from the mature transcript which could then be translated into a slightly shorter but functional dystrophin.

Antisense 2′-O-methyl oligonucleotides directed to the splice sites flanking exon 23 induced 100% skipping of that exon in primary cultured *mdx* myotubes. Induced exon skipping was not as efficient in immortalized mouse myoblast cell lines.

The deletion of exon 23 does not disrupt the reading frame of the dystrophin transcript and should allow the synthesis of a shorter but presumably functional protein. Such an approach has the potential to reduce the severity of a Duchenne mutation to the milder Becker phenotype.

Keywords: antisense oligonucleotides, splicing, induced exon skipping, Duchenne muscular dystrophy

1 INTRODUCTION

Mutations in the human dystrophin gene are responsible for Duchenne muscular dystrophy (DMD) and the less severe, allelic form of Becker muscular dystrophy (BMD). Typically, nonsense or frameshift mutations in the dystrophin gene result in premature termination of protein synthesis and are associated with DMD.[1] Presenting at a frequency of 1 in 3500 live male births, boys affected with DMD show signs of weakness between the ages of 3-5 years and become restricted to a wheelchair by the age of 12 years. The majority will die before their 20th birthday due to respiratory or cardiac failure.[2]

It has been well documented that the milder BMD phenotype generally correlates to in-frame gene deletions which allow the synthesis of shorter but semi-functional dystrophin proteins.[3] In addition, rare dystrophin-positive (revertant) fibres have been detected in at least 50% of DMD patients suggesting that some nonsense or frameshift mutations may not completely preclude the synthesis of an apparently functional dystrophin. These revertant fibres are not thought to be of any significant benefit to the DMD male due to their low incidence, typically less than 1%.[4] It has been proposed that the dystrophin in the revertant fibres has arisen from an exon skipping mechanism, either somatic deletions or alternative splicing, which has by-passed the primary mutation.[5,6]

Although the *mdx* mouse carries a nonsense mutation in exon 23,[7] we have detected alternatively processed dystrophin gene transcripts which skip several exons (including exon 23) but still maintain the reading frame in normal mouse and *mdx* muscle.[8] These "Becker-like" transcripts could be translated into a shorter but presumably functional dystrophin. Regardless of their origin (somatic deletion or alternative splicing), the persistence of these dystrophin positive fibres would imply that these fibres have a selective advantage over the dystrophin-negative fibres.

The huge dystrophin gene consists of 79 exons spanning 2.4 Mb. Extensive pre-mRNA processing is required to produce the mature mRNA of 14000 bases.[9] Although inducing specific genomic deletions does not seem currently feasible, the potential exists for intervention during the processing of the dystrophin pre-mRNA. Previous work by Ryszard Kole and colleagues indicated it was possible to target specific splice sites with antisense oligonucleotides (AOs) and re-program intron removal from the β-globin pre-mRNA.[10] As shown below in Fig 1, targeting AOs to the specific splice sites flanking exon 23 could prevent association with the normal splicing components so that the nonsense mutation is removed from the mature *mdx* dystrophin mRNA.

Normal Gene pre-mRNA mRNA Protein

D..Y..S..T..R..O..P..H..I..N---->D..Y..S..T..R..O..P..H..I..N---->DYSTROPHIN--->DYSTROPHIN

Dystrophic Gene pre-mRNA mRNA Protein

D..Y..S..mdx..R..O..P..H..I.N------>D..Y..S..mdx..R..O..P.H..I..N----->DYSmdxROPHIN---->DYS

Dystrophic Gene Exon skipping Becker-like mRNA and protein

D..Y..S..mdx..R..O..P..H..I.N---->D..Y..S..mdx..R..O..P.H..I..N--->DYSROPHIN--->DYSROPHIN

Fig. 1. Diagrammatic representation of targeted exon skipping in the *mdx* dystrophin gene transcript to allow synthesis of a Becker-like protein. The introns are represented by ".." and the *mdx* mutation causes premature termination of translation of the dystrophin mRNA. Antisense oligonucleotides directed at ".m" and "x." induce specific skipping of the mutated exon during removal of flanking introns from the pre-mRNA.

This AO-based approach could also be used to induce skipping of one or more exons that flank genomic deletions in order to restore the reading frame of a DMD mRNA. We describe preliminary experiments in the application of AOs to specifically re-direct processing of the *mdx* dystrophin pre-mRNA.

2 MATERIALS AND METHODS

Splice sites flanking mouse dystrophin exon 23 were identified after preparing a PCR template by long range amplification across dystrophin introns 22 and 23 (rTth, XL, Roche Molecular Systems, New Jersey, USA). The 4 kb PCR product was partially sequenced with the original PCR oligonucleotides and forward and reverse primers to exon 23 to characterize the splice sites flanking exon 23 (ABI Prism Dye terminator chemistry with the products analysed on a 373A Applied Biosystems DNA Sequencer, USA).

Details of all primers used in this study have been published.[14] 2′-O-methyloligoribonucleotides, synthesized by Hybridon Inc (Cambridge, MA,

USA), were resuspended in sterile water to a final concentration of 1 mg/ml. AO:Lipofectin™ complexes were prepared as described by the manufacturer and used to transfect primary *mdx* myoblasts and an immortalized mouse cell line (C2C12).

Total RNA was prepared from myotube cultures using RNAzol B (Tel-Test Inc, USA) essentially as described by the manufacturer. Total RNA was ethanol precipitated twice, washed with 70% ethanol, air dried and resuspended in water. Single step reverse transcriptase PCR (RT-PCR) was carried out using the Titan Expand Kit (Boehringer Mannheim, Germany) using approximately 500 ng total RNA with an outer primer pair spanning dystrophin exons 20 to 26. An aliquot of the primary RT-PCR was then re-amplified using an inner primer set across exons 20 to 26.

3 RESULTS

3.1 Characterization of Dystrophin Splice Sites and Design of AOs

Amplification of mouse genomic DNA using primers Ex22F and 24R generated a single product of approximately 4.0 kb of which introns 22 and 23 were found to be about 900 and 2,850 bp long respectively. Limited DNA sequencing of this material identified the splice sites flanking introns 22 and 23. Characterization of these splice sites allowed the design of 2′-O-methyl AOs which would anneal to the 3′ splice site of intron 22 (18mer) and the 5′ splice site of intron 23 (20mer) of the mouse dystrophin pre-mRNA. Control AOs included a random sequence or a 20mer targeted to an aberrant splice site in human β-globin pre-mRNA.[10]

3.2 Induced Exon 23 Skipping During Dystrophin Pre-mRNA Processing in C2C12 Cells

The AO-approach was first tested on C2C12 cells, an immortalized mouse myoblast cell line, which can be induced to fuse and express normal dystrophin mRNA. Differentiating cells were transfected with a mixture of dys 3′ and dys 5′ AOs complexed with Lipofectin™. Dystrophin mRNA was examined after reverse transcription-PCR across exons 20 to 26 to detect any skipping of not only exon 23 but also adjacent exons. A dose-dependant accumulation of the exon 23-deleted transcript was observed at 48 and 96 hours after transfection.[14] Identification of the alternatively processed dystrophin transcripts was confirmed by direct DNA sequencing. All C2C12 cultures treated with 50 nM or more AO:Lipofectin complexes consistently generated dystrophin transcripts that skipped exon 23. This particular transcript reached significant levels at 200 nM of both AOs, while at 400 nM additional shorter transcripts became evident. More detailed experiments were then carried out on primary muscle cells isolated from *mdx* mice.

3.3 Induced Exon 23 Skipping During Dystrophin Pre-mRNA Processing in *mdx* Cells

Primary *mdx* myoblast cultures were transfected with AO:Lipofectin complexes at a final concentration of 400 nM. Only the dys 5′ AO was found to induce skipping of exon 23 from the *mdx* dystrophin gene transcript. The dys 3′ AO had no detectable effect on dystrophin splicing. The control AOs (β-globin and random) also had no apparent effect on dystrophin gene transcript processing. Ryszard Kole had previously observed that AOs directed at the 5′ splice sites tended to be more efficient at influencing splicing of pre-mRNA transcripts.

As shown in Fig. 2, exon 23 skipping was readily apparent after 6 hours and was complete after 24 hours.[14]

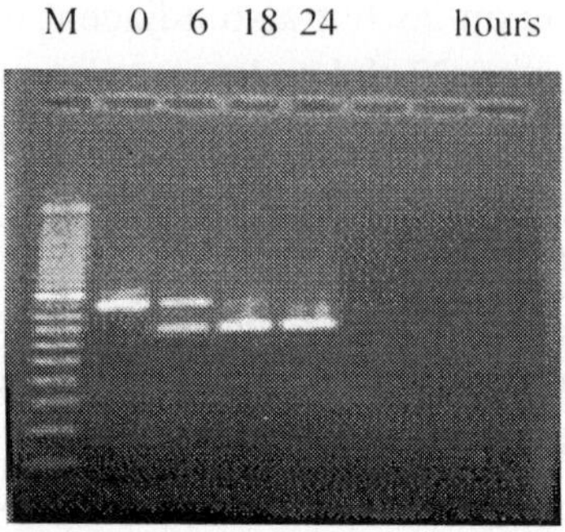

Fig. 2. RT-PCR across dystrophin exons 20-26 from RNA extracted from primary *mdx* cultures transfected with the 5′ dys AO. The shorter fragment was sequenced and, as shown in Fig. 3, resulted from the splicing of exons 22 to 24.

---------------Exon 22----------------] [----------------Exon 23------------
G A C T C G G G A A AT T A C A G G C T C T G C A A A G T T C T T T

---------------Exon 23----------------] [----------------Exon 24------------
A A C T T C G A A A A T T T C A G A A T C A C A T A A A A A C C T T

---------------Exon 22----------------] [----------------Exon 24------------

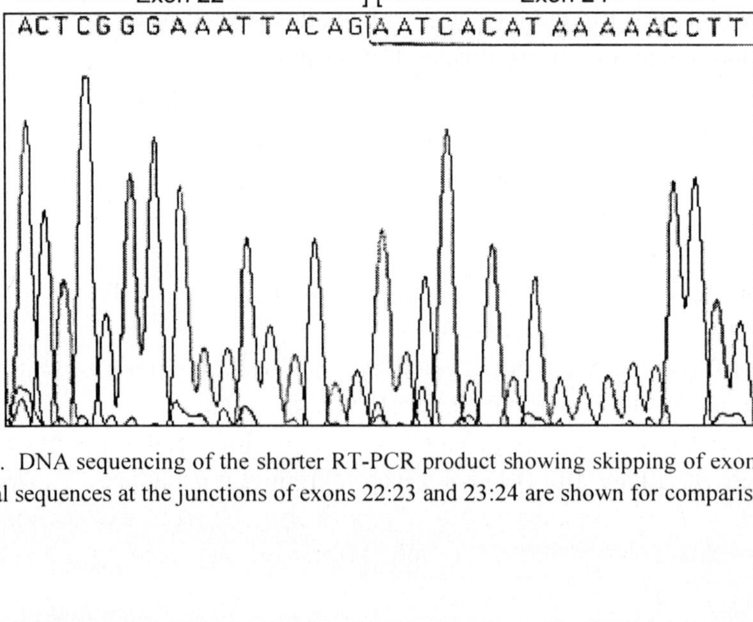

Fig. 3. DNA sequencing of the shorter RT-PCR product showing skipping of exon 23. The normal sequences at the junctions of exons 22:23 and 23:24 are shown for comparison.

4 DISCUSSION

DMD is typically associated with less than 3% of normal dystrophin expression levels while BMD is seen with dystrophin levels of 10% or more. The presence of any dystrophin, albeit of reduced quality and/or quantity, in BMD patients highlights the significance and ability of a shortened but semi-functional dystrophin to reduce the severity of DMD. Louise Nicholson observed that DMD males with no dystrophin entered a wheelchair at a mean age of 7.9 years while those with some detectable dystrophin were restricted to a wheelchair at an average of 9.9 years.[11]

There is a high *de novo* mutation rate in the dystrophin gene where one in three cases has no prior family history. Although prenatal screening can reduce the frequency of DMD in families known to be at risk, new cases will continue to be seen. Developing a therapy for this devastating disorder becomes imperative.

Gene or cell replacement therapies for DMD to date have relied on introducing the missing or defective gene product in tissues where the functional dystrophin is missing. The primary target tissue for cell/gene replacement has been skeletal muscle but it must be noted that dystrophin isoforms are also expressed in a variety of other tissues. The introduction of a muscle-specific isoform into muscle may address the problem of the missing protein in that particular tissue but may not replace missing isoforms expressed in other tissues. Multiple dystrophin isoforms are encoded by this one gene where expression is regulated by at least seven promoters in a cell specific manner.[12] Three distinct promoters at the 5′ end of the gene direct the synthesis of three high molecular weight (427 kDa) isoforms expressed in skeletal and cardiac muscle, Purkinje cells and neurons of the cerebral cortex. Shorter isoforms derive from internal promoters and are expressed in a variety of tissues including the retina, central nervous system, peripheral nerve, liver, brain and testes.[12] Some of these isoforms would also be affected by DMD/BMD mutations, depending upon the nature and position of the particular gene defect.

Several studies have suggested that some DMD mutations have been by-passed via a naturally occurring exon-skipping mechanism.[5,6,8] We had examined *mdx* mouse muscle for alternatively-processed dystrophin gene mRNAs (that skipped exon 23 but maintained the reading frame) and identified at least 4 transcripts which could be translated into shorter but presumably functional dystrophin.[8] Since these dystrophin-positive fibres occur naturally and these revertant fibres have been shown to persist and increase in frequency with the age of the muscle, there appear to be no immunological reactions to this dystrophin. As this revertant dystrophin appears to be correctly localized at the sarcolemma of the muscle fibre, based on immunostaining, the identification of the precise exon re-arrangements could provide natural and proven templates for induced exon-skipping using AOs.

Antisense oligonucleotides have previously been applied to down-regulate gene expression by blocking translation or by targeting an RNA for RNaseH degradation.[13] In another approach, Ryszard Kole used phosphorothioate 2'-O-methyl oligoribonucleotides (which do not induce RNaseH activity when bound to the RNA target) to block an aberrant splice site and restore correct splicing in a thalassemic pre-mRNA.[10]

Our initial AO experiments were undertaken in C2C12 cells, an immortalized normal mouse myoblast line with the aim of inducing the minimal change to by-pass the *mdx* mutation, that is skipping of only exon 23.[14] Targeted skipping of exon 23 was consistently induced in a dose-dependent fashion when both 3′ and 5′ AOs were delivered to those cells. Skipping of only exon 23 was never observed in any untreated C2C12 cultures. It should also be noted that during our search for alternatively-processed dystrophin gene transcripts from skeletal muscle of *mdx* and normal mice, dystrophin transcripts skipping only exon 23 were never observed to occur naturally.[8]

Transfection of primary *mdx* myoblast/tube cultures with both dys 3′ and 5′ AOs also resulted in exon-23-skipping but with a higher efficiency. Not only was exon-23-skipping complete 24 hours after transfection but other alternatively-processed transcripts were not consistently generated. It was subsequently found that exon-23-skipping could be efficiently induced in transfected *mdx* myoblasts using only the dys 5′ AO. Low levels of exon 23 skipping were also detected in C2C12 cells after transfection with only the dys 5′ AO with the dys 3′ AO having no apparent effect on splicing when used on its own. The inability of the dys 3′ AO to induce exon skipping suggests that the splicing factors binding to the 3′ splice site may be more difficult to displace than their 5′ counterpart. The fact that the dys 3′ AO was two nucleotides shorter than that directed to the 5′ splice site may have also contributed to this result. This latter possibility was discounted after a 30mer to the 3′ splice site was found to have no detectable affect on splicing.

A combination of mechanisms may be responsible for this striking improvement of exon-23-skipping in *mdx* cells compared to C2C12 cells: (i) oligonucleotide-Lipofectin complex uptake by primary *mdx* myoblasts may be more efficient than that by immortalized C2C12 cells; (ii) the dys 5′ AO may only be effective at blocking low levels of dystrophin pre-mRNA processing; (iii) the accessibility of the 5' splice site to the oligonucleotide may be increased by the stop codon mutation in exon 23 of dystrophin pre-mRNA.

Masafumi Matsuo and colleagues reported that exon-19-skipping in the human dystrophin mRNA can be induced by targeting an AO to the recognition sequence of that exon in lymphoblastoid cells.[15] A number of similarities between the human exon 19 and *mdx* exon 23 skipping were apparent. In both cases, exon-skipping was detectable after 6 hours and complete after 24 hours. Targeted exon-skipping in both examples appeared to be specific in that rearrangements in other parts of the dystrophin mRNA were not routinely detected. Similar AO concentrations (200 nM) were needed to induce complete skipping, as were the ratios of DNA to liposome. The efficiency of the AOs in inducing exon-skipping was greatly reduced in the absence of a liposome delivery system. These similarities would imply that parameters for efficient exon-skipping in the mouse should be readily extrapolated to human cells and to other regions of the dystrophin gene.

All gene and genetic therapy approaches face the challenge of achieving efficient delivery. This issue will also need to be addressed in the delivery of AOs. However, advantages of an AO-based approach to the suppression of DMD dystrophin gene mutations include the slow turnover of the muscle fibres, the natural ability of these cells to take up DNA and the potential for repeated and systemic delivery. The apparent specificity of the dys 5′ AO would suggest that systemic delivery may be possible with targeted dystrophin exon-skipping only being significant in those cells expressing dystrophin. As long as the AOs do not induce an immune response or become toxic, repeated systemic delivery should be possible. If the antisense sequences are incorporated into vectors that lead to stable expression of antisense RNA, the latter concern would be eliminated.

In summary, AOs may be regarded as highly specific nucleic acid drugs that may be able to re-program existing cellular machinery to minimize the consequences of serious dystrophin gene mutations. This antisense-based therapy has the potential to reduce the severity of the disease so that a boy with a DMD genotype would only develop a milder Becker phenotype.

Acknowledgments

This work was supported by the Muscular Dystrophy Association of Western Australia, the Neuromuscular Foundation of Western Australia and by grants from the Muscular Dystrophy Association of USA. I wish to acknowledge the efforts of Tina Ly, Frances Lloyd and Kylie Carville. In particular, Dr Sue Fletcher for helpful discussions, suggestions and for critically reviewing this manuscript.

References

1 Koenig M, Hoffman E, Moyer M *et al*; The molecular basis for Duchenne versus Becker muscular dystrophy: Correlation of severity with the type of deletion, *Americnl Journal of Human Genetics* 1989; **45**: 498–506.

2 Emery AH, *Duchenne Muscular Dystrophy*, 2, Oxford University Press; New York, 1987.

3 Monaco A, Bertelson C, Liechti-Gallati S, Moser H, Kunkel L, An explanation for the phenotypic differences between patients bearing partial deletions at the DMD locus, *Genomics* 1988; **2**: 90–95.

4 Fanin M, Daniele G, Cadaldini M, Miorin M, Vitiello L, Angelini C, Dystrophin-positive fibers in Duchenne dystrophy: origin and correlation to clinical course, *Muscle & Nerve* 1995; **18**: 1115–1120.

5 Klein C, Coovert D, Bulman D, Ray P, Mendell J, Burghes A, Somatic reversion/suppression in Duchenne muscular dystrophy (DMD): evidence supporting a frame-restoring mechanism in rare dystrophin-positive fibers, *American Journal of Human Genetics* 1992; **50**: 950–959.

6 Wallgren-Pettersson C, Jasani B, Rosser L, Lazarou L, Nicholson L, Clarke A, Immunological evidence for second or somatic mutations as the underlying cause of dystrophin expression by isolated fibers in Xp21 muscular dystrophy of Duchenne-type severity, *Journal of Neurological Sciences* 1993; **118**: 56–63.

7 Sicinski P, Geneg Y, Ryder-Cook A, Barnard E, Darlison M, Barnard P, The molecular basis of muscular dystrophy in the *mdx* mouse: a point mutation, *Science* 1989; **244**: 1578–1580.

8 Wilton SD, Dye D, Laing NG, Dystrophin gene transcripts skipping the *mdx* mutation, *Muscle & Nerve* 1997; **20**: 728–734.

9 Roberts R, Coffer A, Bobrow M, Bentley D, Exon structure of the human dystrophin gene. *Genomics* 1993; **16**: 536–538.

10 Sierakowska H, Sambade M, Agrawal S, Kole R, Repair of thalassemic human beta globin mRNA in mammalian cells by antisense oligonucleotides, *Proceedings of the National Acadamy of Sciences, USA* 1996; **93**: 12480–12484.

11 Nicholson L, Davidson K, Johnson M *et al*; Dystrophin in skeletal muscle: II. Immunoreactivity in patients with Xp21muscular dystrophy. *Journal of Neurological Sciences* 1989; **94**: 137–146.

12 Sadoulet-Puccio HM, Kunkel LM, Dystrophin and its isoforms, *Brain Pathology* 1996; **6**:25–35.

13 Agrawal S, Importance of nucleotide sequence and chemical modifications of antisense oligonucleotides, *Biochimica et Biophysica Acta* 1999; **1489**: 53–68.

14 Wilton SD., Lloyd F, Carville K, Fletcher S, Honeyman K, Agrawal S, R. Kole, Specific removal of the nonsense mutation from the *mdx*

dystrophin mRNA using antisense oligonucleotides, *Neuromuscular Disorders* 1999; **9**: 330–338.

15 Pramono Z, Takeshima Y, Alimsardjono H, Ishii A, Takeda S, Matsuo M, Induction of exon skipping of the dystrophin transcript in lymphoblastoid cells by transfecting on antisense oligodeoxynucleotide complementary to an exon recognition sequence, *Biochemical & Biophysical Research Communications* 1996; **226**: 445–449.

FRONTIERS IN HUMAN GENETICS
Diseases and Technologies
© 2001 by World Scientific Publishing Co. Pte. Ltd.

TREATMENT OF DUCHENNE MUSCULAR DYSTROPHY AT THE mRNA LEVEL

MASAFUMI MATSUO* and YASUHIRO TAKESHIMA†
Division of Genetics, International Center for Medical Research, and †Department of Pediatrics, Kobe University School of Medicine,
7-5-1, Kusunoki-cho, Chuo-ku, Kobe, Hyogo, 650-0017, Japan
**E-mail: matsuo@kobe-u.ac.jp*

Duchenne muscular dystrophy (DMD) is a progressive muscle wasting disease and so far, the treatment of DMD has not yet been established. Here, gene therapy is proposed as a novel treatment for DMD whereby the correction of the translational reading frame transforms severe DMD into the milder Becker muscular dystrophy. Based on the molecular analysis of dystrophin Kobe where the presence of an intra-exon deletion caused exon skipping during splicing, a part of the exon 19 sequence of the dystrophin gene was found to function as a splicing enhancer sequence, a sequence necessary for proper splicing. When oligonucleotides complementary to this sequence were added to the culture medium of lymphoblastoid cells, exon 19 skipping was specifically induced. This raises the possibility of a new therapeutic approach for DMD where dystrophin pre-mRNA splicing can be modulated by an oligonucleotide against a splicing enhancer sequence to produce an in-frame transcript that is able to produce truncated dystrophin. One natural example of this transformation is described.
Keywords: dystrophin, splicing, splicing enhancer sequence, antisense oligonucleotide, Duchenne muscular dystrophy, treatment

1 FRAME-SHIFT HYPOTHESIS

Duchenne muscular dystrophy (DMD) is a common inherited disease with a worldwide incidence of 1 in 3500 male births. DMD patients appear normal until the age of 3-5 years, after which they begin to experience difficulty in rising from the floor, climbing stairs, and other activities involving the large

* Corresponding author

proximal skeletal muscles. The muscular weakness is characteristically progressive. Affected individuals are wheelchair-bound by the age of 12 and succumb to cardiac or respiratory failure in their mid to late 20s. Interestingly, Becker muscular dystrophy (BMD) a milder form of X-linked muscular dystrophy, is distinguished from DMD by its delayed onset, later dependence on wheel-chair support and longer life span: affected boys remain ambulatory beyond the age of 16 years and a few may lead near-normal lives.

Both DMD and BMD are caused by mutations in the dystrophin gene, which extends over 3000 kb on Xp21 and has 70 exons encoded on a 14 kb mRNA. Approximately two-thirds of DMD and BMD patients carry deletions in dystrophin.[1] However, the extent of the deletion does not always correlate with the severity of the disease: some BMD patients with mild symptoms have deletions encompassing numerous exons, whereas some DMD patients with severe symptoms lack only a few exons. According to the frame-shift hypothesis,[2] BMD patients with long deletions may be able to produce a dystrophin mRNA that, despite a portion of its sequence being internally deleted, would still direct the production of an internally truncated semi-functional protein. The shorter deletions harbored by severe DMD patients, on the other hand, would bring together exons that, when spliced, would change the translational reading frame resulting in appearance of stop codon. Subsequent gene analyses have shown that over 90% of the deletion mutations that cause BMD maintain the dystrophin transcript reading frame, whereas frameshifts caused by deletions usually result in DMD.

Immunohistological analyses have demonstrated that dystrophin is present in muscle cell membranes. As expected, this protein is completely missing in boys with DMD, whereas muscle tissue from BMD patients contains reduced amounts of dystrophin. Thus, DMD and BMD represent examples of allelic heterogeneity. Western blot analyses using dystrophin antibodies have revealed a band corresponding to 427 kb, close to the predicted dystrophin size, in extracts of normal muscle tissues. Shorter proteins may sometimes be detected in tissues extracts of patients with BMD.[3]

2 TREATMENT OF DMD

The treatment of DMD is, of course, one of the primary goals of all DMD research work. Even with newly acquired knowledge, the biological role of dystrophin remains to be speculated and our understanding of the disease still remains incomplete. Consequently, the progression of the disease can not yet be slowed by presently known therapeutic treatments. Although much progress has been made in this field of study, we still seem to be a long way from achieving any clinically significant results.

As the injection of dystrophin minigenes into muscle may not be feasible for some time, an alternative strategy for the treatment of DMD might be to retard the progression of its clinical symptoms, i.e. to convert DMD into a BMD phenotype. Theoretically, this therapy can be done by modifying the dystrophin mRNA thereby changing a frameshift mutation causing DMD into an in-frame mutation characteristic of BMD. For this purpose, it may be possible to modify the editing of the mRNA. However, the exact mechanism of mRNA editing is not clear and thus can not yet be employed for therapy. Nonetheless, it may be possible to control the intron splicing during mRNA maturation. Until now, no method has been established to manipulate splicing, although splicing errors have been reported in mutations that affect the cis-acting elements implicated in pre-mRNA splicing, the 5' and 3' splicing site sequences, branch point sequences and their locations.

3 DYSTROPHIN KOBE

Although all of the consensus sequences known to be required for splicing were unaffected in a particular dystrophin gene mutation named dystrophin Kobe, we found that exon skipping during splicing was induced by the presence of an intra-exon deletion mutation in the genome. The deletion was detected by PCR analysis which revealed that the product amplified from the exon 19 encompassing region from the DMD case in question was smaller than normal. Sequence analysis confirmed that the case had a deletion of 52 bp out of the 88 bp exon 19 donor.[4] This 52 bp deletion was considered to result in a frameshift mutation that may cause DMD. The dystrophin transcript of dystrophin Kobe was then analyzed using reverse-transcription PCR (RT-PCR). Surprisingly, the product amplified from the region encompassing exon 19 was smaller than predicted according to the genomic DNA analysis results. Sequence analysis indicated that the whole of exon 19 was missing from the dystrophin cDNA causing an out-of-frame mutation. In particular, this indicated that the deletion mutation within an exon sequence could induce a splicing error during mRNA maturation even though the known consensus sequences at the 5’ and 3’ splicing sites of exon 19 were maintained[5] (Fig. 1).

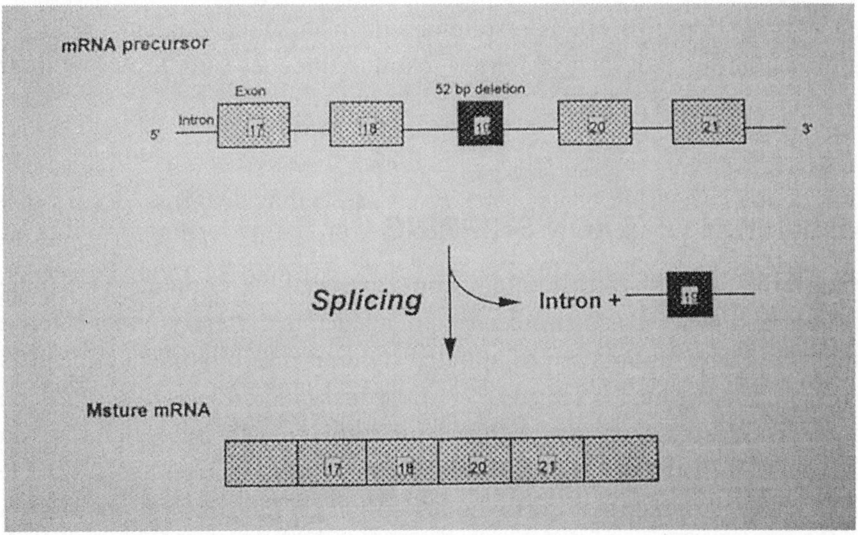

Fig. 1 Schematic representation of exon 19 of dystrophin Kobe and the splicing error that occurred. In the mRNA precursor sequence (upper line), 52bp of the 88bp exon 19 of the dystrophin gene (shaded box) are deleted in dystrophin Kobe. This truncated exon 19 is spliced out together with the intron to produce mature mRNA lacking all of exon 19.

4 IDENTIFICATION OF A SPLICING ENHANCER SEQUENCE

Exon 19 skipping in dystrophin Kobe suggests that the deleted sequence of exon 19 may function as a cis-acting element for exact splicing of the upstream and downstream introns. To investigate this potential role of exon 19, an *in vitro* splicing system using artificial dystrophin mRNA precursors (pre-mRNAs) was established. Pre-mRNA, containing exon 18, truncated intron 18, and exon 19, was spliced precisely *in vitro* whereas the splicing of intron 18 was almost completely abolished when the wild type exon 19 was replaced by the dystrophin Kobe exon 19.[6] In addition an antisense 31 mer 2'-*O*-methylribonucleotide, complementary to the 5' half of the deleted sequence in dystrophin Kobe exon 19, inhibited *in vitro* splicing of the wild type pre-mRNA in a dose- and time-dependent manner.[6]

These results suggest that the deleted region of the exon 19 sequence in dystrophin Kobe is a splicing enhancer sequence that is a likely target for splicing factors to identify exon sequences and promote their inclusion in the mature mRNA.

5 INDUCTION OF EXON SKIPPING

The possibility that oligonucleotides could be used as modulators of gene expression, and hence as chemotherapeutic agents, is currently under intense investigation. The modulation of splicing by antisense oligonucleotides has recently attracted much attention.[7,8] Dominski and Kole described an elegant experiment in which aberrant splicing induced by a thalassemia mutation was corrected by an antisense 2'-*O*-methylribonucleotide (2'-*O*-Me RNA).[8] This prompted us to test whether splicing of dystrophin pre-mRNA could also be modulated by an antisense oligonucleotide as the first step towards evaluating the potential therapeutic use of antisense oligonucleotides to correct aberrant splicing reactions in patients with DMD.

The antisense oligonucleotide was transfected into normal lymphoblastoid cells and the resulting dystrophin mRNA was analyzed. With this treatment, remarkably, skipping of exon 19 started to appear after 6 hours of incubation and complete skipping was observed after 24 hours of incubation[9] (Fig. 2). None of the other 78 dystrophin exons were skipped and exon 19 skipping could not be induced by the sense oligonucleotide. These results showed that antisense oligonucleotides against a splicing enhancer sequence can induce exon skipping even in living cells.

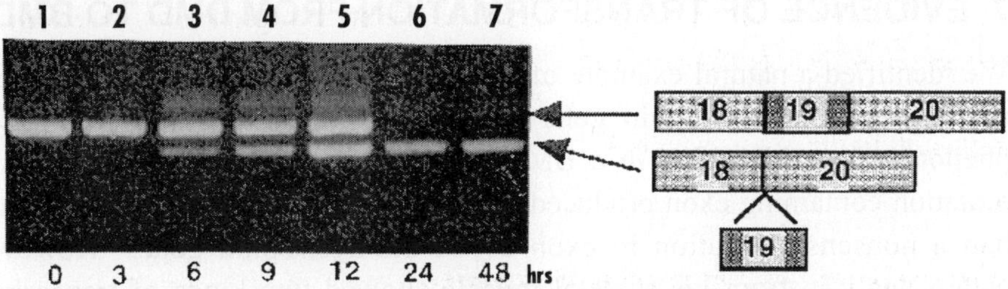

Fig. 2 Effect of the period of incubation with antisense oligonucleotide on splicing. For transcription to proceed, lymphoblastoid cells were incubated for the indicated time. A region encompassing exons 18 to 20, obtained from the dystrophin mRNA extracted from the cells, was then amplified by RT nested PCR. Only one amplified product corresponding to the normal cDNA was amplified after 0 or 3 hours of incubation (lanes 1 and 2). A new product lacking exon 19 started to be produced after 6 hours of incubation (lane 3), and the relative abundance of this product increased thereafter. All products were derived from the product with exon 19 skipping after 24 hours of incubation or longer (lanes 6 and 7). The exon component of the amplified product was schematically described on the right.

6 CORRECTION OF THE TRANSLATIONAL READING-FRAME BY EXON SKIPPING

The first *in vitro* evidence that dystrophin pre-mRNA splicing can be modulated by an antisense oligonucleotide raises the possibility of a new therapeutic approach for DMD. For example, the same antisense oligonucleotide may be used to treat a DMD case with a 242-nucleotide deletion of exon 20. If we are able to induce exon 19 (88 bp) skipping *in vivo*, the dystrophin transcript will lack both exons 19 and 20 but the translational reading frame will be restored. As a result, this modulation of splicing should transform DMD into BMD.[10]

7 EVIDENCE OF TRANSFORMATION FROM DMD TO BMD

We identified a natural example of a DMD to BMD conversion; a nonsense mutation of the dystrophin gene that was expected to result in a DMD phenotype was identified in a BMD case where skipping of the nonsense mutation containing exon produced in-frame dystrophin transcript. The case had a nonsense mutation in exon 27 of the dystrophin gene. However, cDNA analysis from his skeletal muscle showed two kinds of transcripts; one had a normal exon content and the other transcript showed the skipping of exon 27 (Fig. 3).[11] Since exon 27 encoded 183 nt, the resulting dystrophin transcript maintained the translational reading frame thereby producing a truncated dystrophin. This result confirmed our hypothesis that DMD can be treated by the correction of the translational reading frame.

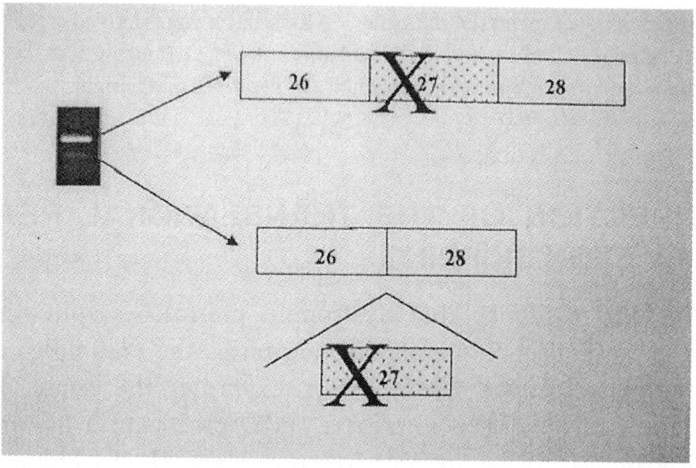

Fig.3 Amplification of dystrophin cDNA encompassing exons 25 to 30 from BMD. Amplified products encompassing exons 25 to 30 were obtained from muscle cDNA by RT-PCR. Two bands were visualized in the index case. The exon component of the amplified product was schematically described on the right and X represents a nonsense mutation.

References

1. Hoffman EP, Kunkel LM, Dystrophin abnormalities in Duchenne/Becker muscular dystrophy. *Neuron* 1989;**2**:1019-1029
2. Monaco AP, Bertelson CJ, Liechti-Gallati S, Moser H, Kunkel LM, An explanation for the phenotypic differences between patients bearing partial deletions of the DMD locus. *Genomics* 1988; **2**:90-95
3. Hoffman EP, Fischbeck KH, Brown RH, Johnson M, Medori R, Loike JD, Harris JB, *et al.*, Characterization of dystrophin in muscle-biopsy specimens from patients with Duchenne's or Becker's muscular dystrophy. *N Engl J Med* 1988; **318**:1363-1368
4. Matsuo M, Masumura T, Nakajima T, Kitoh Y, Takumi T, Nishio H, Koga J, *et al.*, A very small frame-shifting deletion within exon 19 of the Duchenne muscular dystrophy gene. *Biochem Biophys Res Commun* 1990;**170**:963-967
5. Matsuo M, Masumura T, Nishio H, Nakajima T, Kitoh Y, Takumi T, Koga J, *et al.*, Exon skipping during splicing of dystrophin mRNA precursor due to an intraexon deletion in the dystrophin gene of Duchenne muscular dystrophy Kobe. *J Clin Invest* 1991; **87**:2127-2131
6. Takeshima Y, Nishio H, Sakamoto H, Nakamura H, Matsuo M , Modulation of *in vitro* splicing of the upstream intron by modifying an intra-exon sequence which is deleted from the dystrophin gene in dystrophin Kobe. *J Clin Invest* 1995; **95**:515-520
7. Barabino SM, Sproat BS, Lamond AI, Antisense probes targeted to an internal domain in U2 snRNP specifically inhibit the second step of pre-mRNA splicing. *Nucleic Acids Res* 1992; **20**:4457-4464
8. Dominski Z, Kole R, Restoration of correct splicing in thalassemic pre-mRNA by antisense oligonucleotides. *Proc Natl Acad Sci USA* 1993; **90**:8673-8677
9. Pramono ZA, Takeshima Y, Alimsardjono H, Ishii A, Takeda S, Matsuo M, Induction of exon skipping of the dystrophin transcript in

lymphoblastoid cells by transfecting an antisense oligodeoxynucleotide complementary to an exon recognition sequence. *Biochem Biophys Res Commun* 1996; **226**:445-449

10. Matsuo M, Duchenne/Becker muscular dystrophy: from molecular diagnosis to gene therapy. *Brain Dev* 1996; **18**:167-172
11. Shiga N, Takeshima Y, Sakamoto H, Inoue K, Yokota Y, Yokoyama M, Matsuo M, Disruption of the splicing enhancer sequence within exon 27 of the dystrophin gene by a nonsense mutation induces partial skipping of the exon and is responsible for Becker muscular dystrophy. *J Clin Invest* 1997; **100**:2204-2210

FRONTIERS IN HUMAN GENETICS
Diseases and Technologies
© 2001 by World Scientific Publishing Co. Pte. Ltd.

STRATEGY OF GENE THERAPY FOR LIVER CIRRHOSIS AND LIVER CANCER

JIRO FUJIMOTO
First Department of Surgery, Hyogo College of Medicine, Nishinomiya 663-8501, Japan
E-mail: sfujimo@hyo-med.ac.jp

Liver cirrhosis and liver cancer are major cause of morbidity and mortality worldwide with no effective therapy. Our aim is to establish a novel approach for liver cirrhosis and liver cancer utilizing *in vivo* gene therapy. To achieve effective gene expression *in vivo*, we used well established transfection method, hemagglutinating virus of Japan (HVJ)-liposome. Liver cirrhosis induced by dimethylnitorosamine (DMN) in rat is characterized by parenchymal collapse followed by hyper-accumulation of fiber tissue. All cirrhotic rats died of liver dysfunction by 7 weeks after the initial injection of DMN. Muscle directed gene transfer of hepatocyte growth factor (HGF) was performed. After repetitive gene therapy, the massive fibrosis was in nearly total remission and all cirrhotic rats were rescued from fatal cirrhosis. Alpha-fetoprotein (AFP) is highly expressed in most patients with liver cancer. To achieve specific gene expression in AFP-producing tumor, we used AFP promoter. AFP producing HUH7 tumors were established in the liver of SCID mice, vector consisting of HSV-TK gene with AFP promoter (AFP-TK) was introduced to these tumor bearing mice *via* splenic/portal injection. Repetitive *in vivo* transfection of AFP-TK followed by ganciclovir treatment can achieve abrogation of tumors in the liver. Our present results indicate that the gene therapy may hold promise for the treatment of patients with liver cirrhosis/ liver cancer.
Keywords: cirrhosis, liver cancer, gene therapy, hepatocyte growth factor, HVJ-liposome, alpha-fetoprotein

1 INTRODUCTION

Liver cirrhosis is the terminal stage of chronic liver disease with widespread fibrous scar tissue. Various contributing factors induce cirrhosis, including viral hepatitis, alcohol abuse, prolonged biliary obstruction and some genetically transmitted metabolic disorders. Liver cirrhosis is a major cause of morbidity and mortality worldwide with no effective therapy. Complications of liver cirrhosis are often fatal. Hepatic dysfunction, esophageal varices, and liver cancer are the most serious complication.

Hepatocellular carcinoma (HCC) is one of the most common human malignancies in the world, especially in eastern Asia and Africa.[1] Several treatment options are available including surgery, the prognosis of this disease is still poor. Multiple tumors and recurrence after the treatment are common in patients with HCC,[2] gene therapy may be a useful method for treating multiple liver tumors in these patients. We demonstrate a novel approach of gene therapy for liver cirrhosis,[3] and liver cancer.

2 METHODS

To achieve effective gene transfer *in vivo*, we used hemagglutinating virus of Japan (HVJ)-liposome. Plasmid DNA were encapsulated in HVJ-liposome as previously described by Kaneda *et al.*[4] Briefly, phosphatidilserine, phosphatidylcholine, and cholesterol were mixed and dried. This dried lipid and plasmid DNA, which had been complexed with high-mobility group-1 nonhistone chromosomal protein purified from calf thymus, were shaken vigorously and sonicated to form liposomes. UV-irradiated HVJ (3000 hemagglutinating units) was mixed with the liposome suspension, and residual free HVJ was removed from the HVJ-liposome by sucrose density gradient centrifugation.

Dimethylnitrosamine (DMN) was intraperitoneally administrated weekly for three consecutive days during our experimental schedule. All cirrhotic rats died of liver dysfunction by seven weeks after the initial injection of DMN. Beginning after the 4th weekly DMN-administration (when there are serious liver cirrhosis), rats were injected through their skeletal muscles once a week with HVJ-liposome containing hepatocyte growth factor (HGF) vector. To test the feasibility of gene therapy for liver cancer, SICD mice were injected with HUH7 cells, alpha-fetoprotein (AFP)-producing human hepatoma cells. Severe combined immunodeficient (SCID) mice were anesthetized and transverse incision was made in the left flank through the skin and peritoneum to expose the spleen. Then HUH7 cells were injected into portal vein through the splenic hilum using a 30-gauge needle. For natural killer cell depletion *in vivo*, SCID mice were injected intravenously with anti-asialo GM1 rabbit serum. We used reconstructed plasmid vector in which herpes simplex thymidine kinase (HSV-TK) gene was driven under the control of AFP gene promoter (AFP-

TK).[5] AFP-TK was encapsulated in the HVJ-liposome and injected to the portal vein of tumor bearing mice followed by ganciclovir treatment.

3 RESULTS AND DISCUSSION

3.1 Gene Therapy for Liver Cirrhosis

Repetitive transfections of HGF into skeletal muscles induced a high plasma level of HGF and a tyrosine phosphorylation of c-Met/HGF receptor. Recently, evidence has been presented that overproduction of transforming growth factor-β1 (TGF-β1) is a major cause of tissue fibrosis in various organs. In the liver, TGF-β1 induces the phenotypic transition of hepatic stellate cells to proliferating myofibroblast-like cells which enhance production of extracellular matrix components. Furthermore, TGF-β1 induces apoptotic cell death in hepatocytes. We therefore examined expression of TGF-β1 in the cirrhotic rat liver. Immunohistochemical analysis and Northern blot analysis revealed that TGF-β1 expression was markedly reduced by the treatment with HGF-HVJ-liposome. Consistent with the finding, repetitive HGF-HVJ-liposome treatments dramatically decreased fibrous connective tissue components on Glisson's sheath as well as pseudolobule formations. The fibrosis in both peripotal and centrilobular areas were remarkably diminished, liver acinus were reconstructed with hepatocyte regeneration. Quantitative analysis of fibrosis by image analysis techniques demonstrated that HGF-HVJ-liposome treatments produced more than 70% diminution of fibrosis in the liver. Apoptotic cells in the liver was investigated by TUNEL method which detects fragmented DNA *in situ*. Apoptotic cells were found in the cirrhosis liver, HGF gene therapy prevented this DMN-induced hepatocyte apoptosis. All cirrhotic control rats died within 45 days after starting DMN-administration by liver failure. Notably, transfection of rats with 40 mg of HGF DNA rescued all rats from fatal liver cirrhosis and they were free from cirrhosis. Since repetitive *in vivo* transfection using HVJ-liposome is simple and safe, and can be carried out without significant inflammation or immune response, transduction of HGF gene in skeletal muscles presented here may be translated into an useful clinical regiment for gene therapy in the treatments of patients with advanced liver cirrhosis.

3.2 Gene Therapy for Liver Cancer

AFP-TK plasmid vector was encapsulated into the HVJ-liposome and injected to splenic hilus at 7, 21, and 35 days after the tumor injection. Mice were devided into four groups: mice with single injection of AFP-TK, double injection, triple injection and untreated. In the untreated group, the majority of mice injected with HUH7 developed large cannon ball-like tumors in the liver. Triple, double, and a single injection of AFP-TK vector followed by ganciclovir treatment significantly inhibited the outgrowth of 56-day hepatic tumors and improved the survival of the animals. Triple gene transfer of AFP-TK completely inhibited the tumor outgrowth. Expression of *E. coli LacZ* gene was evident by X-gal staining in HUH7 tumors but not in liver tissues in mice injected with AFP-*LacZ* vector. The liver tissues were histologically normal and no lymphocyte infiltration was observed. Thus, specific gene expression in AFP-producing tumor in the liver was obtained by utilizing AFP-promoter *in vivo.* Gene therapy using AFP-TK vector inhibits growth of liver tumor of human hepatoma, and the repetitive treatments boost the effect. This regime may be potentially useful for the treatment of patients with progressive liver cancer, which is otherwise fatal and untreated by conventional therapy.

References

1. Colombo M, Hepatocellular carcinoma, *J Hepatol* 1992; 15: 225-236.
2. Levin B, Amos C, Therapy of unresectable hepatocellular carcinoma, *N Eng J Med* 1995; 332: 1294-1296.
3. Ueki T, Kaneda Y, Tsutsui H, Nakanishi K, Sawa Y, Morishita R, Matsumoto K, Nakamura T, Takahashi H, Okamoto E, Fujimoto J, Curative gene therapy of liver cirrhosis by hepatocyte growth factor in rats, *Nature Med* 1999; 5: 226-230.

4. Kaneda Y, Iwaki K, Uchida T, Increased expression of DNA cointroduced with nuclear protein in adult rat liver, Science 1989; 243: 375-378.
5. Kaneko S, Hallenbeck P, Kotani T, Nakabayashi H, McGarrity G, Tamaoki T, Anderson WF, Chiang Y, *Cancer Res* 1995; 55: 3105-3109.
6. Hirano T, Fujimoto J, Ueki T, Yamamoto H, Iwasaki T, Morishita R, Sawa Y, Kaneda Y, Takahashi H, Okamoto E, Persistent gene expression in rat liver *in vivo* by repetitive transfections using HVJ-liposome, *Gene Ther* 1998; 5: 459-464.

FRONTIERS IN HUMAN GENETICS
Diseases and Technologies
© 2001 by World Scientific Publishing Co. Pte. Ltd.

GENE THERAPY FOR PROSTATE CANCER; DEVELOPMENT OF TISSUE SPECIFIC PROMOTER-BASED GENE THERAPY

AKINOBU GOTOH[*,‡], TOSHIRO SHIRAKAWA[*], YOSHITAKA WADA[*], SONG-CHU KO[†], CHINGHAI KAO[†], LELAND W. K. CHUNG[†] and SADAO KAMIDONO[*]

[*] *Department of Urology, Kobe University School of Medicine, 7-5-2, Kusunoki-cho, Chuo-ku, Kobe 650-0017 Japan*

[†] *Molecular Urology and Therapeutics Program, Department of Urology, University of Virginia, Charlottesville, VA, USA*

[‡] *E-mail: gotoh@med.kobe-u.ac.jp*

We have defined the androgen responsiveness and status of prostate specific antigen (PSA) secretion of prostate cancer cells as well as non-prostatic cells transduced with PSA promoter reporter constructs. Androgen-independent (AI) human prostate cancer remains a lethal phenotype for which there is no effective therapy. AI prostate cancer produces and secretes large amounts of PSA at both primary and metastatic sites. The aim of this investigation is to explore the use of the PSA promoter as a mean of prostate cell specific expression of a toxic gene thymidine kinase (TK) to an AI PSA-producing human prostate cancer cell line. An adenovirus vector carrying human herpes simplex thymidine kinase (TK) gene under control of the PSA promoter (Ad-PSA-TK) was generated to target PSA-producing AI prostate cancer cells. Upon the administration of acyclovir, Ad-PSA-TK efficiently killed the AI PSA-producing human prostate cancer cells. This paper discusses the importance of tissue specific promoter system for using gene therapy of prostate cancer, and summarizes recent gene therapy strategies developed to target this process. This strategy can potentially be used in combination with current therapeutic modalities to achieve more effective tumor cell-kill with much reduced toxicity.

Keywords: prostate cancer, prostate specific antigen(PSA), tissue specific promoter, thymidine kinase(TK), adenovirus, gene therapy

---

[‡] Corresponding author

1 INTRODUCTION

Adenocarcinoma of prostate has become a significant health problem, being the most commonly diagnosed cancer and the second leading cause of death from cancer among males in the United States.[1] An estimated 317100 U.S. men were diagnosed to have clinical prostate cancer and 41400 deaths of the disease were reported in 1996.[2] Approximately one-third of men over the age of 50 years are affected, but not every one of these patients will eventually die from the disease.[3] When prostate cancer becomes advanced and acquires increasing metastatic potential, androgen withdrawal is widely applied to delay the progression of disease. Inevitably, tumors become hormonal refractory and exhibit androgen-independent (AI) characteristics. In these patients, it has been documented that further response to endocrine therapy is rare and chemotherapy is of limited value.[4] For these reasons, we have attempted to develop gene therapy strategies for the treatment of AI prostate cancers. Currently, there are two prototypes of gene therapy protocols that are under clinical or preclinical evaluation for the treatment of cancer. The first strategy is corrective gene therapy. This involves either restoration of defective genes or inactivation of activated genes in preneoplastic or neoplastic cells to restore the loss of growth control pathways.[5] The second is cytoreductive gene therapy which is based upon selective destruction of malignant cells.[5] Cytoreductive gene therapy involves the delivery of gene(s) that will lead to increased sensitivity of cancer cells to drug or toxin therapy,[6] cause cell cycle arrest, or trigger apoptosis.[7,8] Several groups have constructed vectors containing tissue-specific promoters that restrict expression of the transduced cytotoxic genes to specific target cells.[5,9] Prostate cancer is a particularly appropriate target for such an approach.[10] PSA, a 34 kDa chymotrypsin-like serine protease, is secreted almost exclusively by normal and cancerous epithelial cells lining both prostatic acini and ducts.[11,12] Because PSA is highly expressed by most (~ 99%) primary and metastatic prostate cancers,[13,14] it may be possible to use the PSA promoter to drive expression of therapeutic gene(s) specifically in cancerous epithelial cells. To accomplish our goal, we employed an AI PSA-producing LNCaP subline, C4-2,[15] as a target for evaluating the specificity and androgen responsiveness of a 5837 bp PSA promoter luciferase construct, which was found to be highly active in androgen-dependent LNCaP cells.[16] The studies employed both DNA mediated transfections and adenovirus vectors to deliver a therapeutic toxic gene, human herpes simplex (hs) virus thymidine kinase (TK). We evaluated the specific AI human prostate cancer cell killing following bioactivation of the precusor acyclovir by viral TK. The possible clinical implications of this tissue-specific PSA promoter-driven toxic gene therapy are discussed.

2 MATERIALS AND METHODS

2.1 Cell Lines and Cell Cultures

C4-2 was established from a LNCaP tumor grown in a castrated host[15] and WH was established from a human bladder transitional cell carcinoma.[17] All cell lines were maintained in T-medium [80% Dulbecco's modified Eagle's medium (DMEM; GIBCO, Grand Island, NY), 20% F12K(Irving Scientific, Santa Ana, CA), 3 g/L NaHCO3, 100 units/liter penicillin G, 100 mg/ml streptomycin, 5 mg/ml insulin, 13.6 pg/ml triiodothyronine, 5 mg/ml transferrin, 0.25 mg/ml biotin, and 25 mg/ml adenine] with 5% fetal bovine serum (FBS; Sigma Chemical Company, St. Louis, MO).

2.2 Construction of Recombinant Adenoviruses

A recombinant adenovirus containing the PSA promoter and thymidine kinase (TK) was produced by cotransfecting pΔE1sp1BPSA-TK (shuttle vector containing expression cassette) and pBHG-11 (recombinant vector) plasmids as previously described.[10,18] Recombinant virus was cloned from individual plaques, amplified, and purified by $CsCl_2$ centrifugation method. The virus stock was then dialyzed against 10 mM Tris buffer pH 7.5. The plaque-forming unit (PFU) of the viruses was measured by a standard biologic plaque forming assay.[19]

2.3 Thymidine Kinase Activity Assay

Cell extracts containing TK enzyme were prepared from Ad-PSA-TK infected C4-2 cells (passages 35–40) and WH cells (passages of 25–40). Ad-

PSA-TK virus was added at 20 PFU/cell directly to each of the 100 mm dishes containing 60% to 70% confluent cells, and the cell cultures were incubated for an additional 24 hours. The culture medium was then replaced, and after a further 48 hours, the culture medium was removed, infected cells were washed with PBS, and the expressed thymidine kinase enzyme was released from the cells by the freeze and thaw method. The resulting cell suspension was centrifuged at 300 x *g*. for ten minutes and the supernatant was frozen at –80°C. TK activity assays were performed according to the method of Ives *et al.*[20] with modifications.[21] Briefly, supernatant from approximately $1x10^6$ cells containing TK enzyme was mixed with an equal volume of TK assay substrate mixture containing 0.2 mCi (10 mM) [^{3}H]-Ganciclovir (GCV) (Moravek Biochemicals, Brea, CA), 3 mM $MgCl_2$, 3 mM ATP, 10 mg/ml bovine serum albumin, and 50 mM phosphate buffer (pH 6.5). The reaction mixture was then incubated in 36°C water bath. At various time points, the reaction mixture was transferred onto DE-81 disks (Whatman, Hillsboro, OR). The air-dried disks were washed with 50% ethanol, transferred to scintillation vials, and incubated for one hour with 1.0 ml of 0.1 N HCl. Ten milliliters of scintillation fluid was added and the amount of phosphorylated [^{3}H]-GCV was counted in a Beckman Scintillation Counter (Beckman Instruments INC, Schaumburg, IL). GCV, used for TK activity assay here, is mechanistically similar to acyclovir (ACV) and differs from it structurally only by the addition of a hydroxymethyl group. Because of this one side chain difference, GCV is administered only through intravenous route, while ACV can be administered orally, intravenously, and topically to human. ACV and GCV are both commonly used against herpes virus infection.

2.4 Acyclovir Cytotoxicity Assay

Cells were plated in 24-well tissue culture plates and infected with 20 multiplicity of infection (MOI) units of Ad-PSA-TK virus for 24 hours. Cells were allowed to recover for one day from virus infection before acyclovir (10 mg/ml) (Burroughs Wellcome, Research Triangle, N.C.) treatment. Cell numbers were measured by crystal violet staining.[22]

2.5 Androgen-Independent PSA-Producing Human Prostate Cancer Model

Androgen independent C4-2 cell suspension (1 x 10^6) in 50 ml. of media was mixed with 50 ml. of Matrigel (Collaborative Biomedical Products, Inc., Bedford, MA) and injected into the left flank of castrated nude mice (*n* = 21).

This study was initiated at day 0 when the tumor volume (volume = long axis x short axis^2 x 0.523) was at least 45 mm^3. Experimental animals were divided into four groups: Group I, no treatment (*n* = 5); Group II, ACV only (*n* = 5); Group III, Ad-PSA-TK only (*n* = 6), and Group IV, Ad-PSA-TK with ACV (*n* = 5). Animals in groups II and IV received daily ACV (40 mg/kg body weight) intraperitoneally from day 0 to day 20. While animals in Group III and IV received intratumoral injections of 5 x10^8 PFUs of Ad-PSA-TK on days 1, 7, and 14. Tumor volumes were measured weekly and serum PSA was measured every two weeks for a total period of four weeks. Student's *t* test was used to determine significance.[10]

3 RESULTS

3.1 Development of a Recombinant Adenovirus, Ad-PSA-TK, for Selective Inhibition of Growth of AI PSA-Producing Human Prostate Cancer Cell Line

Because the PSA promoter-driven reporter gene shows unique cell specificity and androgen-induced expression, we constructed a replication-defective adenovirus carrying the PSA promoter and a toxic therapeutic gene, human herpes simplex thymidine kinase gene (Ad-PSA-TK). To test whether this therapeutic gene was functional in PSA-producing cells, we compared TK expression in AI C4-2 (PSA-producing) and a control human bladder transitional cell carcinoma line, WH (non-PSA-producing). The infectivities of C4-2 and WH were compared by using a range of Ad-CMV-b-gal as the infecting virus (from 0 to 100 MOI) as described previously.[23] Infectivities of both cell lines were comparable, reaching 90% at a virus titer of 20 MOIs/target cell. When infected with a virus titer greater than 100 MOI, viral toxicity was noted.[10] Time-dependent expression of TK activity was assayed in C4-2 and WH cells 72 hours after infection with Ad-PSA-TK. Figure 1 shows that TK activity, as assayed by the conversion of [^{3}H]-

GCV into phosphorylated form, was detectable in the Ad vector infected C4-2 cell line; whereas, no TK activity was detected in WH cells infected similarly with Ad-PSA-TK virus.[10]

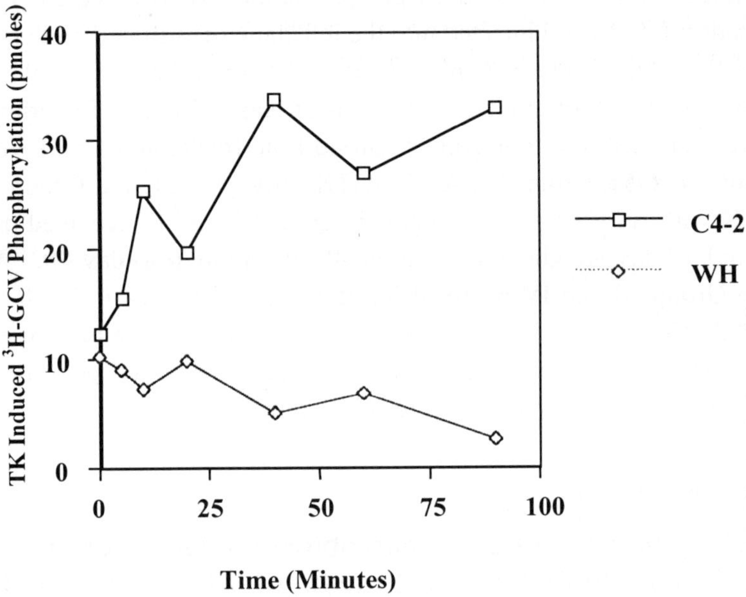

Fig. 1 Thymidine kinase activity assay. WH and C4-2 cells were infected with 20 MOIs of Ad-PSA-TK. The amount of TK enzyme was indirectly determined by measuring the phosphorylation rate of [^{3}H]-ganciclovir mixed with supernatant from lyzed 1×10^6 cells. Significant TK activity was detected in PSA producing C4-2 cells but not in non-PSA producing WH cells.

3.2 ACV-Mediated Cell-Kill in Ad-PSA-TK Infected C4-2 Cells

To examine whether Ad-PSA-TK infection confers ACV-induced cytotoxicity, we first tested the toxicity of ACV (range 0 to 320 mg/ml) in noninfected C4-2 and WH cells and observed that ACV at doses below 80 mg/ml did not appreciably inhibit the growth of C4-2 and WH cells. When infected with Ad-PSA-TK, C4-2 but not WH cells, exhibited sensitivity toward TK-induced cell-kill upon the addition of 10 mg/ml. ACV (Figs. 2A and B) [10]. In comparison to controls, ACV inhibited 70% of the growth of C4-2 (Fig. 2B) but was ineffective in inhibiting the growth of WH cells (Fig. 2A) *in vitro*. Ad-PSA-TK infection alone at 20 PFU/cell did not affect the growth of both C4-2 and WH cells in the absence ACV.

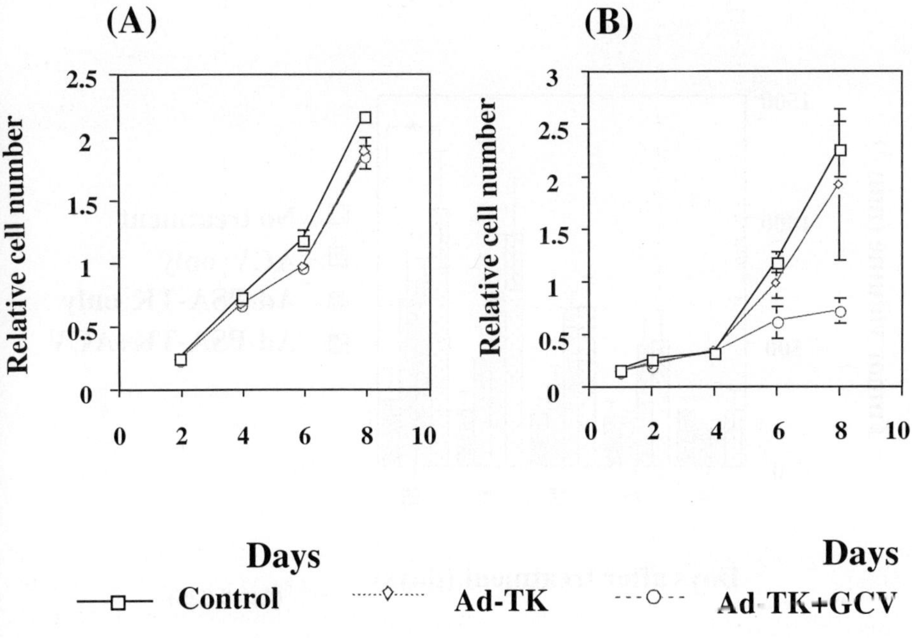

Fig. 2 Acyclovir cytotoxicity assay. (A) WH and (B) C4-2 cells were infected with 20 MOIs of Ad-PSA-TK on day 1. Cell growth was then measured by crystal violet staining with or without the presence of acyclovir (10 mg/ml) in cell culture media. Cell-specific killing was observed only in PSA producing C4-2 cells infected with Ad-PSA-TK followed by acyclovir treatment

3.3 Ad-PSA-TK and ACV Inhibit the Growth of AI Prostate Cancer *In Vivo*

C4-2 tumor-bearing castrated animals were selected as a model to demonstrate the efficacy of Ad-PSA-TK and ACV to inhibit the growth of AI prostate tumor *in vivo*.[10] (Fig. 3) Both tumor size (Fig. 3) and PSA production (Fig. 4) were significantly attenuated in the treated group ($n = 5$) as opposed to the control group ($n = 5$) ($p < 0.025$ and 0.01, respectively).[10]

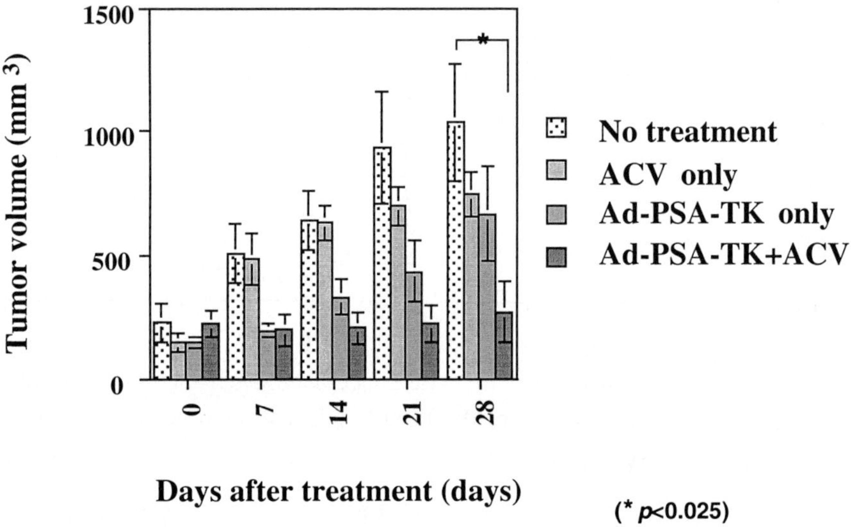

Fig. 3 Attenuation of AI C4-2 tumor growth by Ad-PSA-TK and ACV administration. Animals bearing C4-2 subcutaneous tumors received no treatment or ACV, Ad-PSA-TK, or Ad-PSA-TK/ACV treatment as described in Materials and Methods. Standard error shown below figure. Tumor sizes were monitored weekly. Significance determined by Student's *t* test.

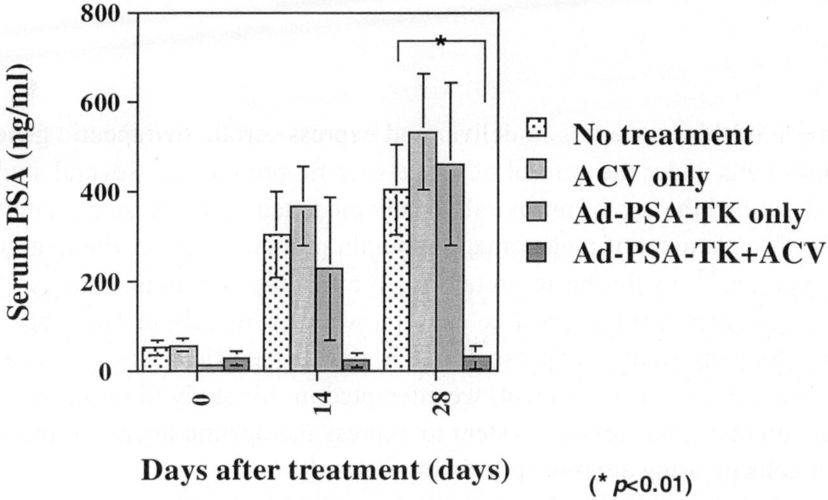

Fig. 4 Inhibition of AI C4-2 tumor PSA production by Ad-PSA-TK and ACV administration. Serum PSA levels of animals described in Fig. 3 were measured biweekly.Note that Ad-PSA-TK/ACV effectively inhibited PSA production by C4-2 tumors.In contrast to inhibition of tumor growth, Ad-PSA-TK alone did not inhibit PSA production by C4-2 tumors.Significance determined by Student's *t* test.

4. DISCUSSION

Adenovirus-based gene therapy for the delivery of both restoration and cytotoxic genes to target tumor epithelial cells has been the subject of intensive study in recent years.[23–25] The success of gene therapy depends on delivering the candidate gene into the target cells with a specificity, and how well the expression of the delivering genes can be directed. At present, the clinical application of gene therapy is limited by the capacities of gene delivery vector; if the vector is likened to a truck, then it can be seen as consisting of the actual body of the truck, representing the delivery vehicle, and the truck's cargo, representing the candidate gene. In order to deliver the candidate gene effectively to its target site, there is a need to develop high-performing delivery vehicles such as the adenovirus vector and to enhance the performance of the promoter, which can be likened to the truck driver.

So far, universal promoters such as CMV and RSV have been used, but these do not allow effective delivery of the candidate gene exclusively to the target site. Universal promoters also do not allow adequate mitigation of the vector's disadvantageous effects. In response to these problems, recent times have seen the development of tissue-specific promoters as a means of ensuring that the vector delivers effectively on the target tissue. A logical extension of this concept is to deliver and express certain therapeutic gene(s) to tumor cells under the control of tissue-specific promoters. Several studies have described the use of retroviral vectors mediated by tyrosinase promoter-TK for the treatment of melanoma,[26] albumin promoter-TK for the treatment of hepatoma,[27] myelin basic protein for the treatment of brain tumors.[28] In contrast to other cancers, prostate cancers, whether in early or late stages, is known to consistently express PSA, a tissue-specific marker. Because retrovirus infection is inefficient, we attempted in this study to make use of a highly infectious adenovirus system to express therapeutic agents in prostate cancer cells by using a tissue-specific promoter, PSA.

Our data support the concept that a PSA promoter is required to confer tissue-specific expression. Under the experimental conditions, applied the PSA promoter can be activated to a level comparable to the well-known strong CMV-promoter raising the promise of the efficacy of this promoter for the delivery of other potential therapeutic gene(s) to both AD and AI PSA-producing human prostate cancer cells (data not shown). High titer, high infectivity and a wide range of target tissues make adenovirus a popular vehicle for cancer gene therapy.[24]

In summary, a promising new adenoviral strategy using PSA promoter mediated transfer of TK gene to AI prostate cancer cells has been developed. The Ad-PSA-TK was shown to be highly effective in inducing cell-kill in vitro in the presence of acyclovir. Our results suggest that a PSA promoter is a powerful and specific promoter for driving the expression of therapeutic gene(s) in human prostate cancer cells. The high level of expression of the PSA promoter reporter in AI PSA-producing prostate cancer cells is particularly noteworthy. This could imply that the PSA promoter has significant advantages over other promoters in delivering therapeutic genes, particularly to AI and PSA-producing human prostate cancer cells.

While the adenovirus vector used in the present study shows excellent rates of success for gene delivery and gene expression, many problems remain in areas such as *in vivo* immune response to the virus. Recently, safety concerns have been raised regarding contamination with replication-competent adenovirus. The successful high titer is expected to improve as further advances in medical engineering technology lead to the development of a next generation of adenovirus and non-virus vectors. The development of a tissue-specific vector discussed here is part of this process.

So far, in animal experiments conducted as pre-clinical trials, it has not been possible to faithfully mimic human pathology in experimental tumors, so that there is a discrepancy between experimental results and actual intended therapeutic effect. The research of gene therapy requires the development of an adequate experimental animal model. The coming years are expected to see the development of numerous enhanced and high-performing vectors; an important concern will be deciding how to adapt these to clinical applications and how to combine them with current methods of therapy. Further advances in the basic research area of gene therapy are required in order to help resolve these problems.

The present paper discusses the potential for clinical application of gene therapy to androgen-independent human prostate cancer. The novel vector created by the authors, which combines an adenovirus with a suicide gene and has tissue-specific promoter, was shown to be useful for gene therapy in hormone-independent prostate cancer.

References

1. Wingo PA, Tong T, Bolden S. Cancer statistics. *CA Cancer J* 1995; **45:** 8–30.

2. Parker SL, Tong T, Bolden S. *et al.* Cancer statistics. *CA Cancer J. Clin.* 1996; **46:**5–30.

3. Fair WR. The natural history of locally confined prostate cancer: A review. *Prostate* 1992; **4**(Suppl)**:** 79–84.

4. Dijkman GA, van Moorselaar RJA, van Ginckel R, et al. Antitumoral effects of liarozole in androgen-dependent and independent R3327-Dunning prostate adenocarcinomas. *J Urol* 1994; **151:** 217–222.

5. Sanda MG, Simons JW. Gene therapy for urological cancer. *Urology* 1994; **44:** 617–624.

6. Obiri NI, Hillman GG, Haas GP *et al.* Expression of high affinity interleukin-4 receptors on human renal cell carcinoma cells and

inhibition of tumor cell growth *in vitro* by interleukin-4. *J Clin Invest* 1993; **91:** 88–93.

7. Liu TJ, el-Naggar AK, McDonnel TJ *et al.* Apoptosis induction mediated by wild-type p53 adenoviral gene transfer in squamous cell carcinoma of the head and neck. *Cancer Res* 1995; **55:** 3117–3122.

8. Yang C, Cirielli C, Capogrossi MC *et al.* Adenovirus-mediated wild-type p53 expression induces apoptosis and suppresses tumorigenesis of prostatic tumor cells. *Cancer Res* 1995; **55:** 4210–4213.

9. Pang S, Taneja S, Dardashti K *et al.* Prostate tissue specificity of the prostate-specific antigen promoter isolated from a patient with prostate cancer. *Human Gene Therapy* 1995; **6:** 1417–1522.

10. Gotoh A, Ko S-C, Shirakawa T *et al.* Development of prostate-specific antigen promoter-based gene therapy for androgen-independent human prostate cancer. *J Urol.* 1998; **160:** 220–229.

11. Wang MC, Valenzuela LA, Murphy GP *et al.* Purification of a human prostate specific antigen. *Invest Urol* 1979;**17:**159–163.

12. Watt KWK, Lee P-J, M'Timkulu T *et al.* Human prostate-specific antigen: structural and functional similarity with serine proteases. *Proc Natl Acad Sci USA* 1986; **83:** 3166–3170.

13. Stamey TA, Yang N, Hay AR *et al.* Prostate-specific antigen as a serum marker for adenocarcinoma of the prostate. *N Engl J Med* 1987; **317:** 909–916.

14. Karr JF, Kantor JA, Hand PH *et al.* The presence of prostate-specific antigen-related genes in primates and the expression of recombinant human prostate-specific antigen in a transfected murine cell line. *Cancer Res* 1995; **55:** 2455–2462.

15. Thalmann GN, Anezinis PE, Chang SM *et al.* Androgen-independent cancer progression and bone metastasis in the LNCaP model of human prostate cancer. *Cancer Res* 1994; **54:** 2577–2581.

16. Cleutjens KBJM, van Eekelen CCEM, van der Korput HAGM *et al.* Two androgen response regions cooperate in steroid hormone regulated activity of the prostate specific antigen promoter. *J Biol Chem* 1996; **271:** 6379–6388.

17. Zhau HE, Hong S-J, Chung LWK. A fetal rat urogenital sinus mesenchymal cell line (rUGM): accelerated growth and conferral of androgen-induced growth responiveness upon a human bladder cancer epithelial cell line in vivo. *Int J Cancer* 1994;**56:**706–714.

18. Bett AJ, Haddara W, Prevec L *et al.* An efficient and flexible system for construction of adenovirus vectors with insertions or deletions in early regions 1 and 3. *Proc Natl Acad Sci USA* 1994; **91:** 8802–8806.

19. de Wet JR, Wood KV, de Luca M *et al.* Firefly luciferase gene: sturucture and expression in mammalian cells. *Mol Cell Biol* 1987; **7:** 725–732.

20. Ives DH, Furham JP, Tucker VS. Rapid determination of nucleotide kinase and nucleotidase activities with tritium-labeled substances. *Anal Biochem* 1969; **28:** 192–205.

21. Ko S-C, Cheon J, Kao C *et al.* Osteocalcin promoter-based toxic gene therapy for the treatment of osteosarcoma in experimental models. *Cancer Res* 1996; **56:** 4614–4619.

22. Sanchez-Prieto R, Vargas JA, Carnero A *et al.* Modulation of cellular chemoresistance in keratinocytes by activation of different oncogenes. *Int J Cancer* 1995; **60:** 235–243.

23. Ko SC, Gotoh A,Thalmann G *et al.* Molecular therapy with recombinant p53 adenovirus in an androgen-independent metastatic human prostate cancer model. *Human Gene Therapy* 1996; **7:** 1683–1691.

24. Zhang W-W, Fang X, Mazur W *et al.* High-efficiency gene transfer and high-level expression of wild-type p53 in human lung cancer cells mediated by recombinant adenovirus. *Cancer Gene Therapy* 1994; **1:** 5–13.

25. Yang C, Cirielli C, Capogrossi MC *et al.* Adenovirus-mediated wild-type p53 expression induces apoptosis and suppresses tumorigenesis of prostatic tumor cells. *Cancer Res* 1995; **55:** 4210–4214.

26. Vile RG, Hart IR. *In vitro* and *in vivo* targeting of gene expression to melanoma cells. *Cancer Res* 1993; **53:** 962–967.

27. Macri P, Gordon JW. Delayed morbidity and mortality of albumin/SV40 T-antigen transgenic mice after insertion of an alpha-fetoprotein/herpes virus thymidine kinase transgene and treatment with ganciclovir. *Human Gene Therapy* 1994; **5:**175–182

28. Shimizu K. Selective gene therapy of malignant gliomas using brain-specific promoters: its efficacy and basic investigations. *Jap J Clin Med* 1994; **52:** 3053–3058.

IV ABSTRACTS

Mitochondrial DNA and disease: Sequence polymorphisms in the mtDNA

Sangkot Marzuki, Herawati Sudoyo and Patcharin Pramoonjago
Eijkman Institue for Molecular Biology, Diponegoro 69, Jakarta 10430, Indonesia

The mitochondrial DNA (mtDNA) evolves at a much higher rate compared to that of the nucleus. Many disease-related mutations in the mtDNA have been identified, underlying a wide range of human disorders, from relatively rare neuro-muscular syndromes to more common neurosensorial deafness and diabetes mellitus. A large number of "normal" sequence polymorphisms have been documented also in the mtDNA and while by themselves these polymorphic sequence variants do not confer any clinical abnormality, there have been suggestions that the cumulative effect of some of these variants might predispose the expression of disease-related mtDNA mutations, or even other non-mitochondrial diseases. We have recently investigated mtDNA sequence polymorphisms in the Southeast Asian populations. A total of 180 polymorphic sites have been identified. In this communication we present evidence for:

(a) two main mtDNA haplogroups which predispose in the Southeast Asian populations the formation/expression of the 11778G→A mtDNA mutation underlying Lebers Hereditary Optic Neuropathy.

(b) a polymorphic site in the replication/transcription control region of the mtDNA as a predisposing factor for type 2 diabetes mellitus.

Emerging technologies for the mapping of complex human traits

Dave Dailey
PE Biosystems, a division of Perkin Elmer, Foster City, USA

Traditional linkage mapping has relied on family-based studies which require microsatellite genotyping of a modest number of individuals at several hundred loci. Recent population-based efforts to map genes which contribute to complex diseases will require scoring thousands to tens of thousands of loci for larger numbers of individuals. The potential usefulness of microsatellites and single nucleotide polymorphisms as well as recent advances in technology for marker scoring will be presented.

Intellectual property protection of genes and gene products

Jacqueline C. Lui
Lawrence Y.D. Ho and Associates, 30 Bideford Rd., #07-01, Singapore 229922

Intellectual property protection has become an important part of research and development in biotechnology. This is particularly true in the area of human genetics, in which the cost of developing and marketing new therapeutics and diagnostics run into billions of dollars each year. Without some form of assurance that they will be given monopoly of the market for a limited period of time, companies cannot justify the high cost of investment needed to complete research, clinical trials and regulatory approval.

Obtaining patent protection is one of the most effective means for attracting financial investment in commercially promising products. To obtain a patent for an invention, an applicant typically has to explain how the invention is novel and inventive over the prior art, and to describe the invention in sufficient details for one in the art to perform the invention. Besides these common criteria, an invention in the area of human genetics must fall within a patentable subject matter and also demonstrate convincing utility. Examples of human genes and gene products which have been given protection include erythropoietin, human tissue plasminogen activator, human growth hormone and the *BRCA1* genes.

Gene-immunotherapy of cutaneous metastases of human carcinoma using allogenic and xenogenic MHC DNA-liposome complexes

S.K. Tay and K.M. Hui†*
**Department of Obstetrics & Gynaecology, Singapore General Hospital,* †*Singapore Cancer Centre and Laboratory of Molecular Immunology, Institute of Molecular and Cell Biology, National University of Singapore*

The generation of strong tumour-specific immunity by *in situ* gene therapy is an attractive approach for the eradication of human cancer lesions. A study was performed to examine the toxicities of employing the human *HLA-A2*, *HLA-B13* and the murine $H\text{-}2K^k$ genes to generate tumour regression in 20 patients with different cancer types via DC-Chol/DOPE cationic liposomes. The patients were given four weekly injections of a DNA-liposome mixture directly into a cutaneous nodule. These procedures resulted in no significant clinical side-effects. The *HLA-A2* gene gave the highest level of expression *in situ*. Although all patients treated had progressive systemic disease and eventually succumbed to their disease, strong local responses were generated in the treated nodules. Of the eight patients whose cutaneous nodules received *HLA-A2* DNA, two completely regressed while five tumour nodules gave a partial local response. All but one of the patients who received *HLA-A2* liposome mixtures and had a subsequent local response were either cervical or ovarian carcinoma patients. This local response, seen in a group of patients who had relapsed systemic metastatic disease and were refractory to all available therapies, demonstrates the generation of a strong local immune response following our in situ gene therapy protocol. Further studies to investigate the use of *HLA-A2* DC-Chol/DOPE cationic liposomes for immunotherapy of cervical and ovarian cancers are warranted.

Generation of tumor-specific immunity through DNA-mediated gene transfer

Kam-Man Hui
Cellular and Molecular Research, National Cancer Centre, c/o Singapore General Hospital, Outram Road, Singapore 169608.

The immune system can recognize and destroy modified, "non-self" cellular components such as cancer cells that arise as a consequence of viral infections or somatic mutations. The recognition and destruction of "foreign, non-self" component is a vigorous immune response and is mostly mediated by the activation of cytotoxic T lymphocytes (CTL). Cancer cells can proliferate and thrive in the presence of such a sophisticated and efficient immune system either actively, by down-regulating the immune response or passively, by having weakly immunogenic antigens on their cell surface so that the host's immune system cannot detect their presence. In our laboratory, we are interested in stimulating the generation of tumor-specific immunity in cancer-bearing animals by manipulating the expression of surface antigens on cancer cells through gene transfer experiments. We have adapted the approach of introducing allogeneic Major Histocompatibility Complex (MHC) genes into cancer cells to render the cells more immunogenic and theoretically, enhancing their detection by the immune system. We have also compared the application of various vectors including gene gun, adenovirus and cationic liposomes for the delivery of DNA *in vivo*. We have observed that the delivery of allogeneic MHC DNA by cationic liposomes *in vivo* will generate strong tumor-specific immunity in cancer-bearing animals.

Advances in real-time quantitative PCR

Mary Grace Brubacher
PE Biosystems, a division of Perkin Elmer, Foster City, USA

Real-time quantitative PCR was made possible in 1996 with the introduction of the ABI Prism 7700 Sequence Detection System. Real-time detection is a quantum leap for quantitative PCR applications. This presentation will summarize how real-time quantitative PCR works as well as further advances made in this area.

Modeling biological function

Dhiraj K. Pathak
Glaxo-Wellcome Inc, North Carolina, USA

Advances in genomics are contributing to an exponential growth in the amount of data available on individual genes and proteins. A critical need is to integrate the individual data in order to understand the overall biological system. One way to integrate the data is to build functional models. Such models help in understanding complex interactions underlying molecular processes. They can also be used to generate hypotheses driving further experimentation. Many approaches are available for modeling function. We describe an approach based on a representation of molecular events and their interactions that can help in understanding function and generating new hypotheses.

The application of FISH (fluorescent *in situ* hybridisation) for the identification of a mosaic pseudodicentric Y-chromosome marker in a Turner patient.

J. Lim, K.Y. Loke*[†]*, K. Prabhakaran*[‡]*, T.M. Chua* and L. Gole**
**Department of Obstetrics and Gynaecology;* [†]*Department of Paediatrics;*
[‡]*Department of Surgery, National University of Singapore*

Approximately 6% to 15% of patients with clinical features of Turner's syndrome have a mosaic karyotype, ie a 45,X cell line and another with a small sex chromosome marker of undetermined origin. It is important to identify the origin of these markers as the presence of a Y or rearranged Y chromosome indicates an increased risk of developing gonadoblastoma.

A 2 year old girl was referred for enlarged clitoris. Chromosomal analysis of peripheral blood revealed one cell line to be hypodiploid (48%), and the other diploid (52%) with a small-sized marker chromosome (45,X/46,X,+mar). G, C and Q banding could not confirm irrevocably that the marker originated from Y chromosome. Using FISH probes for the Y centromere, two pairs of signals were consistently observed at both the distal ends; one pair was always split, suggesting an inactive centromere. The patient's karyotype was subsequently diagnosed as 45,X/46,X,+psu dic(Yp). She underwent clitoroplasty with a bilateral gonadectomy. Intraoperatively, there were streak ovaries on the left. The right gonad consisted of both testicular and ovarian elements.

It is difficult to diagnose a dicentric Yp marker chromosome as the morphology of a dicentric Yp is comparable to a normal Y, having one centromere suppressed and not apparent. This report illustrates the usefulness of FISH in the identification of marker chromosomes. It also supports the observation that there is an increased chance for the patient to be a phenotypic female with the somatic features of Turner's Syndrome when a 45,X cell line is mosaic with a complete/partial Y chromosome.

Allelic association study of a region on chromosome 17q in Singaporean Chinese with essential hypertension

X. Zhou, J. Knight*[†]*, G. Gardner*[†]*, M.J. Caulfield*[†]*, H.M. Wu*[‡]*, E.P.H. Yap*[‡]*, E.A. Taylor* and V.M.S. Oh**

**Dept of Medicine, National University of Singapore, Singapore;* [†]*Dept of Clin. Pharmacology, St Bartholomew's Hospital, London;* [‡]*Defence Medical Research Institute, Singapore.*

The several genes responsible for cryptogenic ('essential') hypertension are mostly unknown. One of the main blood pressure loci is near the angiotensin- converting enzyme (ACE) gene on rat chromosome 10. The homologous region on human chromosome 17q was linked to hypertension in a European sib-pair study (Julier *et al,* 1997). In the NUS Genetics of Hypertension (NUSIGHT) Study, we recently acquired about 90 hypertensive families, most of whom contain affected sibling-pairs. We enrolled persons whose pre-treatment resting blood pressure was ≥150/95 mmHg by standardised manometry, or ≥140/90 mmHg on 24-h ambulatory recording. Disease onset was ≤60 years. In this preliminary study, we genotyped 47 hypertensive Chinese patients and 100 population controls matched for ethnicity, using three microsatellite markers D17S951, D17S934 and D17S806. These markers reside in this particular region but are distinct from the ACE gene locus (17q23). Polymerase chain reaction (PCR) amplification of genomic DNA was carried out using fluorescent-labelled primers, and PCR products were sized on an automated sequencer. We applied the Markov-chain Monte Carlo simulation to test statistical significance. There was no significant difference in allele frequencies between hypertensive and control subjects for all three markers ($p > 0.05$, chi-square). From the results of this limited study, we conclude that we could not detect any allelic association between the microsatellite markers D17S951, D17S934 and D17S806 on human chromosome 17q with essential hypertension in Singapore Chinese.

Association studies of candidate genes for severe sporadic myopia

R.Y.Y. Yong, H.M. Wu and E.P.H. Yap
Defence Medical Research Institute, NUS Clinical Research Centre MD11, Singapore 119260

Myopia, the most common ocular disorder in man, is highly prevalent in Singapore. Data of familial clustering, inter-ethnic differences in prevalence rate and twin studies demonstrate the contribution of genetic factors. The aim of our project is to identify the genes that predispose to severe sporadic myopia. Currently we have screened 17 candidate genes by allelic association studies in myopes and controls. The candidate genes selected consist of 8 retinal genes (4 involved in phototransduction and 4 retinal structural genes), 2 scleral genes (Collagen IIA1 & Myocilin) and 7 neural genes (genes involved in the synthesis and transduction of the neurotransmitter dopamine). A total of 33 short tandem repeat (STR), 2 variable number of tandem repeat (VNTR) and 1 restricted fragment length polymorphism (RFLP) markers were studied. High throughput genotyping of STR markers were achieved by performing PCR in 96-well plates and multiplexing panels of up to 10 markers on an automated sequencer. Genotyping of VNTR and RFLP were done using 96-lane diagonal gels and manual scoring of alleles. 93 unrelated Chinese with severe myopia (worse than – 8.0 D) and 95 emmetrope controls were recruited as study subjects. A total of 6768 genotypes were carried out. Statistical analysis of significance was done by standard and Monte Carlo simulation tests. Significant allelic associations were detected for an intragenic marker in the rhodopsin gene (p=0.009), a flanking marker of the dopamine receptor D4 gene (p=0.006), and the TIGR gene (p<0.002), suggesting that these genes or loci nearby may be involved in myopia susceptibility.

The glaucoma gene trabecular meshwork inducible glucocorticoid response (TIGR) or myocilin gene is associated with severe sporadic myopia

Hui Min Wu, Xiang Hua Yu, Rita Yong and Eric Yap
Defence Medical Research Institute, NUS Clinical Research Centre MD11, Singapore 119260.

The aetiological mechanisms for myopia in humans are unknown. Increased intra-ocular pressure accompanied by weakening of the scleral wall is hypothesized. The risk of developing open angle glaucoma is higher in myopes than in non-myopes. Furthermore, intra-ocular pressure in myopes is reportedly higher than that in non-myopic controls. Recently mutations in the TIGR or MYOC (myocilin) gene were identified in pedigrees with juvenile-onset open angle glaucoma. We sought to determine if TIGR was a candidate gene for severe early-onset sporadic myopia. A case control study was designed to detect linkage disequilibrium and allelic association. 97 unrelated myopes (spherical equivalence worse than -8.0 dioptres) and 92 matched emmetropic controls were identified by population-based screening of 18-20 year old males of Chinese ethnicity. Four short tandem repeat (STR) markers closely linked to TIGR and 3 intragenic markers (2 STR and 1 novel SNP) were selected. The microsatellites were genotyped by PCR with fluorescent end-labeled primers, and sized on a fluorescent sequencer. Significant allelic association was found between myopia and a dinucleotide STR marker in the 5' proximal promoter region. Among the 4 common alleles, the relative risk for myopia increased from 0.7 to 4.3 as repeat length decreased. With a subset of severe myopia worse than 12D, there was an even greater change in relative risk from 0.4 to 6.6. No disease association was found with the other intragenic markers. Allelic association between myopia and TIGR suggests that TIGR is involved in the development of severe sporadic myopia. The possibility that microsatellite polymorphisms in the promoter may affect TIGR function needs to be investigated further.

cDNA characterisation and chromosomal mapping of human Golgi SNARE GS27 and GS28 to chromosome 17

Thuan D. Bui, Elaine R. Levy*[†], *V. Nathan Subramaniam*, Stephen L. Lowe*, Lei Lu* and Wanjin Hong**

**Membrane Biology Laboratory, Institute of Molecular and Cell Biology, 30 Medical Drive, Singapore 117609, Singapore,* [†]*Cytogenetics Group, Wellcome Trust Centre for Human Genetics, Windmill Road, Oxford OX3 7BN, United Kingdom.*

Transport of proteins along the exocytotic pathway is primarily achieved by vesicular intermediates. Two proteins, Golgi SNARE of 27 kD, GS27 and of 28kD, GS28 (approved nomenclature GOSR2 and GOSR1, respectively), are.important trafficking membrane proteins between the endoplasmic reticulum (ER) and the Golgi, and between Golgi subcompartments. Here, we present the human GS27 and GS28 cDNA sequences. They encode a predicted protein of 212 and 250 amino acids, respectively. Chromosomal mapping analyses reveal that human GS27 is located on chromosome 17q21 and GS28 approximately on 17q11. The chromosomal location of GS27 near a locus implicated in familial essential hypertension and its known function in trafficking place it to be a potential candidate gene for this disease.

Analysis of ectopic dystrophin mRNA

H.H. Khng[*], *P.S. Low*[*], *W.L. Lee*[*,†] *and P.S. Lai*[*]

[*]*Department of Paediatrics, National University of Singapore*

[†]*Department of Neurology, National Neuroscience Institute, Singapore*

The dystrophin gene is encoded by a large RNA transcript processed from 79 exons covering 2 million base-pair sequences. This large size and complexity of the gene presents a challenge to direct identification of mutations. One approach to this problem is to analyse the expression of ectopic dystrophin mRNA transcripts. Although the levels of such transcripts are low, with one mRNA copy per 500-1,000 cells, we have found that it was possible to identify the splicing events of pathogenic gene lesions using lymphoblastoid cell lines. Aberrant sized transcripts have been found by amplifying the entire coding region of the dystrophin mRNA by reverse transcription and nested PCR. These overlapping transcripts amplified from 10 reactions ranged approximately from 0.7 kb to 1.3 kb. Sequencing of the altered transcripts as compared to wild type revealed that exons were spliced out in four patients. Disruption of translational reading frame occurs due to production of downstream nonsense codons. The clinical phenotypes manifested by these patients are expected to be due to these truncated transcripts.

Germline β-catenin mutation is absent in familial adenomatous polyposis patients without APC mutation.

P.Y. Cheah, X. Cao, K.W. Eu and F.C. Seow
Dept of Colorectal Surgery, Singapore General Hospital, Singapore 169608

Inactivation of the adenomatous polyposis coli (APC) gene has been shown to initiate the majority of colorectal cancer (CRC), including a familial form called familial adenomatous polyposis (FAP). One consequence of the APC mutation is the activation of the β-catenin/T-cell transcription factor (Tcf) pathway. A recent study has shown that about half of the sporadic CRC lacking APC mutation has β-catenin mutation suggesting that β-catenin mutation can substitute for APC mutation in the initiation of colorectal tumorigenesis. However, the frequency of β-catenin germline mutation in FAP has not been reported. Using the Protein Truncation Test (PTT), we screen the entire coding region of APC in 26 unrelated FAP kindreds and found germline mutations in twenty families. We then screened for β-catenin germline mutations in the rest of the families lacking detectable APC mutations. No missense mutations at GSK-3β phosphorylation sites or interstitial deletion of exon 3 of β-catenin was found. Our results indicate that APC germline mutations are frequent but β-catenin germline mutations are rare in FAP patients suggesting that β-catenin mutation can not substitute for APC mutation in the initiation of FAP.

The detection of *Entamoeba histolytica* DNA in amoebic liver abscess pus using polymerase chain reaction

J. Khoo, S.W. Ng*, S. Zaman*, R. Hussain†, M. Ahmed†, V. Zaman†, and R. Ahmed‡*
**Biotechnology Department, Ngee Ann Polytechnic, Singapore; †Aga Khan University, Karachi, Pakistan; ‡Liaquat Medical Hospital, Karachi, Pakistan*

An important and serious complication of intestinal infection with *E. histolytica* is the involvement of the liver (hepatic amoebiasis). In hepatic amoebiasis, abscess formation in the liver is usually diagnosed by the clinical picture (pain in the right upper quadrant and fever), ultrasound examination and positive serology. However, none of these tests are definitive and amoebae are rarely seen in the aspirated pus. It is for this reason that the feasibility of using polymerase chain reaction for the detection of *E. histolytica* DNA in liver abscess pus was investigated.
A comparative study was done to verify the sensitivity of 15 pairs of primers specific for detecting *E. histolytica* in liver abscess, using polymerase chain reaction. Liver abscess aspirates from 20 serology-positive patients were collected under ultrasound observation. Of the 15 pairs of previously published primers investigated (for references refer to poster), only one set gave consistently reliable results. Based on these results, we recommend that this pair of primers can be successfully used to confirm the diagnosis of amebic liver abscess.

Identification of differential gene expression between pelvic endometriotic tissue and uterine endometrium by using cDNA subtractive hybridization.

Weiping Hu and Sun Kuie Tay
Department of Obstetrics & Gynaecology, Singapore General Hospital

Pelvic endometriosis is the presence of ectopic endometrium in the pelvis. It affects a large proportion of women at reproductive age. It is also one of the major causes of infertility. There is yet no conclusion on the etiology and pathogenesis of endometriosis. Epidemiological studies involving familial endometriosis and endometriosis in twins suggest a genetic basis of this disease. It is hypothesized in this study that specific genetic alterations and/or aberrant expression of specific genes in endometrial cells is the underlying molecular mechanism for successful implantation and growth of the disseminated endometrial cells to form endometriosis. We here present some preliminary results.

A new protocol, based on magnetic bead assisted cDNA subtraction, was recently developed to identify differential gene expression. It was employed to make a systematic search for the differentially expressed genes between pelvic endometriotic tissue and uterine endometrium. The subtracted cDNAs were PCR-amplified and cloned into TA vectors, then used to transform competent cells to generate a subtractive cDNA library. Colony lift hybridization was used for differential screening of the subtracted library and about 50 colonies were selected. The selected colonies were directly analyzed with PCR and proved to contain inserts with size from 300 to 1500 bp. They were also confirmed by restriction enzyme digestion. DNA sequencing and hybridization analysis of the isolated genes is now being performed.

Transfection of urothelial cells using non-viral vectors

Carmel Lawrencia, Ratha Mahendran and Kesavan Esuvaranathan
Department of Surgery, National University of Singapore

Clinical trials using BCG have shown that superficial bladder cancer can be cured by the induction of an immune response. To date, the most efficient gene transfection systems are viral but the use of viruses have raised many safety questions. Our objective was to develop a rapid non-viral system for the transfection of urothelial cells *in vitro* and *in vivo.*

MB49 cells were transfected with reporter gene *pCMVlacZ* using a number of commercial transfecting agents. Transfection efficiency was determined either by staining or assaying for β-galactosidase activity. *In vivo,* mice were transfected intravesically. After 2 days the bladder, lungs, kidneys, heart, liver and spleen were stained for β-gal activity. Exposure and durability of expression were also evaluated.

In vitro transfection efficiency was in the order of Dotap > Superfect > Fugene. The addition of cholesterol to Dotap and Superfect further improved efficiency by 3.8 fold and 2.6 fold respectively. Both *in vitro* and *in vivo* expression was observed 15 minutes after exposure of the Dotap+ cholesterol/DNA complex to cells. Protein production although detectable after I hour, peaked only after 48 hrs. Interestingly expression was sustained up to 30 days *in vivo* and was localised in the bladder. Nuclear and cytoplasmic analysis showed that with the Dotap+cholesterol complex DNA was found in both the nucleus and cytoplasm. TEM analysis *in vivo* showed no accumulation of agent or lipid.

From our data, it appears that Dotap+ cholesterol is the best agent for the transfection of urothelial cells *in vitro* and *in vivo.*

Progestins inhibit the growth of MDA-MB-231 cells transfected with progesterone receptor cDNA

Valerie C.L. Lin, Eng Hen Ng†, Swee Eng Aw*, Michelle G.K. Tan*, Esther H.L. Ng†, Vivian S.W. Chan* and Gay Hui Ho**
**Department of Clinical Research, and †Department of General Surgery, Singapore General Hospital Republic of Singapore 169608*

Since progesterone exerts its effects mainly via estrogen-dependent progesterone receptor (PgR), the expression of progesterone's effects may be overshadowed by the priming effect of estrogen. PgR expression vectors were transfected into ER-α- and PgR-negative breast cancer cells MDA-MB-231 so that the functions of progesterone can be studied independent of estrogens and ER. Eight stable transfectant clones expressing both PgR isoform A and B were studied for their growth response to progesterone and its analogues. While progesterone had no effect on growth in the control transfectant, the hormone markedly inhibited DNA synthesis and cell growth in all PgR-transfectants dose-dependently from 10^{-12}-10^{-6} M. This growth inhibition was associated with an arrest of cells in the G0/G1 phase of the cell cycle. Progestins medroxyprogesterone acetate, Org2058, R5020 also strongly inhibited DNA synthesis and their doses required for maximal inhibition of 60 - 70% were 10^{-17} M, 10^{-13} M and 10^{-7} M, respectively. Antiprogestin ZK98299 alone had no effect, but the compound was capable of counteracting the inhibitory effect of progesterone. In contrast, RU486 inhibited DNA synthesis and it showed no further effects when it was used concurrently with progesterone. These results indicate that progestins are per se antiproliferative via a PgR-mediated mechanism in breast cancer cells. More importantly, we have shown that progestins may exert effective inhibitory control over the cell growth if the PgR expression is reactivated in ER- and PgR-negative breast cancer cells.

Strategies for EBV vaccine design using computational tools

P. Kangueane, M.K. Sakharkar*, K.S. Lim*, Lin Kui*, E.C. Ren†, V. Brusic‡ and P.R. Kolatkar**

**BioInformatics Centre, †Department of Microbiology, ‡Kent Ridge Digital Laboratory, National University of Singapore, Singapore*

The Epstein Barr virus (EBV) is a human γ herpes virus associated with a number of clinical manifestations in humans. These include nasopharyngeal carcinoma (NPC), Burkitt's lymphoma (BL), pseudo-lymphomatous lung carcinoma and gastric carcinoma. EBV has also been associated with 30% of Hodgkin's disease and 20% of breast cancer tumors. Worldwide, EBV has infected 50-90% of human population. An effective vaccine is lacking; hence there is an urgent need for EBV vaccine development.

Advances in molecular immunology and biotechnology show that short peptides that are targets of immune recognition are potent vaccine candidates. The mapping of different regions of the viral antigen to a specific immune response is crucial for the design of synthetic peptide vaccines. The high polymorphism of HLA among the human population and the allelic variation between individuals has made the identification of peptidic targets a complex task. To date, nearly 500 HLA class-I and more than 400 HLA class-II allelic variants have been assigned. Several synthetic peptides corresponding to a number of different epitopes have to be incorporated in such a vaccine.

Computational methods provide means to determine specific vaccine components efficiently. The relevant techniques for determining EBV vaccine components include protein sequence alignment, matrix models, artificial neural networks, homology modeling and protein threading. We have used Kleisli Related Integration System (KRIS) for data integration and sequence information retrieval. This poster highlights the strategies defined for EBV vaccine design using computational tools.

Rare mutations in low density lipoprotein receptor gene detected by denaturing gradient gel electrophoresis and direct sequencing in two Malay women with familial hypercholesterolaemia

Seng Tzer Jing and Evelyn Koay
Department of Pathology, National University of Singapore

Familial hypercholesterolaemia (FH) is an autosomal dominant inherited disorder caused by mutations in the low density lipoprotein receptor (LDLR) gene that lead to elevated plasma levels of LDL cholesterol, accelerated atherosclerosis and premature coronary artery disease. Molecular methods have an advantage over serum lipid profiling in providing an accurate diagnosis of FH by defining the mutation involved and allowing early detection of asymptomatic carriers of the mutation within the same family. Eight patients were recruited for this study on the basis of high plasma LDL cholesterol levels. Initial mutation screening was performed by denaturing gradient gel electrophoresis (DGGE). Each of the 18 exons and the promoter region was amplified by polymerase chain reaction (PCR) using optimised pairs of primers with a GC-clamp in either the 5' or 3' end. An aberrant DGGE pattern in the exon 9 region was detected in two female Malay patients. Cycle sequencing of exon 9 was performed using the Dye Terminator Ready Reaction Mix (Perkin Elmer). A GγC sequence variation in codon 1284 of the LDLR gene was found, resulting in a Asnγ Lys change in the LDLR protein sequence. This mutation has been recently reported in South Africans [1, 2], but it has not been previously found in Asians. Our results indicate that DGGE is a reliable screening method for detecting the presence of sequence variation in the LDLR gene, defining the specific region of the gene that requires sequence analysis, thereby facilitating the rapid molecular diagnosis of FH.

Genetic analysis of four short tandem repeats loci for three ethnic groups in Singapore

W.F. Tan-Siew[*], *S. Y. Chuah*[*], *Violet P.E. Phang*[†], *S.T. Chow*[*]
[*]*Department of Scientific Services, Institute of Science and Forensic Medicine, 11 Outram Road, Singapore 169078.* [†]*Department of Biological Sciences, National University of Singapore, Kent Ridge, Singapore 117600.*

Allele and genotype frequencies for four tetrameric short tandem (STR) loci were determined for three population groups – Chinese, Malays and Indians – in Singapore. The technique uses multiplex polymerase chain reaction (PCR) and electrophoresis of the PCR products in denaturing polyacrylamide gels coupled with fluorescent-based detection. The loci are HUMTHO1, HUMfes, HUMvWA and HUMF13A. Statistical evaluations on these four loci indicated that the samples met the Hardy-Weinberg. In addition, there was no evidence for association of alleles between the four loci. The product of allele frequencies from the data from the sample populations in this study can be used in forensic analyses to estimate the frequency of the STR DNA genotype.

Non-parametric linkage analysis of dopamine D2 receptor and essential hypertension in Singaporean Chinese sib-pairs

Yap E.P.H.[*], *Wu H.M.*[*], *Zhou X.*[†], *Taylor E.A.*[†] *and Oh V.M.S.*[†]
[*]*Defence Medical Research Institute, Singapore;* [†]*Dept of Medicine, National University of Singapore, Singapore*

Abnormalities in the dopaminergic system have been demonstrated in rat models of hypertension and are implicated in the pathogenesis of hypertension in humans. Dopamine has a diverse range central and peripheral effects (including the regulation of renal natriuresis and vasomotor tone, and the control of catecholamine, vasopressin and renin-angiotensin-aldosterone secretion). These are mediated by two classes of receptors (D1/D5 and D2/D3/D4). We investigated the role of the dopamine D2 receptor (DRD2) in this genetic linkage study of essential (idiopathic) hypertension. Forty-nine sib-pairs concordant for hypertension and of Chinese descent were recruited as part of the NUSIGHT study, with a further 190 ethnic-matched samples serving as population controls. Two polymorphisms in the coding region (Ser311, NcoI) and a 3' marker (TaqIA) were genotyped by PCR-RFLP. Non-parametric single-point linkage analysis of each of these markers was carried out. There was an increased sharing of alleles identical by state at the TaqIA locus (p=0.0025). This interesting finding suggests that DRD2 could be involved in hypertension in this population. Further studies using more polymorphic markers and larger samples sizes are warranted.